Wireless Engineering: From Theory to Practice

Wireless Engineering: From Theory to Practice

Edited by **Dravin Rager**

CWILLFORD PRESS

New York

Published by Willford Press,
118-35 Queens Blvd., Suite 400,
Forest Hills, NY 11375, USA
www.willfordpress.com

Wireless Engineering: From Theory to Practice
Edited by Dravin Rager

© 2016 Willford Press

International Standard Book Number: 978-1-68285-107-4 (Hardback)

Contents

Preface

The field of wireless engineering is rapidly expanding because of the growing demand for pervasive communication networks for both business as well as individual networks. It has widespread applications for mobile and large scale networks. This book attempts to provide a detailed overview of the current status of wireless engineering - from its theories to applications at present. It is a compilation of up-to-date information on the different components, control algorithms and programs, and routing protocols involved in mobile and wireless communication networks. Industry experts and scientists from different parts of the globe have contributed their precious inputs and data in the chapters of this book. It would be an excellent source of reference for all the graduate and postgraduate students, individual researchers and engineering professionals.

All of the data presented henceforth, was collaborated in the wake of recent advancements in the field. The aim of this book is to present the diversified developments from across the globe in a comprehensible manner. The opinions expressed in each chapter belong solely to the contributing authors. Their interpretations of the topics are the integral part of this book, which I have carefully compiled for a better understanding of the readers.

At the end, I would like to thank all those who dedicated their time and efforts for the successful completion of this book. I also wish to convey my gratitude towards my friends and family who supported me at every step.

Editor

Automatic security assessment for next generation wireless mobile networks

Francesco Palmieri[a], Ugo Fiore[b] and Aniello Castiglione[c,*]

[a]*Dipartimento di Ingegneria dell'Informazione, Seconda Università degli Studi di Napoli, Aversa (CE), Italy*

[b]*Università degli Studi di Napoli "Federico II", Napoli, Italy*

[c]*Dipartimento di Informatica "R. M. Capocelli", Università degli Studi di Salerno, Via Ponte don Melillo, I-84084 Fisciano (SA), Italy*

Abstract. Wireless networks are more and more popular in our life, but their increasing pervasiveness and widespread coverage raises serious security concerns. Mobile client devices potentially migrate, usually passing through very light access control policies, between numerous and heterogeneous wireless environments, bringing with them software vulnerabilities as well as possibly malicious code. To cope with these new security threats the paper proposes a new active third party authentication, authorization and security assessment strategy in which, once a device enters a new Wi-Fi environment, it is subjected to analysis by the infrastructure, and if it is found to be dangerously insecure, it is immediately taken out from the network and denied further access until its vulnerabilities have been fixed. The security assessment module, that is the fundamental component of the aforementioned strategy, takes advantage from a reliable knowledge base containing semantically-rich information about the mobile node under examination, dynamically provided by network mapping and configuration assessment facilities. It implements a fully automatic security analysis framework, based on AHP, which has been conceived to be flexible and customizable, to provide automated support for real-time execution of complex security/risk evaluation tasks which depends on the results obtained from different kind of analysis tools and methodologies. Encouraging results have been achieved utilizing a proof-of-concept model based on current technology and standard open-source networking tools.

Keywords: Dynamic access control, active networks, analytic hierarchy process, multiple criteria decision analysis, nomadic computing, security audit, security assessment, ubiquitous networking

1. Introduction

The networking world is going mobile and the heterogeneous wireless communication infrastructures enabling ubiquitous connectivity are now an important part of our daily life and their fields of application are rapidly increasing. Most of us take for granted the ability to be connected from anywhere, at any time and at a reasonable cost, so that the widespread adoption of low-cost wireless technologies, such as Wi-Fi and 3G, makes nomadic networking more and more common. Simultaneously, the increasing number of possible malicious/hostile behaviour, exploiting available known vulnerabilities or design flaws may lead to dramatic consequences. In fact, as mobile users migrate from a wireless access point to another, by moving between home, office, airport and hotel, they take with them a large number

*Corresponding author: Aniello Castiglione, Dipartimento di Informatica "R.M. Capocelli" – Università degli Studi di Salerno, Via Ponte don Melillo, I-84084 Fisciano (SA), Italy.
E-mail: castiglione@ieee.org.

of networked devices such as mobile phones, tablet PCs, handhelds and other "electronic hitchhikers", probably with unpatched or misconfigured applications. The continual afflux of vulnerable/compromised nodes threatens the integrity of the environments, as well as that of other mobile nodes operating within the same environments. As corrupted machines move from network to network, they will be able to quickly spread offending code to more network resources and to new users. In situations where transit routing is used [27], internal traffic between mobile nodes may even cross the network border.

The traditional paradigm based on the separation between a secure area "inside" and a hostile environment "outside" can no longer be effective, because authenticated users can bring uncontrolled machines in the network. A user may unwittingly bring in active threats such as viruses, Trojan Horses, Denial-of-Service daemons, or even create a hole for a human intruder. Alternatively, they may bring in passive threats such as vulnerable packages or poorly configured software. Particularly, worms could use nomadic trends to augment their spreading activity in dense urban centers or large campuses, without promptly spread to the Internet. This is a fundamental security threat that must be addressed. To contain or at least mitigate the impact and spread of these attack "vectors", wireless access control environments must be able to examine and evaluate clients on their first access to the network for potential threats or vulnerabilities.

Accordingly, a new third party authentication, authorization and audit/assessment paradigm in which once a device enters an environment it is subjected to active analysis by the infrastructure (*Admission Control*) is proposed. Moreover, if its security level is found to be not adequate it is immediately taken out from the network and denied for any further access until its problems and vulnerabilities have been fixed. The essence of the proposed architecture is allow or deny access to the network to systems based on their "perceived" threat level, i.e., before they can become infected or attack other systems. Hence, to evaluate their "health" status, an active *vulnerability assessment* is carried out, followed by a continuous passive monitoring against further suspect activities such as port scans, broadcasts or other hostile activities. The above assessment is based on a fully automated methodology for deterministically evaluating the "security level" of a networked object/node by observing it from the outside without requiring any information about its equipment, structure, security policies and services offered. The associated security analysis process takes advantage from a reliable dynamically built knowledge base containing semantically-rich information about the node under examination. This information is provided by several network mapping and security assessment tools relying on different service/application discovery, vulnerability analysis, operating system and protocol fingerprinting techniques, integrated within a common framework. It provides a complete set of metrics aiming at representing in a concise and reliable way the most significant security-related properties characterizing the mobile node itself. The above knowledge base is then exploited by using a structured decision-making methodology such as AHP (Analytic Hierarchy Process) [29] to properly combine the specific metrics and evidences collected and reliably guess the fundamental properties that are needed to explicitly rank, according to standardized risk assessment schemes and methodologies, its overall security degree. This results in a set of quantitative security indicators that could be used as a reference for the numerical evaluation of the node security. In order to take the decision on the node access admission or refusal, a numerical comparison between the resulting security levels and a set of preconfigured thresholds defining the "minimal" acceptance security profile can be performed.

The proposed approach seems to be promising for several reasons. Firstly, it does not rely only on global network traffic monitoring. Secondly, it does not require the use of specific agents to be installed on the mobile hosts, implying an undesired strong administrative control on the networked systems, so it is very suitable for the untrusted mobile networking environment. Thirdly, the whole security arrangement is easily applicable at any network location and does not depend from specific software packages. In fact, the underlying assessment methodology is flexible, customizable, extensible, and can easily integrate other off-the-shelf analysis tools to provides automated support for the execution of complex risk evaluation tasks that require the results obtained from several different analysis processes. Finally, this type of infrastructure would strongly encourage active and timely patching of vulnerable and exploited systems, increasing the overall network security. It would also benefit users by protecting their systems, as well as keeping them up to date, and benefit local providers by protecting their infrastructure and reducing theft of service. Its deployment would also protect the Internet as a whole by slowing the spread of worms, viruses, and dramatically reducing the available population of denial-of-service agents.

2. Related work

Self-defending networks, supporting dynamic access control mechanisms have received a significant attention in literature to face the continuously evolving security challenges that require high degrees of autonomy, scalability, adaptability, and robustness. Generally, in all the environments where no reliance can be made on the presence of an experienced administrator, automated methods are called for in order to detect and block malfunctioning or malicious nodes. Nomadic wireless environments, as well as Mobile Ad-hoc NETworks (MANETs) and sensor networks, clearly belong to this class. Dynamic access control can be traced back to an early work by Thomas and Sandhu [36]. Knorr [15] used Petri net workflows to achieve dynamic access control. In [12] a vulnerability scanner is used to keep an updated record of their current vulnerabilities. The authors in [1] use the historical record of vulnerabilities and their frequency to assess the overall security for each service, by estimating the probability that new vulnerabilities will be discovered and form the basis for new attacks. In [35] a system to manage network level access based on threat information is presented. To each node and service is associated an access threshold, which is checked against the threat level of a potential access requestor to grant or deny access to it.

In [8] the authors assess the risk associated with granting a given access request and derive a corresponding level of trust required. In parallel, the trust level of the actual requesting subject is calculated and compared against the established level of admission. The work in [17] presents a Budget-Based Access Control Model aiming at mitigating the threats from insiders whose risk level may be unknown to the involved organization. A more innovative approach in self-defending networks is related to the research area of Artificial Immune Systems (AIS) [24], covering the development of biologically-inspired security systems that focus on the monitoring and detection of improper or dangerous behaviour [2] to trigger the artificial immune system reaction. The majority of research works on AIS-based intrusion/Anomaly Detection in wireless networks [14] is centered on passive detection of danger signals. Recent works have been focused on fault reconnaissance agents [31], while other research directions have integrated in this context several nonlinear techniques such as nonlinear classifiers [21] and phase-space reconstruction [32]. Similar approaches have also led to improvements in the modeling of mobility in wireless networks [9,10]. These ideas have also been adapted to Anomaly Detection [25,33] and offer a promising perspective for the identification of signals that linear systems fail to detect. In contrast, the work in [23] emphasized the role of preventive auditing and access control, coupled with active monitoring. This proposal builds upon the previous one and extends it, by adding a more robust and versatile self-defending framework and by detailing the key vulnerability assessment issues that constitute its foundation.

3. The admission control subsystem

Network administrators are highly interested in controlling which devices are allowed onto their networks, and the associated access admission decision is always driven by security concerns. In mobile wireless LAN environments this is even more important, because these networks are usually deployed to grant ubiquitous access to the Internet and are therefore popular targets for hackers. In addition, many systems/devices requiring connection to these environments are owned by people nearly completely unaware of security chores like keeping their operating systems or anti-virus definitions up-to-date. The fundamental concept of the proposed active access control strategy is that once a client mobile device enters a new environment it has to be analyzed by the infrastructure to check for potential vulnerabilities and contaminants and its security degree is evaluated in order to decide if network access can be granted without compromising the target environment. This enhanced security facility can be implemented as an additional access control phase immediately following the traditional IEEE 802.11 authentication and authorization paradigm. Here, the involved wireless access point (playing the *Authenticator* role) performing successful IEEE 802.1x authentication of the supplicant against the RADIUS or DIAMETER server (which is called *Authenticator* or *AS*) when the mobile node receives full network access and is reachable through an IP address, requires its complete security analysis by a new network entity called the *Auditor*. To ensure the proper model scalability in huge and complex network environments there may be many *Auditors* associated to the different access points, each one (eventually more than one) dedicated to the coverage of a specific area. This also provides fault tolerance and implicit load distribution where necessary. There is a large set of possible types of analyses that could be performed by the *Auditor*, including external network scans and probes, virus scans, or behaviour monitoring. For instance, a simple examination might determine the versions of installed OS and software along with appropriate security patches, verifying the existence of open vulnerabilities where applicable. During the different examination phases, the *Auditor* computes a complex multi-dimensional score, which represents the resulting client insecurity degree, that is a vector of different indexes associated to each security category, where each index is calculated as the weighted combination of individual scores assigned to the different analysis results.

The task of specifying the individual scores to be assigned to the different categories (or security criteria) is handled through a complex decision-making strategy, known as AHP, that structures the available choices concerning the different criteria into a hierarchy and assigns a relative importance weight to each one of these criteria by comparing and ranking all the possible alternatives. That is, a numerical weight or priority is derived for each element of the hierarchy, allowing diverse and often incommensurable elements to be compared to one another in a rational and consistent way. Since the target access network environments and their security requirements may vary widely, the above comparison process is driven by the experience and knowledge of several security experts (maybe involved in the specific administration task), with the role of decision makers, that systematically evaluate the various elements by comparing them to one another, two at a time, with respect to their impact on an element above them in the hierarchy. This results in a judgments matrix, defined at the initial system configuration time, expressing in terms of pairwise comparisons, the relative values of a set of security attributes. For instance, it may define the relative importance to the node operating system security degree as opposed to the presence of a certain type of vulnerability or to the lack of a required security update/patch. The choice is whether the former security criterion is extremely more important, rather more important, as important, and so on down to extremely less important, than the other one.

Once the security analysis terminates, the resulting degree is compared with an acceptance threshold configured at the *Auditor* level and, if the computed security score does not fall below the threshold,

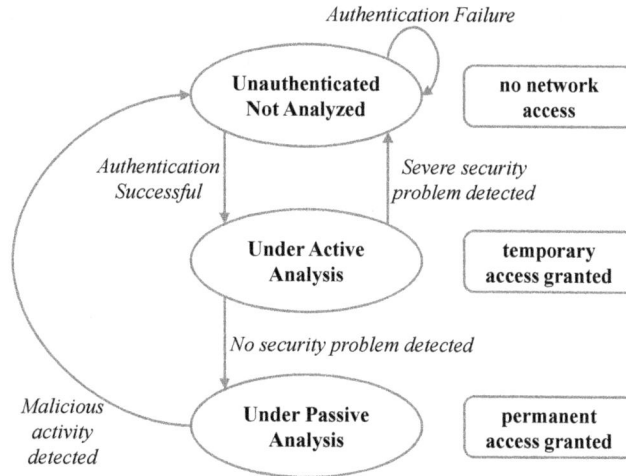

Fig. 1. The admission control process status diagram.

the *Auditor* returns a reject response to the access points that immediately denies any further network access to the examined device. Otherwise, the mobile client does not lose its network access, but the examination for the involved node will transition from the active to the passive state, which means that using standard intrusion detection techniques, the *Auditor*, on behalf of the local infrastructure, could continuously examine network traffic to determine if any entity is trying to launch a port scan of any other form of attack or take over other machines. In this case the *Auditor* quickly notifies the above event to the access point and consequently the node is immediately taken off from the network. The whole process is depicted in Fig. 1.

The issue of mobile clients possibly equipped with anti-scan mechanisms deserves a specific discussion. These clients will resist the scanning, thus not triggering any vulnerability detection alert. The presence of anti-scan systems can result from a legitimate intent of raising the security of the client, but it can also be part of a malicious plan to evade surveillance and enter the network. If the second case holds, the client would likely be granted access, but subsequent active behavior infringing the security policy would be detected by the Auditor though passive monitoring. There is, though, the threat of a malicious system crafted to appear as a secure client, whose purpose is to enter the network and grab information via passive monitoring. The gathered data could then be released after disconnection.

The proposed architecture can scale quite easily. It can be adapted to a large network by simply replicating its components. All that is needed is to properly set up relationships between each *Auditor* and its controlled access points. The resulting system is flexible, customizable, extensible, and can easily integrate other common off-the-shelf tools. In addition, it provides automated support for the execution of complex tasks which require the results obtained from several different tools.

3.1. Implementing access control

Available technologies implement wireless security and authentication/access control mechanisms such as IEEE 802.11i and IEEE 802.1x to prevent the possibility of unauthorized users gaining undue access to a protected network. Unfortunately, this arrangement falls short as a tool to avoid the inside spread of hostile agents, since in the modern nomadic networking environment authentication and authorization are no longer sufficient and need to be completed with a strong security assessment aiming at clearly

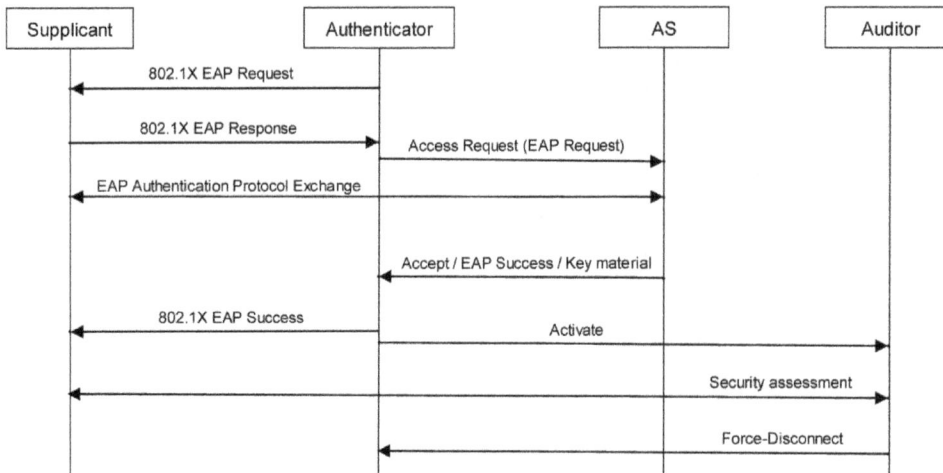

Fig. 2. Simplified access control scenario.

	Octet Number
Protocol Version	1
Packet Type	2
Packet Body Length	3-4
Identifier	5-8
Client MAC Address	9-14
Client IP Address	15-18

Fig. 3. The *Auditor* <-> access point message interface.

highlighting any potential menace. Therefore our proposed security strategy operates according to the scenario described in Fig. 2.

Four entities are involved, called the Supplicant (the wireless station), the Authenticator (the access point), the Authentication Server, and the *Auditor*. We assume, as in IEEE 802.11i, that also the access point and *Auditor* have a trustworthy UDP channel between them that can be used to exchange information and action triggering messages, ensuring authenticity, integrity and non-repudiation. After successful authentication, when a wireless device is granted access to the network, it usually issues a "DHCP Request" for an address. The local DHCP server then hands an IP address to the wireless device passing through the access point which has been modified to detect the DHCP response message (message type "DHCP ACK") and acquire knowledge about the layer-3 availability of a new device together with its IP address. Now the access point has to ask the *Auditor* for the needed security analysis.

Basically, the communication between the *Auditor* and the access point is twofold: the access point signals the *Auditor* that a new client has associated and that the assessment should begin; the Auditor may request that the access point disconnect the client in case the computed score exceed the admissibility threshold. So only two messages are necessary, respectively the "Activate" message and the "Force-Disconnect" message, whose layout is shown in Fig. 3. Port UDP/7171 is used on both the *Auditor* and the access point to exchange messages with the other party. The message interface, as shown in Fig. 3, has been carefully designed by leaving room for further extensions. For instance, message authentication may necessary to avoid potential DoS attacks based on forging such messages. The detailed message structure is reported in Fig. 3, whose involved fields are described below:

- *Protocol Version.* This field is one octet in length and represents an unsigned binary number. Its value identifies the version of protocol supported by the sender.
- *Packet Type.* This field is one octet in length and represents an unsigned binary number. Its value determines the type of packet being transmitted. The following types are defined:

    1. *Activate.* A value of 0000 0000 indicates that the message carries an Activate packet.
    2. *Force-Disconnect.* A value of 0000 0001 indicates that the message is a Force-Disconnect packet.

- *Packet Body Length.* This field is two octets in length and represents an unsigned binary number. The value of this field defines the length in octets of the packet body: a value of 0 indicates that there is no packet body.
- *Identifier.* The Identifier field is one octet in length and allows matching of responses with requests.
- *Client MAC Address.* This field is six octets in length and specifies the MAC address of the client.
- *Client IP Address.* This field is four octets in length and indicates the IP address of the client.

No acknowledgement messages have been provided because the communication between the access point and the *Auditor* is asynchronous and no one of the involved entities need information about the completion of task by the other. In addition, synchronous communication would have introduced unacceptable latency.

3.2. Passive monitoring

The *Auditor* also acts as a passive malicious activity detector. This is a necessary facility, because in some cases the analysis can take too much time for mobile users, and so the decision adopted has been to grant immediate access to users, without waiting for the assessment completion. This may open a security hole, because a mobile client can immediately be active to propagate malware, but this activity would be immediately noticed by the passive detector. In fact, if, at any time, some hostile activity such as scanning, flooding or known attacks is discovered from an associated client, the *Auditor* requests the access point to disconnect it. To perform passive monitoring the *Auditor* must be equipped with two network interfaces. The first one is used to communicate with the connected access points, while the other will be dedicated to passively listen (to discover illegitimate activity) all the traffic generated by its associated Authenticators/access points. This mechanism, that is implemented through port mirroring on the underlying network infrastructure, allows the *Auditor* to continuously analyze, for each connected node, some properly chosen traffic health parameters, directly reflecting the network behaviour in presence of various malicious/hostile activities (massive scans, or DoS attacks), outbreaks (viruses and worms propagation) or other botnet-related services, and checking them against a "sanity" per-time limit threshold.

The most significant parameters used in the prototypal passive monitoring facility are the outgoing flow and the connection failure rate. The outgoing flow of each node represents the number of outbound nodes that an internal machine may contact per time interval. This choice is motivated from the observation that during normal operation the volume of outbound flows to unique machines is relatively small, and that this volume generally greatly increases in presence of an outgoing DoS attack or when a machines is performing an aggressive port or address space scan by seeking for susceptible nodes to be exploited for security vulnerabilities. By choosing a good threshold value and using some well-crafted heuristics to discriminate some classical hostile behaviour, such as trying access, eventually according to a locality principle, to the same set of ports on many hosts on a target network, the above detection technique may reach a satisfactory level of success by minimizing the number of false positives that may cause guiltless nodes to be abruptly taken out from the network. The rate of failed connection requests is another significant traffic health parameter that can be measured by monitoring the failure replies that are sent to each specific node. The failure rate measured for a normal host is likely to be low. For most Internet applications (www, telnet, ftp, etc.), a user normally types domain names instead of raw IP addresses to identify the servers. Domain names are resolved by Domain Name System (DNS) for IP addresses. If DNS cannot find the address of a given name, the application will not issue a connection request. Hence, mistyping or stale web links do not result in failed connection requests. An ICMP host-unreachable packet is returned only when the server is off-line or the DNS record is stale, which are both uncommon for popular or regularly-maintained sites (e.g., Yahoo, Google, E-bay, CNN, universities, governments, enterprises, etc.) that attract most of Internet traffic. Moreover, a frequent user typically has a list of favorite sites (servers) to which most of his/her connections are made. Since those sites are known to work most of the time, the failure rate for such a user is likely to be low. If a connection fails due to network congestion, it does not affect the measurement of the failure rate because no "ICMP host-unreachable" or "RST" packet is returned. On the other hand, the failure rate measured for a compromised host, performing scanning activities, is likely to be high. Unlike normal traffic, most connection requests, initiated in such a situation, are doomed to fail because the destination addresses are randomly picked, and in addition, the associated ports may not be in use, without any service listening on it. Accordingly, by passively monitoring the outgoing flow and outgoing failure rate from the traffic flow counters associated to all the admitted nodes, the *Auditor* can easily identify potential offending hosts, such those belonging to BotNets and participating to Distributed DoS attacks, also after successfully passing the security assessment phase. Thus, if any of the measured activity indicators for one of these host exceeds a pre-configured alarm threshold (set at the system configuration time), the active *Auditor* immediately sends the "Force-Disconnect" message to the Authenticator, causing the involved mobile node to be immediately dropped from the wireless network.

The passive monitoring facility can be extended by integrating within the *Auditor* a very simple and flexible network intrusion detection system (NIDS) such as **snort** (open-source intrusion detection system) [28] to perform in a more sophisticated way real-time traffic and protocol analysis, signature matching against an up-to-date signature database and packet logging. Anyway, since passive monitoring is essentially based on the on-line analysis of outgoing traffic patterns, it can also be effective in identifying any "noisy" (i.e. volume-based) unknown (a.k.a. "zero-day") attacks originated by the mobile nodes under observation.

4. The security assessment process

Security assessment is a complicated, expensive and time-consuming process that consists in many steps and provides input to the *Auditor*'s risk evaluation needs in form of valid and reliable data about several security aspects of the involved node. Because of the impossibility to directly measure the risk associated to the admission of an unknown mobile node, different factors that may affect its security degree have to be measured, combined and correlated in an effective security assessment methodology. Thus, in order to examine the adequacy of the current security degree of a specific host, several different aspects and security-relevant properties within the scope of the assessment process have to be considered.

All the following tasks produce several results that are often measured on different scales, which makes them difficult to be compared. Depending on the targets of security measurements and the needed security level granularity several different approaches to security assessment can be identified:

- *Observation*: implies that the target to be analyzed is only viewed from the outside, while its internal characteristics are not taken into consideration at all.
- *Security testing*: describes a security assessment approach based on the extensive use of vulnerability scanners and penetration testing. The number of detected vulnerabilities in a target system under test is used to determine several security metrics. While could be many questions about the relevance of these metrics, this approach is really effective in capturing valuable information on the security strength of the examined nodes [16]. However, this methodology is highly dependent on the tools used in the penetration testing process and on the skills of the IT experts choosing the specific tests to define the needed metrics. Hence, security testing alone is a useful and effective technique for identification of vulnerabilities but lacks the "total picture" view of formal evaluation [4].
- *Security functionality structure*: in order for the assessment to stress security relevant characteristics of the particular node under examination, the meaning of the security functionalities and defense countermeasures implemented on the target network has to be well defined. This applies to the verification of the compliance to some specific security policies and controls through the correlation of specific security metrics [34]. Such security metrics monitor the accomplishments of the policies goals and objectives by quantifying their possible degree of violation by the target host. Magnitude, scale and interpretation are three important elements of a security metric that need to be decided for a successful correlation [26]. Most of the methods developed to approach security assessment have enormous difficulties to capture the complexity of the process because of the quantity of different measurable properties that are involved.

The automatic assessment performed by the *Auditor* has the objective of characterizing the security of a nomadic node when requiring access to the network, by using several metrics whose combined observation results in different clearly quantified security indexes, each one associated to a specific security perspective. Such indexes, combined into a multi-dimensional score in the form of a vector representing the specific security profile of the involved node, can be easily compared against the minimal security requirements of the target infrastructure to perform admission control. The underlying methodology takes into account the specific security properties and criteria associated to the individual evaluation metrics and combines them within a structured hierarchical decision process which assess the relative importance of these criteria, comparing the alternatives for each of them, and determining an overall ranking associated to the confidence according to which each criterion will provide correct information about the security index describing a particular security aspect. The information about the most important properties will be put together to form a relevance matrix which is then used to generate a vector of "fundamental" security indexes that is the final multi-dimensional security score. In order to automatically estimate and analyze the associated metrics values, the aforementioned security criteria must be described in a rigorous, controllable way, according to a formal model containing a possibly complete information about the node Operating System, connection properties and deployed services (identified through port or vulnerability detection scans, or assumptions based on some configuration evidences). Such a security analysis model has to be designed by taking into account the models used by the existing remote management and vulnerability scanning tools.

The key for a successful and effective assessment is the availability of quantitative data, as complete and accurate as possible, for the generation of the above security analysis model. This implies the

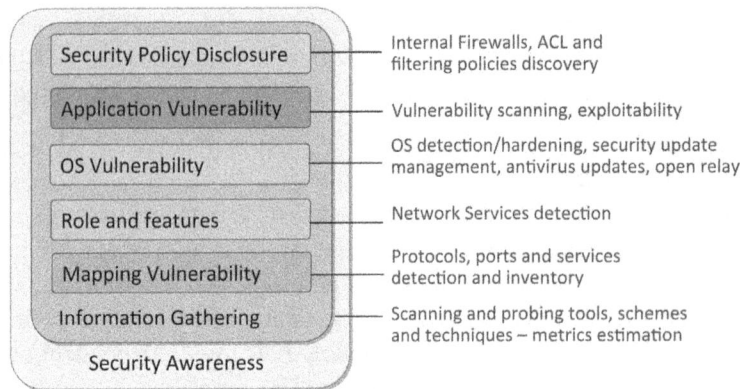

Fig. 4. Basic metrics definition model.

assignment of quantitative values to the individual security criteria and the mathematical combination of them in a way that demonstrates their relative or absolute effects on the overall system security. These effects can be generic numbers that only have bearing on each other or they can be converted into specific cost or risk values. Such a quantitative analysis can be viewed as just a refining or a partitioning of a more complex qualitative security analysis problem. It breaks down the qualitative issues into smaller factors that can have easily obtainable quantities ascribed to them. The inherent hierarchical structure of the above analysis, starting from the principle of breaking down the process into individual measurable components, requires a method to accurately estimate the weight, and hence the relevance, of each component for the determination of an accurate and quantified network security degree. AHP is an ideal decision making methodology to do this, since it allows an easier and efficient identification of the security evaluation criteria, their weighting and analysis.

4.1. Information gathering

The assessment process has been designed to be carried out according to a two phase scheme. The first phase, needed to generate a quantifiable scoring index for each fundamental network security aspect/component, requires the definition and the evaluation of a set of metrics associated with the fundamental security criteria to be considered in the context of the analysis process. The set of metrics used has been determined according to a layered approach which combines several well-known techniques and evaluation methods into a general security awareness consolidation model (see Fig. 4).

The above approach starts from the basic information gathering phase focusing on the determination of the fundamental properties of the target node in terms of service/application availability and role. It also aims at acquiring a basic visibility about the security characteristics of the node, specifically searching for the presence of potential vulnerabilities, lack of controls, weaknesses or misconfiguration problems. The above task takes advantage from a reliable and possibly complete knowledge base, which contains semantically-rich information provided by network mapping and configuration assessment facilities. Accordingly, client profiling will be accomplished through the use of specialized network analysis tools that can be used to examine the mobile node and the services running on it in a fairly non-intrusive manner. For example, the detailed information about the services active on the hosts could be determined by scanning the ports of such hosts. In addition, the results obtained from the execution of one tool are often used as the basis for additional analysis and possibly as input for the execution of other tools. That is, starting from the obtained information on the known hosts and services, more sophisticated vulnerability assessment tools can automatically perform checks on them by looking for vulnerable applications, misconfigured services, and flawed operating system versions. This analysis can thus result in the detection of services that are unnecessary or undesirable in the target environment or in the identification of explicit anomalies and system vulnerabilities. For example, if a port scan was run and if it is determined that a normally unused port, e.g., TCP port 31337, was open on a scanned host, a penalty would be added to the security score indicating that the machine had been potentially exploited and may represent a security problem. In more detail, all the information gathering activities are accomplished by applying the following techniques:

- *OS Detection.* It is of fundamental importance to assess the OS security degree and to continuously monitor its evolution with respect to system configuration or operation changes, along with application and environment evolution. For this sake, both the OS type and its current version and patching level must be reliably identified by remotely assessing the new node requiring network access. Most of the available techniques for remotely detecting the OS running on a system rely on implementation differences between OSes to identify the specific software variant. They work by using a set of network queries and a classification model. Detection is performed by issuing the queries against the networked host, typically by sending carefully crafted network packets, collecting response packets, and feeding these responses into the classification model. If the different implementations of the software predictably generate different responses to the queries, the classification model can use them to reliably identify the remote OS. Several very common OS detection techniques have been used. The first one, called *"IP stack fingerprinting"*, allows the determination of remote OS type by comparison of variations in OS IP stack implementation behaviour. Ambiguities in the RFC definitions of core Internet protocols coupled with the complexity involved in implementing a functional IP stack, enable multiple OS types (and often revisions between OS releases) to be identified remotely by generating specifically constructed packets that will invoke differentiable but repeatable behaviour between OS types, e.g. to distinguish between Linux RedHat and Microsoft Windows 7. Additionally, the pattern of listening ports discovered using service detection techniques may also indicate a specific OS type; this method is particularly applicable to "out of the box" OS installations. By determining the presence of older and possibly unpatched OS releases the *Auditor* can increase the overall insecurity score according to the locally configured policy. For example, if network administrators know that all of the legitimate clients are running Windows boxes, they will likely raise the score of an OS fingerprint indicating Linux, or the reverse. In those cases when OS fingerprinting is able to determine the operating system version, besides the type, higher scores can be assigned to the older versions. The available fingerprinting tools provide a database of thousands of reference summary data structures for known OSs that are continuously kept up-to-date by finding new probes and re-examining the existing ones.

- *Service Detection.* Also known as *"port scanning"* performs the availability detection of a TCP, UDP, or RPC service, e.g. HTTP, DNS, NIS, etc. Listening ports often imply the existence of associated applications/services running, e.g. a listening port 80/tcp often implies an active apache web server or a listening 53/udp can flag the presence of an active DNS server. There are certain vulnerabilities which can be reliably inferred from the status of a port. Many of these vulnerabilities relate to policy violations, i.e. the presence of deprecated services, from the security point of view, or, at worst ports associated to possible worm infection. Here, for each dangerous service or suspicious port detected the insecurity score is increased according to a configurable policy. Service detection will be accomplished with the nmap tool that can provide a very fast and complete assessment of the running services and in some cases their versions. Several different port scanning techniques (i.e.

traditional, sweep, half, xmas and stealth scan [7]) have been used in order to bypass internal security controls and hence to obtain a more and accurate result regarding the ports which are opened, closed, or blocked. More sophisticated techniques inherited from the *Traceroute-Like Analysis* of "Time Exceeded", "Host Unreachable" and "Network Unreachable" packet responses can be used to easily determine the list of local controls/rule sets implementing the internal access control policies of the host.

– *Vulnerability Detection.* Based on the results of the previous probes, that attempt to determine what services are running, more sophisticated scanners (e.g. [5]) with semantic knowledge about up-to-date security issues, try to exploit known vulnerabilities of each service to test the overall system security. Here, the concept of vulnerability is used in a loose sense which includes not only software design errors and implementation flaws but also misconfigurations and questionable user decisions such as using a weak password. More specifically, vulnerability detection techniques work by searching for the presence of default accounts, form validation errors, insecure **CGI-BIN** files, test Web pages, lack of bound checking in service implementation code and other known vulnerabilities. This can be accomplished through a variety of means operating both at the application-level (service behaviour and protocol compliance probing, banner grabbing or exploiting stack smashing buffer overflow for malicious code execution), or at the network-level (packet forgery, hijacking TCP connections, port diversion and ARP or IP spoofing). Testing by exploit involves using a script or program designed to take advantage of a specific vulnerability. However, there is a well-known problem associated with such kind of vulnerability assessment that often prevents it from being used extensively as a basic security practice: the safety of the scanning tools used for information gathering. That is, many scanners can cause adverse effects on the network systems being tested by crashing the involved systems/services (e.g. due to the unpredictable results of exploiting a buffer overflow vulnerability), or even worse leaving permanent damaging side effects and/or undesirable modifications to the system state (e.g., by putting a plus in the "**rhosts**" file). Consequently, the analysis must be accomplished by using a "weakened" exploit code with the only sake of probing the target system to demonstrate the presence of a vulnerability, that should not leave the system itself in a vulnerable or damaged state.

– *Security Policy Discovery.* Some useful information about the implementation of internal security restrictions through personal firewalls and/or packet filters (e.g. IP tables) can be obtained by using several sophisticated ICMP and SNMP-based probing techniques. These techniques, combined with massive scanning activities (e.g. idle scanning) are really effective in guessing the list of controls/rule sets implementing the above access control policies.

The result of the aforementioned operations are retained for a configurable time interval, in order to prevent a client continuously trying to get access from consuming inordinate amounts of resources. The optimal value of this amount of time is a tradeoff between the quest for accurate and up-to-date security assessment as well as the compelling need to save resources.

The adopted philosophy is to implement all the above techniques with the tool best suited for the specific task. This choice is more effective if a tool performs one specific task instead of implementing many different functionalities as a monolithic application. Also, tools that can perform many tasks are, normally, able to limit their operation to exactly what is needed. For example, **nmap** [18] is able to perform ping scans, port scans, OS fingerprinting, and RPC scans in a flexible and very effective way, so that it can be finely tuned to suit the specific needs of each initial discovery activity. On the other hand the Metasploit framework [22], which is an advanced open-source platform providing a set of application programming interfaces (APIs) for packaging vulnerability exploit code in a fully automated fashion, has been used, along with the **OpenVAS** [5] open-source remote security scanner, for detecting vulnerable applications and services running on the available hosts and providing a warning level (to be used for quantitative assessment) for each possible vulnerability. Both the above tools provide very comprehensive and up-to-date vulnerability exploit databases, covering a large variety of architectures, operating systems and services. As a general rule, only open-source applications will be used for that purposes, in order to avoid licensing costs and legalities.

4.2. Security metrics and indexes

Another fundamental step in the security assessment of an host entering the network is to define significant variables that can be used to quantitatively characterize and evaluate its security. However network and security management objectives may substantially change between different target access infrastructures, as well as their basic security requirements and usage policies. Therefore, the security metrics considered by the proposed scheme are associated to a customizable set of security indexes, intended to be easily modified. The choice of such a set of indexes should be based on plausible assumptions in order to obtain significant results. Any choice should be an approximation of the real security degree and associated risk level, with some unavoidable trade-offs between metric plausibility and usability. As a demonstrative example, that might also be used as a starting point, what follows are the basic security indexes:

- *Network Mapping Vulnerability* Reports the host responsiveness to different known scanning techniques.
- *Operating System Vulnerability* Reports the degree of vulnerability associated to the OS used together with its patching/security upgrade status.
- *Application Disclosure, Vulnerability and Threat* These indexed are associated to the number of different vulnerable applications/services running on the hosts as well as to the type (i.e. leading to confidentiality, integrity, availability impacts) and associated risk level of different vulnerabilities.
- *Security Policies Disclosure* Number and types of security and filtering rules (e.g. access controls, anti-spoofing practices etc.).
- *Unexpected Findings* Results from the preceding analyses indicating the presence of unexpected/unauthorized OS or services.

To estimate the *Network Mapping* vulnerability index is necessary to model the basic scan responsiveness features of a node by using a vectors of binary variables $M = \{m_1, \ldots, m_7\}$ where the individual elements are defined as:

- $m_1 = 1$ if the node responds to ICMP echo requests;
- $m_2 = 1$ if the node is responsive to UDP port scans;
- $m_3 = 1$ if the node is responsive to "half scan" techniques;
- $m_4 = 1$ if the node is responsive to "stealth scan" techniques;
- $m_5 = 1$ if the node responds to TCP port scans;
- $m_6 = 1$ if the node shows its full network path when solicited with a traceroute (TTL-expiration triggered) technique;
- $m_7 = 1$ if the node shows an incomplete path when solicited as above.

Starting from the above variables the *Network mapping* vulnerability index Ψ_{NM} can be determined from the linear combination of the above variables m_i, each one weighed with its relative risk level, or security relevance μ_i to be determined through AHP as described in Section 4.3.

$$\Psi_{NM} = \sum_{i=1}^{7} \mu_i m_i \tag{1}$$

After performing a portscan, some open ports will be found to belong to well-known applications, while some others will be unknown. The *Application Disclosure* vulnerability index Ψ_{AD} can be determined by taking the weighted sum of the relative number of the well-known and of the unknown listening ports. The index can thus be defined as follows:

$$\Psi_{AD} = \lambda_1 \frac{W}{P_T + P_U} + \lambda_2 \frac{(P_T + P_U) - W}{P_T + P_U} \tag{2}$$

where W is the count of well-known applications and P_T and P_U are respectively the total number of listening ports resulting from TCP and UDP portscans, and λ_1 and λ_2 are, as usual, the ranking values associated to the relative security relevance of the metric.

As far as *OS Vulnerability* together with the *Application Vulnerability and Threat* categories are concerned, the auditing process will refer to a standardized list of known vulnerability tests such as those provided by **OpenVAS**. This approach will ensure that the vulnerability checklist will be kept up to date as new vulnerabilities are discovered, their details are published, and the corresponding tests are designed and added to the **OpenVAS** archive. Additional, personalized checks may also be added if needed, provided that the actual scanner is instructed on how to perform them.

Analogously, instead of specifying the scores manually for these classes of vulnerabilities, the associated scores (or at least a good starting value) can be advantageously derived by referencing the CVSS (Common Vulnerability Scoring System) [20] scoring for the vulnerability under consideration. CVSS is an open, jointly developed, standard providing a universal method for rating vulnerabilities and quantifying the associated risk. OpenVAS supports the CVSS scoring, that is included in the new vulnerability tests, which are dispatched from the OpenVAS database to the application instances. Most vulnerability checks in OpenVAS are associated to a CVE (Common Vulnerabilities and Exposures) [6] code and any of such codes has a CVSS. CVE is a list or dictionary that provides common names for publicly known information security vulnerabilities. CVSS consists of 3 groups: Base, Temporal and Environmental. For each group, a numeric value in the range 0–10 is given, complemented by a compressed textual representation reflecting the values used to derive the score. While the Base group is associated to intrinsic and unchanging characteristics of a vulnerability, the Temporal group reflects those qualities that change over time, and the Environmental group describes the characteristics that are unique to any given environment. The calculation of Base metric involves many aspects including the complexity of a successful exploit, the possibility of a remote exploit, details about authentication and the security impact, both in terms of confidential data exposure and in terms of reduced system integrity or availability. These details describe intrinsic qualities that do not change over time. The threats generated by a vulnerability, instead do change over time, as they depend on the confidence in the vulnerability description, and on the availability of exploits and remediation solutions.

The mapping from CVSS to the internal assessment score is an important parameter that should be customized to reflect the differences and the priorities of the various environments. In this example, a weighted combination of the Base B_j and the Temporal T_j metrics (ρ_B and ρ_T are the normalized weights for the Base and Temporal metrics, respectively) has been used to obtain a score for the j-th vulnerability belonging to the set of analyzed vulnerabilities V. Since a chain is as strong as its weakest link, the highest value is then selected.

$$\Psi_V = \max_{j \in V} \left(\rho_B B_j + \rho_T T_j \right) \tag{3}$$

Here, multiple sets V may be associated to several different indexes referring to explicit classes, e.g. *Application Vulnerabilities* (Ψ_{AV}) or *OS Vulnerabilities* (Ψ_{OV}), etc.

To calculate the *Security Policies Disclosure* index Ψ_{SD} it is necessary to combine several filtering rules-related metrics calculated from the following counters, that are updated during the initial scan operations, and reflect the number of active access controls presumably configured on internal firewalls or IP filters to protect the host:

- the number of incoming ICMP controls c_1 is incremented for each scanning action resulting in an "Admin prohibited filter" response message to an "ICMP Echo Request" or an "ICMP Info Request";
- the number of generic ICMP controls c_2 is incremented for each scanning action not resulting in any response to an "ICMP Echo Request" solicitation;
- the number of incoming TCP controls c_3 is incremented for each scanning action resulting in an "Admin prohibited filter" response message to a TCP port scan with the "SYN" or "ACK" TCP flags set;
- the number of generic TCP controls c_4 is incremented for each scanning action not resulting in any response to any TCP port scan solicitation;
- the number of incoming UDP controls c_5 is incremented for each scanning action resulting in an "Admin prohibited filter" response message to an UDP port scan solicitation;
- the number of generic UDP controls c_6 is incremented for each scanning action not resulting in any response to any UDP port scan solicitation.

Let $C = \sum_{i=1}^{6} c_i$ be the total number of detected controls, as in the previous cases, the linear combination of the individual rules/controls counters is weighted by using their relative vulnerability/risk priorities ω_i, resulting in the following equation:

$$\Psi_{SD} = \frac{1}{C} \sum_{i=1}^{6} \omega_i c_i \tag{4}$$

Finally, the *Unexpected Findings* index is computed by comparing the results from the previous analyses against a security policy specifying what the expected parameters (e.g. OS or open ports/accessible applications) are like in the environment under consideration. This metric is heavily dependent on the operating environment. For it to be effective, network administrators are requested to compile a list of OSes, applications, and open ports which they consider appropriate in their environment. Anything outside that list will be considered suspicious or harmful, that is each detected problem heavily taxes the computed insecurity score since it typically may be a very dangerous symptom of an open vulnerability. However, the presence of open ports outside an admissible range specified by the network administrator could be evaluated differently, in accordance with locally defined policies. In this context, the metrics chosen are the following:

- the u_1 metric is computed by evaluating, upon detection of an unexpected OS, the reliability of the detection itself. OS fingerprinting tools usually return a percentage which expresses the confidence in their findings;

- the number of unexpected fingerprinted applications u_2 is incremented for each detected application not found in the list of allowed/expected applications;
- the average reliability score of the unexpected fingerprinted applications u_3 is computed by taking the average of the single reliability percentages involved;
- the number of unexpected TCP open ports u_4 is incremented for each TCP port found to be responsive;
- the number of unexpected UDP open ports u_5 is incremented for each TCP port found to be responsive.

The overall index is computed by taking a linear weighted combination (with weights being γ_i) of the above metrics:

$$\Psi_{UF} = \sum_{i=1}^{5} \gamma_i u_i \tag{5}$$

4.3. The ranking methodology

In order to measure the network security indexes correctly and accurately, it is essential to explicitly quantify the importance of each component metric associated to the previously defined fundamental security criteria for the sake of the overall node security. Furthermore, when scoring the node security by combining several individual criteria, care must be taken to ensure that the method used is both simple and consistent. The chosen methodology decomposes the overall security scoring problem into simpler and more manageable sub-problems, each one specialized in ranking the criteria/metrics associated to each individual index, to form a multi-level, multi-target and multi-factors structure. The AHP allows these factors to be compared, with the importance of individual factors being relative to their effect on the problem solution. Ranking decisions, represented by priorities between individual factors, are established using pairwise comparisons determining the relative weights of the basic security evaluation metrics. Different weights will lead to different evaluation results. Such decisions, to be taken at the initial system configuration time (or "training" phase), should rely on the experience and knowledge of several competent security professional or decision makers performing pairwise comparisons between the various performance metrics associated to each criterion, and between the various criteria. Each decision maker has to indicate, according to the security objectives and policies of the target network, his/her own pairwise comparison matrices and the resulting opinions are then aggregated, i.e. for each pairwise comparison the average value is computed. The essence is to construct a matrix expressing the relative values of a set of attributes. For instance, an individual element may represent the relative importance of the presence of exploitable vulnerabilities ensuring super-user access as opposed to the possibility of remotely crashing the host. Each of these judgments is assigned a number on a scale. A commonly used scale (see [3]) is given in Table 1.

In detail, the aforementioned pairwise comparisons matrix E, where the generic element E_{ij} represents the quantitative judgment e_{ij} between the security criterion (or metric) X_i compared with the criterion X_j, is developed as follows:

$$
E = \begin{array}{c} \\ X_1 \\ X_2 \\ \vdots \\ X_n \end{array}
\begin{array}{c} X_1 \quad X_2 \quad \cdots \quad X_n \\ \left(\begin{array}{cccc} 1 & e_{12} & \cdots & e_{1n} \\ \frac{1}{e_{12}} & 1 & \cdots & e_{2n} \\ \vdots & \vdots & \vdots & \vdots \\ \frac{1}{e_{1n}} & \frac{1}{e_{2n}} & \cdots & 1 \end{array} \right) \end{array}
\quad
\begin{cases} E_{ii} = 1 \quad \forall i \\ \\ E_{ji} = \frac{1}{E_{ij}} \quad \forall E_{ij} \neq 0, j > i \end{cases}
\tag{6}
$$

Table 1
Individual judgments scale

Judgment	Rating
Extremely preferred	9
Very strongly to extremely	8
Very strongly preferred	7
Strongly to very strongly	6
Strongly preferred	5
Moderately to strongly	4
Moderately preferred	3
Equally to moderately	2
Equally preferred	1

Starting from the pairwise comparisons matrix, the relative weight ω_i of each security criterion X_i can be easily estimated according to a stepwise process. At first, the total of each column in the matrix E has to be calculated. Then each element in E must be divided by its column total. The resulting matrix $\widehat{E}$ is the normalized pairwise comparison matrix, defined as:

$$
\widehat{E} =
\begin{array}{c}
 \\
X_1 \\
X_2 \\
\vdots \\
X_n
\end{array}
\begin{array}{cccc}
X_1 & X_2 & \cdots & X_n \\
\left(\dfrac{1}{\sum\limits_{i=1}^{n} E_{i1}} \right. & \dfrac{E_{12}}{\sum\limits_{i=1}^{n} E_{i2}} & \cdots & \dfrac{E_{1n}}{\sum\limits_{i=1}^{n} E_{in}} \\
\dfrac{E_{21}}{\sum\limits_{i=1}^{n} E_{i1}} & \dfrac{1}{\sum\limits_{i=1}^{n} E_{i2}} & \cdots & \dfrac{E_{2n}}{\sum\limits_{i=1}^{n} E_{in}} \\
\vdots & \vdots & \vdots & \vdots \\
\dfrac{E_{n1}}{\sum\limits_{i=1}^{n} E_{i1}} & \dfrac{E_{n2}}{\sum\limits_{i=1}^{n} E_{i2}} & \cdots & \left. \dfrac{1}{\sum\limits_{i=1}^{n} E_{in}} \right)
\end{array}
\tag{7}
$$

The next step is to carry out the calculation of the individual weights ω_i. A priority vector $\vec{\omega}^T = (\omega_1, \ldots, \omega_n)$, that represents the relative weight of each performance indicator, can be directly determined from the average of the elements in each row of the normalized pairwise comparison matrix $\widehat{E}$:

$$
\omega_i = \frac{\sum\limits_{j=1}^{n} \widehat{E}_{ij}}{n}
\tag{8}
$$

Using these weights, it is possible to estimate the value of each security index S as a linear combination of the associated security criteria/metric (X_i), each contributing with its relative importance to the overall evaluation (ω_i):

$$
S = \sum_{i=1}^{n} \omega_i X_i
\tag{9}
$$

Once obtained the needed ranking values the last step consists in measuring the consistency of the pairwise comparison judgments used as inputs of the whole ranking process (the so called *Consistency Ratio* or shortly C_R). In other words, this step evaluates the quality of the judgments established in the training phase and reported within the pairwise comparison matrix. Since the final weights strongly depend on the pairwise comparison matrix, it is important to guarantee that the degree of inconsistency among the pairwise comparisons is relatively small. Typically a ratio exceeding 0.10 indicates inconsistent judgments, whereas, C_R values lower than 0.10 are considered reasonable. When the C_R is above 0.10, the pairwise comparisons matrix needs to be re-evaluated to reduce the degree of inconsistency.

To determine the C_R, the weighted sum vector $W = \{w_1, \ldots, w_n\}$ has to be estimated by multiplying each column element of the pairwise comparison matrix E by the relative weight of each item reported in $\vec{\omega}^T$ and summing the resulting values across the rows to obtain:

$$\vec{W} = E \times \vec{\omega} \tag{10}$$

The calculation method is based on Eigenvalue Analysis [13] [11]. More precisely, the inconsistency for a judgment matrix can be estimated as a function of its largest eigenvalue $\lambda_{\max}$ and the order n of the matrix. According to [29] the maximum eigenvalue $\lambda_{\max}$ would be

$$\lambda_{\max} = \frac{\sum_{j=1}^{n} \left(\frac{w_j}{\omega_j} \right)}{n} \tag{11}$$

The aforementioned relative weight (or priority) vectors $\vec{\omega}^T$ can be also viewed as the most relevant eigenvectors of the normalized pairwise comparison matrix. They are represented in their distributive form, that is, they must be normalized by dividing each element of the eigenvector by the sum of its elements so that they sum to 1. The $\vec{\omega}^T$ vectors can be transformed to their idealized form by selecting the largest element in the vector and dividing all the elements by it, so that the largest element assumes value 1, whereas the others assume proportionally lower values. The above relevant eigenvectors $\vec{V}$ can be determined by resolving:

$$(E - \lambda_{\max} I) \times \vec{V} = 0 \tag{12}$$

The consistency ratio C_R is then calculated from another quantity known as the *Consistency Index* C_I [30] defined as:

$$C_I = \frac{(\lambda_{\max} - n)}{n - 1} \tag{13}$$

by dividing it by the value R_I, or *Random Index*, that is the consistency index of a randomly generated pairwise comparison matrix, depending on the number of elements being compared.

$$C_R = \frac{C_I}{R_I} \tag{14}$$

The overall hierarchical structure of the AHP decision process is depicted in Fig. 5, where some example ranking values for the metrics associated to the aforementioned security indexes have been reported.

4.4. Performing the final comparison

The information gathering, metric estimation and ranking processes result into a number m of different security indexes $\{S_1, \ldots, S_m\}$ (that is $\{\Psi_{NM}, \Psi_{AD}, \Psi_{OV}, \Psi_{AV}, \Psi_{SD}, \Psi_{UF}\}$ in the case of the previously presented metrics framework), each one associated to a specific security perspective or point

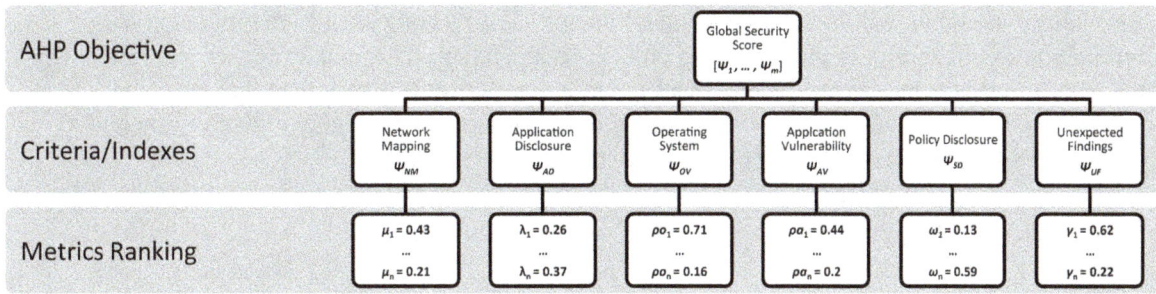

Fig. 5. The AHP decision process schema.

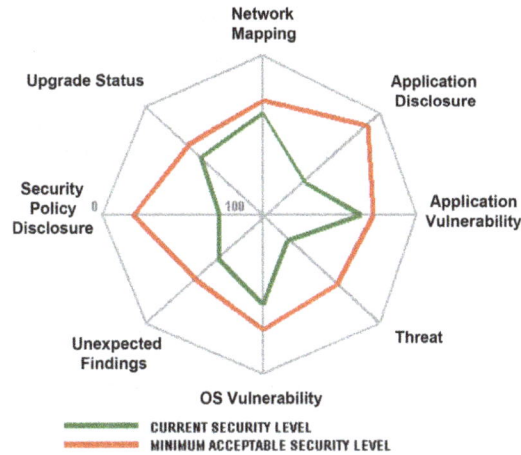

Fig. 6. Kiviat diagram representation of a comparison case.

of observation, combined into a vector $\vec{S}$ that can be viewed as the multi-dimensional score representing the actual security degree of the assessed mobile node.

In the last phase, to take the final decision for admission control, the obtained security score must be compared with the acceptance threshold configured at the *Auditor* level and, if it does not fall below the tolerable insecurity limit defined by the threshold, the node must be taken off from the network and denied further access.

However, both the security score and the threshold are multi-dimensional values and an element-by-element strict comparison against the threshold vector may be too simplistic and may lead to incorrect results. To clarify this issue, the comparison process can be modeled through the Kiviat diagram reported in Fig. 6, where the key idea is to keep the intersection between the current measurement area and the acceptance limit at its highest possible value but ensuring, however, a certain tolerance to very small noncompliance areas.

To cope with this problem, and be more confident in the comparison results, an alternative distance metric, such as the Mahalanobis distance [19] can be used to summarize the multi-dimensional characteristics into a single scale. The Mahalanobis distance is known as an effective and scale-invariant method for determining the similarity between an unknown sample vector and a known one. Such distance metric is based on an outlier detection scheme which considers the variance and covariance of the measured elements rather than just their average values. It calculates the distances, in units of standard deviation, from the group mean and weights the differences within their range of variability accounting for natural variations within the data to be compared. It also compensates the interactions between different elements that are not guaranteed to be fully independent. This makes it more general and preferable to a traditional Euclidean distance that is more appropriate for variables that have the same variance and are uncorrelated. In detail let $\vec{x}$ and $\vec{y}$ be the vectors to be compared and C their covariance matrix, the Mahalanobis distance D_M can be expressed as:

$$D_M(\vec{x}, \vec{y}) = \sqrt{(\vec{x} - \vec{y})^T C^{-1}(\vec{x} - \vec{y})} \tag{15}$$

5. Proof of concept implementation

A simple proof-of-concept implementation of the proposed assessment-based admission control system for Wi-Fi networks has been developed to test the effectiveness of the aforementioned security enforcement strategy, with an emphasis on the use of currently available devices and open-source components. The Authenticator has been implemented on a modified Linksys access point, starting from its publicly available Linux-based firmware. Accordingly, the following new functions have been introduced:

- accept, decode and verify the new protocol messages;
- detect the "DHCP ACK" message and send the "Activate" message;
- disconnect the client upon reception of a "Force-Disconnect" message.

On the other side, the *Auditor* has been implemented on a simple Intel-based laptop PC running Linux with **nmap**, **OpenVAS**, **Metasploit** and **snort** applications installed. Although this implementation it is only a "proof of concept", the simple testbed implemented demonstrated the correct operation of the proposed security framework. In this testbed, the operations of the admission control mechanism has been verified, since the continuous vulnerability monitoring can be viewed as a subordinate case. Unless vulnerability checks are updated, the same tests are performed in both cases, so we tested them only once. An admission test has been performed with 70 wireless devices, where 54 of them were running Windows and 16 some flavors of Linux. The devices were, then, manually analyzed to verify their state of vulnerability, and the results were given to a panel of 5 experts who classified the devices according to their opinion about whether access should have been granted or not. The devices were then graded on the basis of the experts ranking, who deemed them acceptable according to the following scale:

- 5 out of 5: Safe – the device should be admitted;
- 4 out of 5: Reasonably Safe – unsure, but the device should be admitted;
- 2, 3 out of 5: Unsafe – the device should not be admitted;
- 0, 1 out of 5: Dangerous, the device should definitely be dismissed.

Of the 70 devices, 13 were graded as "Safe" and 11 as "Reasonably Safe". Thus, 24 of them received a grade $\geqslant 4$ and would thus have been admitted. The *Auditor* conceded access to 22 devices, none of them was graded "Unsafe" or "Dangerous". In addition, all 13 devices graded "Safe" were admitted. Although this experiment does not represent a conclusive test, its results are encouraging.

6. Conclusions

As millions of users migrate between home, office, airports or railway stations and bookstore, they take with them not only their computer, but also other electronic hitchhikers such as fast propagating malware they picked up elsewhere, threatening the integrity of all the network environments they access in. This problem will only be exacerbated as wireless coverage expands and nomadic behaviour becomes more and more common. This is a fundamental security threat that must be effectively coped with. Accordingly, this work addresses the problem of authenticated users bringing vulnerable or infected machines into a secure network. In a nomadic wireless environment where machines are not under the control of the network administrator, there is no guarantee that operating system patches are correctly applied and anti-malware definitions are updated.

The paper proposes an architecture based on an automatic security assessment framework aiming at evaluating the security characteristics of the inspected host by identifying its vulnerabilities from the outside without any prerequisite information undertaken on it. The proposed architecture can scale quite easily. It can be adapted to a large network by simply replicating its components. All that is needed is to properly set up relationships between each *Auditor* and its controlled access points. The ability to dynamically estimate and compare the security of each new node at its first network access time may lead to increased awareness and improved possibility to apply specific admission control policies where they are needed. Areas for future research and development include better user interaction. Disconnection, and its reason, should be explicitly notified to users so that they would know what happened and the appropriate actions required to fix the problem. After receiving such notification, users should also be given the possibility to download a complete report about the detected vulnerabilities. If a downright disconnection is felt to be too draconian and some restricted environment should be created, a tighter integration with DHCP would also be called for.

References

[1] M.S. Ahmed, E. Al-Shaer and L. Khan, A Novel Quantitative Approach For Measuring Network Security, In *INFOCOM*, pages 1957–1965. IEEE, 2008.

[2] U. Aickelin, P.J. Bentley, S. Cayzer, J. Kim and J. McLeod, Danger Theory: The Link between AIS and IDS? in: *ICARIS*, J. Timmis, P.J. Bentley and E. Hart, eds, volume 2787 of *Lecture Notes in Computer Science*, pages 147–155. Springer, 2003.

[3] D.R. Anderson, D.J. Sweeney and T.A. Williams, *Quantitative methods for business*, South-Western Pub, Cincinnati, Ohio, 2000.

[4] M.A. Bishop, *The Art and Science of Computer Security*, Addison-Wesley Longman Publishing Co., Inc., Boston, MA, USA, 2002.

[5] T. Brown, OpenVAS – The Open Vulnerability Assessment System (OpenVAS). http://www.openvas.org/.

[6] T.M. Corporation, CVE – Common Vulnerabilities and Exposures. http://cve.mitre.org.

[7] M. de Vivo, E. Carrasco, G. Isern and G.O. de Vivo, A review of port scanning techniques, *Computer Communication Review* **29**(2) (1999), 41–48.

[8] N. Dimmock, A. Belokosztolszki, D.M. Eyers, J. Bacon and K. Moody, Using trust and risk in role-based access control policies, In Trent Jaeger and Elena Ferrari, editors, *SACMAT*, pages 156–162. ACM, 2004.

[9] P. Fülöp, S. Imre, S. Szabó and T. Szálka, Accurate mobility modeling and location prediction based on pattern analysis of handover series in mobile networks, *Mobile Information Systems* **5**(3) (2009), 255–289.

[10] A. Gaddah and T. Kunz, Extending mobility to publish/subscribe systems using a pro-active caching approach, *Mobile Information Systems* **6**(4) (2010), 293–324.

[11] C. Godsil and G.F. Royle, Algebraic Graph Theory, http://www.amazon.com/exec/obidos/redirect?path=ASIN/ 0387952209, April 2001.

[12] A. Hess and N. Karowski, Automated protection of end-systems against known attacks. In *Proceedings of IEEE/IST Workshop on Monitoring, Attack Detection and Mitigation,(Tuebingen, Germany)*, 2006.

[13] R.A. Horn and C.R. Johnson, *Matrix analysis*, Cambridge University Press, 1990.

[14] J. Kim, P.J. Bentley, U. Aickelin, J. Greensmith, G. Tedesco and J. Twycross, Immune system approaches to intrusion detection – a review, *Natural Computing* **6**(4) (2007), 413–466.

[15] K. Knorr, Dynamic Access Control through Petri Net Workflows, In *ACSAC*, pages 159–167. IEEE Computer Society, 2000.

[16] D. Levin, Lessons learned in using live red teams in ia experiments, In *DISCEX (1)*, pages 110–119. IEEE Computer Society, 2003.

[17] D. Liu, L.J. Camp, X. Wang and L. Wang, Using Budget-Based Access Control to Manage Operational Risks Caused by Insiders, *Journal of Wireless Mobile Networks, Ubiquitous Computing, and Dependable Applications* **1**(1) (2010), 29–45.

[18] G.F. Lyon, NMAP – Free Security Scanner For Network Exploration and Hacking. http://nmap.org.

[19] P.C. Mahalanobis, On the generalised distance in statistics, In *Proceedings National Institute of Science, India*, volume 2, pages 49–55, April 1936.

[20] P. Mell, K. Scarfone and S. Romanosky, A Complete Guide to the Common Vulnerability Scoring System Version 2.0, http://www.first.org/cvss/cvss-guide.html.

[21] M.E. Mohamed, B.B. Samir and A. Abdullah, Immune Multiagent System for Network Intrusion Detection using Non-linear Classification Algorithm, *International Journal of Computer Applications* **12**(7) (2010), 7–12.

[22] H.D. Moore, Metasploit – free, open source penetration testing solution, http://www.metasploit.com/.

[23] F. Palmieri and U. Fiore, Audit-Based Access Control in Nomadic Wireless Environments, in: *ICCSA (3)*, volume 3982 of *Lecture Notes in Computer Science*, M.L. Gavrilova, O. Gervasi, V. Kumar, C.J.K. Tan, D. Taniar, A. Laganà, Y. Mun and H. Choo, eds, Springer, 2006, pp. 537–545.

[24] F. Palmieri and U. Fiore, Automated detection and containment of worms and viruses into heterogeneous networks: a simple network immune system, *Int J Wire Mob Comput* **2** (May 2007), 47–58.

[25] F. Palmieri and U. Fiore, Network anomaly detection through nonlinear analysis, *Computers & Security* **29**(7) (2010), 737–755.

[26] S.C. Payne, A Guide to Security Metrics, Technical report, SANS Security Essentials GSEC Practical Assignment Version 1.2e, June 2006.

[27] V. Pham, E. Larsen, Ø. Kure and P. Engelstad, Routing of internal MANET traffic over external networks, *Mobile Information Systems* **5**(3) (2009), 291–311.

[28] M. Roesch, SNORT – A free lightweight network intrusion detection system for UNIX and Windows. http://www.snort.org.

[29] T.L. Saaty, *Decision Making for Leaders: The Analytic Hierarchy Process for Decisions in a Complex World*, RWS Publications, Pittsburgh, Pennsylvania, 1999.

[30] T.L. Saaty, Decision making – The Analytic Hierarchy and Network Processes (AHP/ANP), *Journal of Systems Science and Systems Engineering* **13** (2004), 1–35. 10.1007/s11518-006-0151-5.

[31] E. Shakshuki, X. Xing and T.R. Sheltami, Fault reconnaissance agent for sensor networks, *Mobile Information Systems* **6**(3) (2010), 229–247.

[32] Y.Q. Shi, T. Li, W. Chen and Y.M. Fu, Network Security Situation Prediction Using Artificial Immune System and Phase Space Reconstruction, *Applied Mechanics and Materials* **44–47** (2011), 3662–3666.

[33] J. Sun, Y. Wang, H. Si, J. Yuan and X. Shan, Aggregate Human Mobility Modeling Using Principal Component Analysis, *Journal of Wireless Mobile Networks, Ubiquitous Computing, and Dependable Applications* **1**(2/3) (2010), 83–95.

[34] M. Swanson, N. Bartol, J. Sabato, J. Hash and L. Graffo, Security Metrics Guide for Information Technology Systems, Technical report, National Institute of Standards and Technology (NIST) Special Publication 800-55, 2007.

[35] L. Teo, G.-J. Ahn and Y. Zheng, Dynamic and risk-aware network access management. In *SACMAT*, pages 217–230. ACM, 2003.

[36] R.K. Thomas and R.S. Sandhu, Towards a task-based paradigm for flexible and adaptable access control in distributed applications, In *Proceedings on the 1992-1993 workshop on New security paradigms*, NSPW '92-93, pages 138–142, New York, NY, USA, 1993. ACM.

Francesco Palmieri is an assistant professor at the Engineering Faculty of the Second University of Napoli, Italy. His major research interests concern high performance and evolutionary networking protocols and architectures, routing algorithms and network security. Since 1989, he has worked for several International companies on networking-related projects and, starting from 1997, and until 2010 he has been the Director of the telecommunication and networking division of the Federico II University, in Napoli, Italy. He has been closely involved with the development of the Internet in Italy as a senior member of the Technical-Scientific Advisory Committee and of the CSIRT of the Italian NREN GARR. He has published a significant number of papers in leading technical journals and conferences and given many invited talks and keynote speeches.

Ugo Fiore leads the Network Operations Center at the Federico II University, in Naples. He began his career with Italian National Council for Research and has also more than 10 years of experience in the industry, developing software support systems for telco operators. His research interests focus on optimization techniques and algorithms aiming at improving the performance of high-speed core networks. He is also actively pursuing two other research directions: the application of nonlinear techniques to the analysis and classification of traffic; security-related algorithms and protocols.

Aniello Castiglione joined the Dipartimento di Informatica ed Applicazioni "R. M. Capocelli" of the University of Salerno in February 2006. He received a degree in Computer Science and his Ph.D. in Computer Science from the same university. He serve as a reviewer for several international journals (Elsevier, Hindawi, IEEE, Springer) and he has been a member of international conference committees. He is a Member of various associations, including: IEEE (Institute of Electrical and Electronics Engineers), of ACM (Association for Computing Machinery), of IEEE Computer Society, of IEEE Communications Society, of GRIN (Gruppo di Informatica) and of IISFA (International Information System Forensics Association, Italian Chapter). He is a Fellow of FSF (Free Software Foundation) as well as FSFE (Free Software Foundation Europe). For many years, he has been involved in forensic investigations, collaborating with several Law Enforcement agencies as a consultant. His research interests include Data Security, Communication Networks, Digital Forensics, Computer Forensics, Security and Privacy, Security Standards and Cryptography.

2

IEEE 802.15.4 modifications and their impact

M. Goyal, W. Xie and H. Hosseini
Department of Computer Science, University of Wisconsin Milwaukee, Milwaukee WI 53201, USA
E-mail: {mukul,wxie,hosseini}@uwm.edu

Abstract. IEEE 802.15.4 is a popular choice for MAC/PHY protocols in low power and low data rate wireless sensor networks. In this paper, we suggest several modifications to beaconless IEEE 802.15.4 MAC operation and evaluate their impact on the performance via stochastic modeling and simulations. We found that the utility of these modifications is strongly dependent on the traffic load on the network. Accordingly, we make recommendations regarding how these modifications should be used in view of the prevalent traffic load on the network.

1. Introduction

IEEE 802.15.4 [1] provides *physical* (PHY) and *medium access control* (MAC) layer functionality in low power and low data rate wireless sensor networks (WSN). Wireless communication among sensor devices, enabled by IEEE 802.15.4 technology, is increasingly replacing the existing *wired* technology in a wide range of monitoring and control applications in home, urban, building and industrial environments [2–7,10,12,13,15–23]. IEEE 802.15.4 MAC operation is based on *carrier sense multiple access* with *collision avoidance* (CSMA/CA). Thus, an IEEE 802.15.4 node competes with all the nodes in its radio range for access to the transmission channel.

The CSMA/CA based networks are known to suffer from performance deterioration with increase in the number of nodes competing for channel access at any given time, which in turn depends on both the total number of nodes in each other's radio range and their packet generation rates. Future IEEE 802.15.4 networks may consist of several thousand nodes distributed over a large area and an individual node may possibly have hundreds of nodes in its radio range [3]. Also, it is not uncommon for IEEE 802.15.4 nodes to have high packet generation rates over some time intervals. A combination of large number of nodes in each other's radio range and high packet generation rates may have a severe impact on the performance of the wireless sensor network in terms of the packet loss probability and the packet latency.

In [9], we developed a stochastic model for popular *beaconless* operation of IEEE 802.15.4 MAC protocol. Given the number of nodes competing for channel access and their packet generation rates, the model can very accurately predict the packet loss probability as well as the packet latency and thus serve as a useful tool in the design of large scale wireless sensor networks based on IEEE 802.15.4. In this paper, we use this model to investigate the efficacy of various proposals to improve the performance of becaonless IEEE 802.15.4 MAC protocol.

The rest of the paper is organized as follows. Next two sections present background information regarding IEEE 802.15.4. Section 2 presents a brief overview of beaconless IEEE 802.15.4 MAC operation and describes *channel access failure* and *collision failure* that a node may encounter in the process of sending a packet. Section 3 describes different ways a packet collision may occur in the operation of beaconless IEEE 802.15.4 network. In Section 4, we describe the stochastic model presented in [9]. Section 5 describes several modifications to IEEE 802.15.4 beaconless MAC operation and use the stochastic model to evaluate their impact on performance. Based on this evaluation, we make recommendations regarding the modifications to be used for a given traffic load on the network. We demonstrate that these recommendations, derived using the stochastic model, result in better performance in simulations as well. Finally, Section 6 concludes the paper.

2. About IEEE 802.15.4

As mentioned before, IEEE 802.15.4 protocol provides PHY and MAC layer functionality in low power and low data rate wireless sensor networks. Typically, IEEE 802.15.4 constitutes the PHY/MAC layer of a larger protocol suite (e.g. Zigbee [24]), where the upper layers provide multi-hop routing and other functionality to allow formation of large-scale wireless sensor networks. In the following discussion, we assume the popular 2.4 GHz range operation of IEEE 802.15.4 PHY layer, where the information is sent 4 bits, or a *symbol*, at a time and maximum data rate is 250 kbps (or 62500 symbols per second). Various time durations are expressed in terms of symbols.

IEEE 802.15.4 MAC operation has two modes – *beacon-enabled* and *beaconless*. The beacon-enabled mode allows splitting of time into multiple *active* durations with a *cluster* [1] having exclusive access to the transmission channel during its active duration. The coordinator broadcasts a *beacon* to inform other nodes in the cluster about the beginning of the cluster's active duration. The cluster nodes compete for channel access during their active period using a *slotted* CSMA/CA algorithm. In the beaconless operation, there is no division of time into active durations and a node competes for channel access with other nodes in its radio range using an unslotted CSMA/CA algorithm. In this paper, we have focussed on the beaconless operation of IEEE 802.15.4 MAC layer.

As per the unslotted CSMA/CA algorithm, the source node begins a transmission attempt with a CSMA wait for a random number of *backoff* periods (= 20 symbols each) between 0 and $2^{BE} - 1$. Here, *BE* refers to a variable called the *backoff exponent* that is initially set to the value of *macMinBE* parameter (by default 3). After the CSMA wait, the source node determines if the channel is available for transmission. This *clear channel assessment* (CCA) is performed over a time duration of 8 symbols. If the CCA fails (i.e. the channel is found to be busy), the node increments *BE* (up to the value of *macMaxBE* parameter, which is 5 by default) and repeats the procedure. If the CCA fails even after *macMaxCSMABackoffs* (by default 4) CSMA waits, a *channel access failure* is declared and any further attempt to transmit the packet is abandoned. If the CCA succeeds, the source node performs an *RX-to-TX* turnaround [2] and transmits the packet. The propagation delay for the packet is expected to be negligible. On receiving the packet, the destination node performs an *RX-to-TX* turnaround and sends the acknowledgement (ACK) if required by the source. No CSMA wait is performed for ACK transmission. As discussed in Section 3, the transmitted packet or its ACK may suffer a collision. In this case, the source node waits for the *macAckWaitDuration* (typically 54 symbols for 2.4GHz operation) for the ACK to arrive and then proceeds with next attempt to transmit the packet. The source node can make upto *macMaxFrameRetries* (3 by default) further attempts to transmit the packet and receive the ACK. The failure to receive an ACK

---

[1] consisting of a *coordinator* and its *associated* nodes

[2] The IEEE 802.15.4 nodes are typically *half-duplex* in nature, i.e. they can not perform both the *transmit* (TX) and *receive* (RX) operations at the same time. The IEEE 802.15.4 specification [1] requires the *RX-to-TX* or *TX-to-RX* turnaround time to be 12 symbols or less.

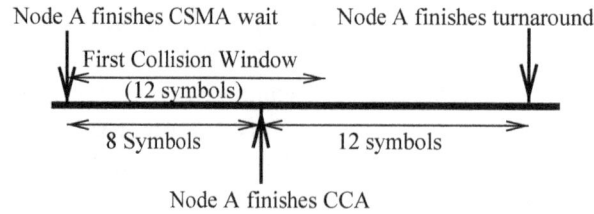

Fig. 1. First Collision Window.

even after *macMaxFrameRetries* +1 attempts causes the IEEE 802.15.4 MAC layer to accept failure in sending the packet. Such a failure is referred to as *collision failure* in the following discussion.

3. Packet collision scenarios in beaconless IEEE 802.15.4 networks

Even though the IEEE 802.15.4 nodes use a *collision avoidance* algorithm to compete for channel access, collisions do occur for reasons described next.

Hidden nodes: Some nodes in a WSN may not be in the hearing range of a node (say node X) and hence may transmit a packet at the same time as node X. Such nodes are called *hidden* nodes for node X. However, if node Y, the destination of node X's transmissions, can hear these hidden nodes, any concurrent transmission by a hidden node would cause node Y to drop node X's transmission.

Collisions due to turnaround time: As mentioned earlier, an IEEE 802.15.4 node may take upto 12 symbols to turn around from *RX* mode to *TX* mode and vice-versa. This non-negligible turnaround time may cause packet collisions to take place in the following situations:

- Suppose, a number of nodes, all in each other's hearing range, are competing for channel access and all of them are doing the CSMA wait at a certain time, hence the transmission channel is idle. Suppose, node A is the first node to wake up at time t. Node A performs a CCA till time $t + 8$, which is guaranteed to succeed, and then performs an *RX-to-TX* turnaround that finishes at time $t + 20$. The transmission channel would continue to be idle until time $t + 20$ when node A begins its packet transmission. Thus, if another node finishes its CSMA wait between times t and $t + 12$, its CCA would succeed and its subsequent packet transmission would collide with that of node A. Figure 1 refers to this 12 symbol duration as the *first collision window*. Note that the first collision window is actually equal to the *RX-to-TX* turnaround time.
- A destination node (say B) needs to complete an *RX-to-TX* turnaround before it can send the ACK for a packet. If another node finishes its CSMA wait during the first 4 symbols of this turnaround, its CCA would succeed and its packet transmission would collide with node B's ACK. Figure 2 refers to this 4 symbol duration as the *second collision window*. Note that the second collision window is the result of CCA duration being less than the *RX-to-TX* turnaround time. Also, note that the second collision window exists only if no collision takes place in the first collision window.
- A destination node would ignore a packet transmission if it begins before the destination has completed the *TX-to-RX* turnaround after sending the ACK for the previous transmission. Even though this situation does not involve a collision, its impact is same as that of a collision.

Corruption due to PHY noise/interference: A packet transmission may get incorrigibly corrupted due to PHY level noise or interference from sources like microwave ovens or WiFi transmissions. The consequent discarding of the packet transmission by the destination has the same impact as a collision.

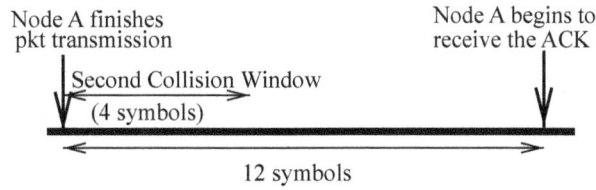

Fig. 2. Second Collision Window.

4. A stochastic model for beaconless IEEE 802.15.4 MAC operation

In [9], we developed a stochastic model to predict the packet loss probability and latency for a group of n source nodes in a network that are in each other's radio range and hence compete with each other for channel access to send packets to one or more destination nodes. The time interval, t, between two consecutive packet send events at each node is assumed to be exponentially distributed with rate $1/T$, i.e. $T = E(t)$. The model further assumes that the *clear channel assessment* (CCA) fails if there is a transmission by any node in the radio range during any part of the CCA duration; otherwise, the CCA succeeds.

The stochastic model takes n and T values as input and generates the expected values for the packet loss probability, $L(n, T)$, and the packet latency, $D(n, T)$, at steady state. Let m be the random variable denoting the number of *active* nodes at any given time. A node is considered *active* only while it has a packet to send and hence is competing for channel access with other active nodes. We denote the corresponding probability of CCA failure as $\alpha(m)$ and the corresponding probability of collision for a transmission as $\beta(m)$. The $\{\alpha, \beta\}$ values for a given m can be used to determine the corresponding packet loss probability $\lambda(m)$ and the corresponding packet latency $\delta(m)$.

The number of active nodes vary with time depending on the total number of source nodes (n) and their packet generation behavior characterized by average inter-packet interval T. The stochastic model is based on the assumption that the probability that m nodes are active at any given time is same as the probability that $m-1$ nodes get a new packet to send while an active node is sending its current packet, i.e. during a time interval equal to the steady state packet latency $D(n, T)$. Thus, for given values of n and T, m is a function of packet latency $D(n, T)$, which in turn is a function of $\delta(m)$ and hence m. The model exploits this cyclic relationship between m and $D(n, T)$ to determine $D(n, T)$, which can readily be translated to the expected value for packet loss probability $L(n, T)$.

4.1. Modeling the CSMA wait duration

During a packet transmission attempt, a node does a CSMA wait before performing the CCA. The CCA failure causes *BE* value to be incremented (up to *macMaxBE*, by default 5) and the CSMA wait to be repeated (up to *macMaxCSMABackoffs* times, by default 4). The CSMA wait duration is a randomly selected number of *backoff periods* ($= 20$ symbols each) in the range $[0, 2^{BE} - 1]$, where each number in the range is equally likely to be selected. With α as the probability of CCA failure, a node would perform on average $1 + \alpha + \alpha^2 + \alpha^3 + \alpha^4$ CSMA waits during a packet transmission attempt. Let w denote the random variable corresponding to a CSMA wait duration. With *macMinBE* and *macMaxBE* parameters at their default values (3 and 5 respectively), i.e., with 3 and 5 as the initial and maximum values of *BE*, the probability that a CSMA wait is equal to i backoff periods, where $0 \leqslant i \leqslant 31$, is:

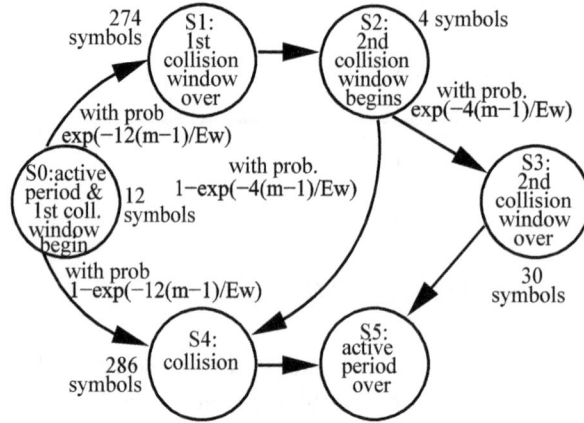

Fig. 3. Different states during the life time of an *active period*.

$$p(w = i) = \begin{cases} \frac{\frac{1}{8} + \frac{1}{16}\alpha + \frac{1}{32}\alpha^2 + \frac{1}{32}\alpha^3 + \frac{1}{32}\alpha^4}{1 + \alpha + \alpha^2 + \alpha^3 + \alpha^4} & 0 \leqslant i \leqslant 7, \\ \frac{\frac{1}{16}\alpha + \frac{1}{32}\alpha^2 + \frac{1}{32}\alpha^3 + \frac{1}{32}\alpha^4}{1 + \alpha + \alpha^2 + \alpha^3 + \alpha^4} & 8 \leqslant i \leqslant 15, \\ \frac{\frac{1}{32}\alpha^2 + \frac{1}{32}\alpha^3 + \frac{1}{32}\alpha^4}{1 + \alpha + \alpha^2 + \alpha^3 + \alpha^4} & 16 \leqslant i \leqslant 31, \\ 0 & i > 31 \end{cases}$$

The expected value of a CSMA wait duration (in terms of backoff periods) can be expressed as the following continuous and monotone increasing function of α, the probability of CCA failure:

$$E_w(\alpha) = \frac{3.5 + 7.5\alpha + 15.5\alpha^2 + 15.5\alpha^3 + 15.5\alpha^4}{1 + \alpha + \alpha^2 + \alpha^3 + \alpha^4} \tag{1}$$

Even though the CSMA wait duration is discrete in nature, we model it as a continuous random variable, $w\prime$, exponentially distributed with rate $1/E_w$, i.e. $w\prime \sim \mathrm{EXP}(1/E_w)$.

4.2. The probability of CCA failure

Consider m active nodes competing for channel access at a certain time. Suppose all the active nodes are performing their CSMA waits initially. Suppose node A is the first node to finish its CSMA wait. We mark this event as the beginning of an *active period*. In other words, node X *triggers* an active period. Node X would find the channel idle and proceed with its packet transmission. Thus, node X would have zero probability of CCA failure. The active period ends when node X, and any other node that begins its transmission during the active period, have finished their transmissions.

If another node, say node Y, finishes its CSMA wait during the active period, its CCA would fail unless node Y's CSMA wait ends during one of the *collision windows* (as identified in Section 3) associated with node X's transmission. Figure 3 shows different states during the life time of an active period under the following assumptions:

– $w\prime \sim \mathrm{EXP}(1/E_w)$ represents the individual CSMA wait time for all the nodes;
– the nodes send 133 byte long packets (including the 6 byte IEEE 802.15.4 PHY header);

- in case of a collision, the last colliding node finishes its CSMA wait just before the end of the collision window.

The *sojourn* times in various states could be explained as follows. State *S0* corresponds to the beginning of the active period as well as that of the first collision window and has a sojourn time equal to 12 symbols. These 12 symbols consist of 8 symbols of CCA by node *X* and the first 4 symbols of node *X*'s *RX-to-TX* turnaround. If at least one active node finishes its CSMA wait during the first collision window, which happens with probability $1 - e^{(-12(m-1)/E_w)}$, a collision is guaranteed. In this case, there is a transition to the *collision* state, *S4*. For simplicity, we assume that the last colliding node finishes its CSMA wait just before the end of the collision window and completes its transmission over next 286 symbols (8 symbols of CCA + 12 symbols of turnaround + 266 symbols of packet transmission), which constitute the sojourn time in state *S4*. The completion of sojourn in state *S4* concludes the active period. On the other hand, if no active node finishes its CSMA wait during the first collision window, which happens with probability $e^{-12(m-1)/E_w}$, node *X* would complete its turnaround (additional 8 symbols) and transmit the packet (266 symbols). These 274 symbols correspond to the sojourn in state *S1*.

The second collision window begins as soon as the destination successfully receives the transmitted packet and starts its *RX-to-TX* turnaround to send the ACK. This corresponds to a transition to state *S2*, where the sojourn time is 4 symbols. If at least one active node finishes its CSMA wait during the second collision window, which happens with probability $1 - e^{(-4(m-1)/E_w)}$, there is a transition to the *collision* state, *S4*. Otherwise, following the completion of 4 symbol long collision window, there is a transition to state *S3*. The sojourn time in state *S3* is 30 symbols during which the destination node completes its *RX-to-TX* turnaround that started in state *S2* (8 symbols) and transmits the 11 byte long ACK (22 symbols). The completion of the ACK transmission completes the active period.

Next, we determine different possible durations of an active period as well as the probability of occurance for each duration. We also determine the probability that an active node finishes its CSMA wait during the active period of a particular duration and the probability of CCA failure for such a node:

- If one or more nodes finish their CSMA waits during the first collision window, which happens with probability $p_1 = 1 - e^{-12(m-1)/E_w}$, the active period would sojourn over state sequence *S0-S4* and last for 298 symbols. The probability that an active node finishes its CSMA wait during such an active period is $q_1 = 1 - e^{-298/E_w}$. Since, the CCA would not fail during the collision window, the probability of CCA failure for such a node would be $\alpha_1 = (298 - 12)/298 = 286/298$.

- If no node finishes its CSMA wait during the two collision windows, which happens with probability $p_2 = e^{-12(m-1)/E_w} \times e^{-4(m-1)/E_w}$, the active period would sojourn over state sequence *S0-S1-S2-S3* and last for 320 symbols. Given that no node finishes its CSMA wait during the two collision windows ($= 12 + 4 = 16$ symbols), the probability that an active node finishes its CSMA wait during such an active period is $q_2 = 1 - e^{-304/E_w}$ and the probability of CCA failure for such a node would be $\alpha_2 = 1$.

- If no node finishes its CSMA wait during the first collision window but some nodes do so during the second collision window, which happens with probability $p_3 = e^{-12(m-1)/E_w} \times (1 - e^{-4(m-1)/E_w})$, the active period would sojourn over state sequence *S0-S1-S2-S4* and last for 576 symbols. Given that no node finishes its CSMA wait during the first collision window ($= 12$ symbols), the probability that an active node finishes its CSMA wait during such an active period is $q_3 = 1 - e^{-564/E_w}$. Since, the CCA does not take place during the first collision window but may take place during the second collision window, the probability of CCA failure for such a node would be $\alpha_3 = (564 - 4)/564 = 560/564$.

Clearly, an active node may finish its CSMA wait either by triggering a new active period or inside an ongoing active period. However, it is not necessary that all $m-1$ active nodes (excluding the one that triggered an active period) would finish their CSMA waits during an ongoing active period. Suppose q_i is the probability that a node finishes its CSMA wait during an ongoing active period, i.e. $q_i \in \{q_1, q_2, q_3\}$. Then the expected number of nodes that would finish their CSMA waits during an active period (including the node that triggers the active period) is $1 + (m-1)q_i$. Therefore, the probability that an active node triggers an active period is $1/(1 + (m-1)q_i$ and the probability that an active node finishes its CSMA wait during an ongoing active period is $(m-1)q_i/(1 + (m-1)q_i)$. Since the node triggering an active period has zero probability of CCA failure, the overall probability of CCA failure can be expressed as a function of m, the number of active nodes, and E_w, the average CSMA wait duration as follows:

$$\alpha(m, E_w) = \sum_{i=1}^{3} p_i \frac{(m-1)q_i}{1 + (m-1)q_i} \alpha_i \tag{2}$$

Note that, for a given value of m, α is a continuous function of E_w, which in turn is a continuous, monotone increasing function of α Eq. (1). This relationship can be exploited to determine the unique value of α for a given m value.

4.3. The probability of collision for a transmission

Out of $m-1$ active nodes still in the middle of their CSMA waits when an active period starts, the probability that i $(0 \leqslant i \leqslant m-1)$ nodes finish their CSMA waits during the first collision window and hence transmit during this active period is $p_{\text{coll1}}(i) = \binom{m-1}{i}(1 - e^{-12/E_w})^i (e^{-12/E_w})^{m-1-i}$. The second collision window comes into picture if no active node finishes its CSMA wait during the first collision window. The probability that i $(0 \leqslant i \leqslant m-1)$ nodes finish their CSMA waits during the second collision window and hence transmit during this active period is $p_{\text{coll2}}(i) = \binom{m-1}{i}(1 - e^{-4/E_w})^i(e^{-4/E_w})^{m-1-i}$.

An active period would consist of just one packet transmission, the one by the node that triggers the active period, if no other node finishes its CSMA wait during the two collision windows. Similarly, an active period would consist of i transmissions, including one by the node that triggers the active period, in two cases: 1) if $(i-1)$ active nodes wake up during the first collision window; 2) if no active node wakes up during the first collision window and $(i-1)$ active nodes wake up during the second collision window. Thus, the probability that i transmissions take place during an active period, including one by the node that triggers the active period, is given by:

$$p_{\text{trans}}(i) = \begin{cases} p_{\text{coll1}}(0)p_{\text{coll2}}(0) & i = 1 \\ p_{\text{coll1}}(i-1) + p_{\text{coll1}}(0)p_{\text{coll2}}(i-1) & 2 \leqslant i \leqslant m \end{cases}$$

Thus, the expected number of transmissions during an active period is $\sum_{i=1}^{m} i \times p_{\text{trans}}(i)$. Clearly, two or more transmissions during an active period imply collision for all these transmissions. Thus, the expected number of transmissions during an active period that witnesses a collision is $\sum_{i=2}^{m} i \times p_{\text{trans}}(i)$. Thus, the fraction of transmissions that experience a collision, or in other words the probability of collision for a transmission, can be described as follows:

$$\beta(m, E_w) = \left(\sum_{i=2}^{m} i \times p_{\text{trans}}(i) \right) / \left(\sum_{i=1}^{m} i \times p_{\text{trans}}(i) \right) \tag{3}$$

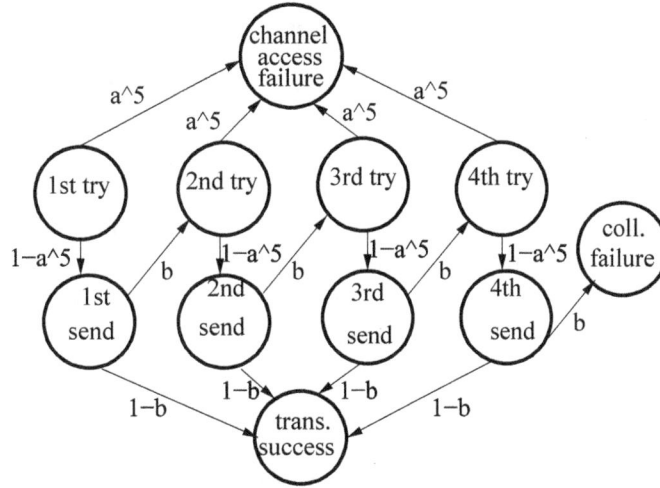

Fig. 4. A state diagram for packet transmission process in IEEE 802.15.4 MAC

As mentioned earlier, the cyclic relationship between α and E_w Eqs (2) and (1) can be used to determine the unique value of α, and hence E_w, for a given m value. Then, we can use Eq. (3) to determine the corresponding β value. Clearly, α and β are functions of m.

4.4. The probability of packet loss in IEEE 802.15.4 MAC

As discussed earlier, the IEEE 802.15.4 MAC layer declares failure in sending a packet if it fails to receive the acknowledgement for the packet even after *1+macMaxFrameRetries* (i.e. 4, by default) transmission attempts. The failure could be declared sooner if, during a transmission attempt, the MAC layer suffers a *channel access failure*, which happens when *1+macMaxCSMABackoffs* (i.e. 5, by default) back-to-back CCA failures take place. Let α be the probability of CCA failure and β be the probability of collision for a packet transmission (or its acknowledgement). Thus, the probability of a *channel access failure* is α^5. Figure 4 shows a state diagram for the packet transmission process followed by IEEE 802.15.4 MAC layer. Clearly, the probability of packet loss, λ, is given by:

$$\lambda(\alpha, \beta) = \alpha^5 + (1 - \alpha^5)\beta(\tag{4}$$
$$\alpha^5 + (1 - \alpha^5)\beta($$
$$\alpha^5 + (1 - \alpha^5)\beta($$
$$\alpha^5 + (1 - \alpha^5)\beta)))$$

Since α and β are functions of m, the probability of packet loss λ is also a function of m and can be represented as $\lambda(m)$.

4.5. The packet latency

Next, we determine the packet latency as a function of the probabilities $\{\alpha, \beta\}$. The packet latency refers to the time required by IEEE 802.15.4 MAC layer to report back the fate of a packet to the upper layer. As discussed earlier, IEEE 802.15.4 MAC layer performs up to *1+macMaxCSMABackoffs* (5 by default) CCAs to find the channel idle during a transmission attempt. Each CCA is preceded by a CSMA wait. With default values for *macMinBE* and *macMaxBE* (3 and 5 respectively), the average duration of the five CSMA waits are 3.5, 7.5, 15.5, 15.5, 15.5 *backoff periods* (= 20 symbols) respectively. Each CSMA wait is followed by 8 symbols of CCA. Thus, after including the CCA duration, the average durations of five CSMA waits are 78, 158, 318, 318 and 318 symbols respectively. If α is the probability of CCA failure, the probability that a transmission attempt involves exactly $i, 1 \leqslant i \leqslant 4$, CSMA waits is $\alpha^{i-1}(1 - \alpha)$. The probability that a transmission attempt involves 5 CSMA waits with success in the 5th CCA is $\alpha^4(1 - \alpha)$. The probability that a transmission attempt involves 5 CSMA waits, all of which end in CCA failures (i.e. there is a channel access failure), is α^5. Thus, total time spent (in symbols) in CSMA waits and CCAs during a transmission attempt can be described as follows:

$$d_{\text{CSMA}} = \begin{cases} 78 & \text{w.p.} 1 - \alpha \\ 78 + 158 & \text{w.p.} \alpha(1 - \alpha) \\ 78 + 158 + 318 & \text{w.p.} \alpha^2(1 - \alpha) \\ 78 + 158 + 2 \times 318 & \text{w.p.} \alpha^3(1 - \alpha) \\ 78 + 158 + 3 \times 318 & \text{w.p.} \alpha^4(1 - \alpha) + \alpha^5 \end{cases}$$

Clearly, the expected duration of a transmission attempt that ends in a channel access failure is $d_{\text{CAF}} = 78 + 158 + 318 + 318 + 318 = 1190$ symbols (or 19.04ms) and the expected duration in symbols, if the transmission attempt does not end in a channel access failure, is $d_{\text{noCAF}} = \frac{1-\alpha}{1-\alpha^5}(78 + 236\alpha + 554\alpha^2 + 872\alpha^3 + 1190\alpha^4)$.

If a CCA is successful, the node would proceed with *RX-to-TX* turnaround ($d_{\text{TA}} = 12$ symbols) and transmit the packet ($d_T = 266$ symbols for a 133 byte packet) and wait for the acknowledgement. The acknowledgement should be received in $d_A = 34$ symbols (12 symbols *RX-to-TX* turnaround for destination + 22 symbols to transmit 11 byte acknowledgement) unless the packet (or acknowledgement) transmission ends up in collision, in which case the node waits for $d_W = 54$ symbols for acknowledgement and then proceeds with next transmission attempt. Let $d' = d_{\text{noCAF}} + d_{\text{TA}} + d_T + d_A$. Following the state diagram for the packet transmission process (Fig. 4), the packet latency, δ, in symbols can be expressed as a function of α and β as follows:

$$\delta(\alpha, \beta) = \alpha^5 d_{\text{CAF}} + (1 - \alpha^5)(d' + \beta(d_W - d_A + \tag{5}$$
$$\alpha^5 d_{\text{CAF}} + (1 - \alpha^5)(d' + \beta(d_W - d_A +$$
$$\alpha^5 d_{\text{CAF}} + (1 - \alpha^5)(d' + \beta(d_W - d_A +$$
$$\alpha^5 d_{\text{CAF}} + (1 - \alpha^5)(d' + \beta(d_W - d_A))))))))$$

Since α and β are functions of m, the packet latency δ is a function of m as well and can be represented as $\delta(m)$.

4.6. Tying it all together

Consider n source nodes, in each other's radio range, with the inter-packet generation (or *arrival*) interval at each node being exponentially distributed with rate $1/T$. Thus, the packet arrivals at each source node can be modeled as a Poisson process with rate $1/T$.

Suppose $D(n, T)$ represent the expected packet latency for the source nodes. For a given m value, the corresponding packet latency, $\delta(m)$, can be calculated using Eq. (5). We assume that, for given values of n and T, m is in turn a function of $D(n,T)$ as described below. This relationship between $\delta(m)$ and $D(n,T)$ can be exploited to determine $D(n,T)$.

Suppose node X is in the process of sending a packet. During the service time of this packet, some active nodes would finish sending their packets (and thus become inactive) while some inactive nodes would get a new packet to send (and thus become active). We assume that the probability that m nodes are active is same as the probability that $m-1$ *inactive* nodes get a new packet to send during the time node X is servicing its packet (equal to the expected packet latency $D(n,T)$). Since the number of active nodes at any given time is small and an active node is unlikely to receive another packet while it is servicing one, we calculate the probability that m nodes are active simply as the probability that $m-1$ nodes, out of $n-1$ source nodes, get a new packet to send during time duration $D(n,T)$.

Since the combined packet arrivals across $n-1$ source nodes can be modeled as a Poisson process with rate $(n-1)/T$, the probability that m nodes are active, i.e. $m-1$ nodes get a new packet to send during time interval $D(n,T)$, is given by:

$$p(m) = \frac{\left(\frac{n-1}{T} D(n,T)\right)^{m-1} e^{-\frac{n-1}{T} D(n,T)}}{(m-1)!} \tag{6}$$

Thus, for given n and T values, the expected packet latency, $D(n,T)$, can be obtained by solving the following recursive equation:

$$D(n,T) = \sum_{m=1}^{n} \delta(m) \times p(m) \tag{7}$$

Once we have determined the expected packet latency $D(n,T)$ for given n and T values, we can determine the probability, $p(m)$, that m nodes are active at any given time using Eq. (6) and hence the expected values of the probability of CCA failure, $A(n,T)$, the probability of collision for a transmission, $B(n,T)$, and the probability of packet loss, $L(n,T)$, as follows:

$$A(n,T) = \sum_{m=1}^{n} \alpha(m) \times p(m) \tag{8}$$

$$B(n,T) = \sum_{m=1}^{n} \beta(m) \times p(m) \tag{9}$$

$$L(n,T) = \sum_{m=1}^{n} \lambda(m) \times p(m) \tag{10}$$

4.7. Evaluating the Model's accuracy

To verify the accuracy of the stochastic model developed above, we performed simulations with a significantly improved version [8] of IEEE 802.15.4 protocol implementation in NS2 simulator [11]. IEEE 802.15.4 MAC layer operated in beaconless mode while IEEE 802.15.4 PHY layer operated in 2.4GHz range. All the IEEE 802.15.4 MAC and PHY parameters had their default values as listed in IEEE 802.15.4 specification [1]. Particularly, the *macMinBE, macMaxBE, macMaxFrameRetries* and *macMaxCSMABackoffs* parameters had values 3, 5, 3 and 4 respectively. The CCA duration was 8

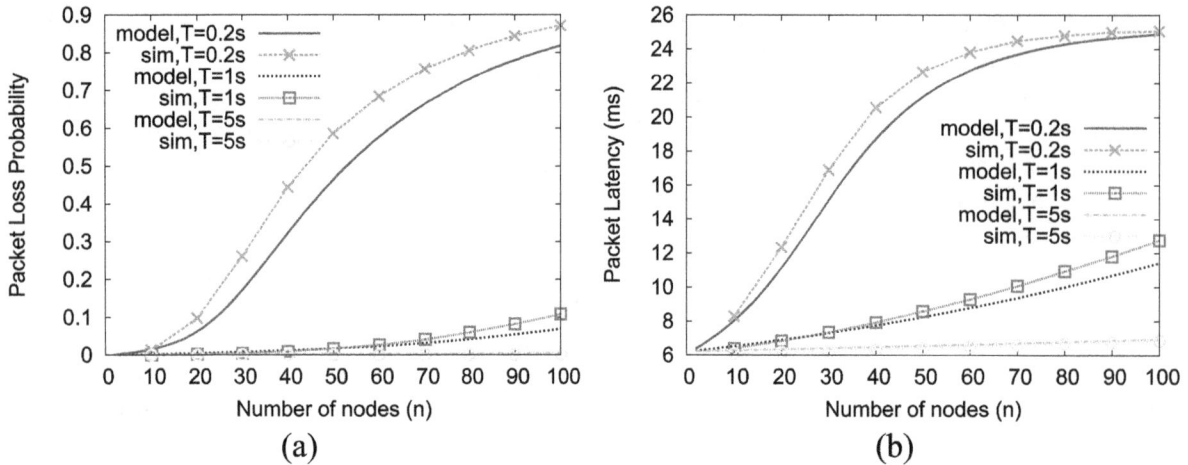

Fig. 5. Comparing the model predictions with NS2 simulation results.

symbols and the packet size used in these simulations was 133 bytes, including 6 bytes of IEEE 802.15.4 PHY header.

Each simulation begins with a certain number of source nodes, $n \in \{10, 20, 30, ..., 100\}$, coming up in a randomly determined sequence within first 100 seconds of the simulation. The source nodes generate packets for a common destination with the time interval between two packet send events at each node being exponentially distributed with average $T \in \{0.2s, 1s, 5s\}$. A successful packet delivery requires the receipt of MAC-level acknowledgement from the destination. All the nodes, including the common destination node, are always in each other's radio range. The simulation time is set such that each node sends close to 10000 packets to the common destination during the simulation. Each node keeps track of the average values of its probability of packet loss (L), i.e. the fraction of packets the MAC layer reports as lost due to channel access or collision failures, and the packet latency (D), i.e. the time difference between the instants when a packet is sent to the MAC layer for transmission and when the MAC layer reports the fate of the packet (successfully delivered or lost) back to the higher layer. The node-level values are then used to calculate the average values of the performance measures, and their confidence intervals, across all the nodes in the network. The 95% confidence intervals were always observed to be within a small range around the average value.

Figure 5 shows the comparison between the average values of the performance measures, calculated across all the nodes, obtained from the simulations and the model predictions. Clearly, there is quite a good match between the model predictions and simulation results for the probability of packet loss (Fig. 5a) and the packet latency (Fig. 5b) for all simulated (n, T) scenarios.

5. Modifications to beaconless IEEE 802.15.4 MAC operation and their evaluation using the stochastic model

It can be observed from Figs 5a and 5b that the packet loss probability and the packet latency increase rapidly as the traffic load in the WSN increases. While the packet latency remains acceptable even for high traffic loads, the probability of packet loss becomes very large even under moderate traffic loads. With 133 byte packet size, a packet transmission takes 300 symbols (266 symbols for packet transmission + 12 symbols for destination's *RX-to-TX* turnaround + 22 symbols for ack transmission) and hence a

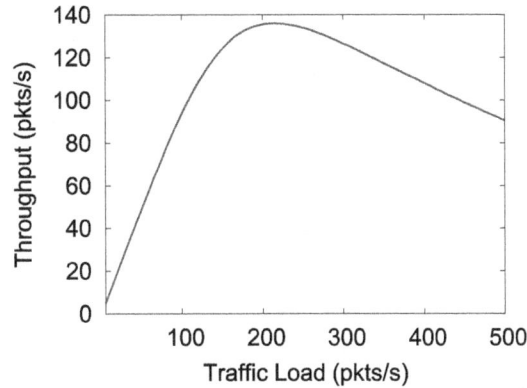

Fig. 6. Throughput versus Traffic Load as predicted by the Model.

maximum packet rate of 208.33 (= 62500/300) packets per second can be supported over a 2.4 GHz channel, assuming all the channel capacity can be utilized. Figure 6 shows how the throughput achieved, as predicted by the model using $n = 100$ and $T = 0.2s$, changes as the traffic load increases. As per Fig. 6, the maximum throughput achieved is approximately 136 packet per second when the offered traffic load is 215 packets per second corresponding to a packet loss probability of 36.72%. Any further increase in traffic load causes almost linear decrease in throughput. Clearly, there is a need to reduce the packet loss probability as the traffic load in the WSN increases. In this section, we investigate the efficacy of various proposals to modify beaconless IEEE 802.15.4 MAC operation in reducing the probability of packet loss while keeping the packet latency under control.

5.1. Eliminating the second collision window

In the earlier sections, we observed that if any node finishes its CSMA wait during one of the two possible collision windows during an active period, all the packet transmissions in the active period would end up in a collision. The first collision window is 12 symbols long and is caused by 12 symbol *RX-to-TX* turnaround time. Thus, the first collision window can be reduced by reducing the *RX-to-TX* turnaround time, which is clearly a hardware issue. The second collision window lies across first 4 symbols of 12 symbol *RX-to-TX* turnaround by a destination node as it finishes receiving the packet and prepares to send the acknowledgement back. If a node finishes its CSMA wait during the second collision window, its CCA would succeed because the transmission channel would be idle over the 8 symbol long CCA duration. Any reduction in the *RX-to-TX* turnaround time would also reduce the second collision window as well. Alternatively, we could eliminate the second congestion window by increasing the CCA duration such that it is larger than the *RX-to-TX* turnaround time. In this section, we analyze the impact of increasing the CCA duration from 8 symbols to 16 symbols. For this purpose, we need to modify the calculations of the probability of CCA failure (α), the probability of collision for a transmission (β) and the packet latency (δ) in the stochastic model of IEEE 802.15.4 MAC operation developed earlier.

5.1.1. The probability of CCA failure

Figure 7 shows different states in an active period when CCA duration is 16 symbols. State *S0* corresponds to the beginning of the active period as well as that of the first (and only) collision window and has a sojourn time equal to first 12 symbols of 16 symbol long CCA by the node that triggers the active period (say node *A*). If at least one active node finishes its CSMA wait during the collision window,

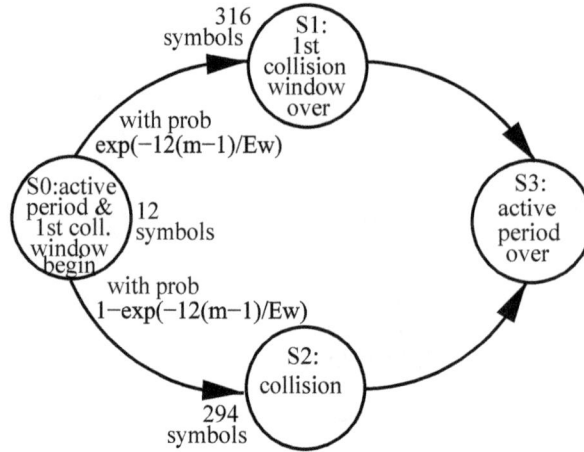

Fig. 7. Different states during the life time of an *active period* when CCA duration is 16 symbols.

which happens with probability $1 - e^{(-12(m-1)/E_w)}$, a collision is guaranteed. In this case, there is a transition to the *collision* state, *S2*. As before, we assume that the last colliding node finishes its CSMA wait just before the end of the collision window. This node would complete its transmission over next 294 symbols (16 symbols of CCA + 12 symbols of turnaround + 266 symbols of packet transmission, assuming 133 byte long packets), which constitute the sojourn time in state *S2*. The completion of sojourn in state *S2* concludes the active period. On the other hand, if no active node finishes its CSMA wait during the collision window, which happens with probability $e^{-12(m-1)/E_w}$, node *A* would complete its CCA (additional 4 symbols), the *RX-to-TX* turnaround (12 symbols), transmit the packet (266 symbols) and receive the acknowledgement (12 symbols of *RX-to-TX* turnaround by destination + 22 symbols for transmission of 11 byte long acknowledgement). These 316 symbols correspond to the sojourn in state *S1*, which also completes the active period.

Thus, different possible durations of an active period as well as other probabilities required to calculate the probability of CCA failure under modified circumstances are as follows:

- If no node finishes its CSMA wait during the collision window, which happens with probability $p_1 = e^{-12(m-1)/E_w}$, the active period would sojourn over states *S0* and *S1* and last for $12 + 316 = 328$ symbols. Given that no node finishes its CSMA wait during the collision window ($= 12$ symbols), the probability that an active node finishes its CSMA wait during such an active period is $q_1 = 1 - e^{-316/E_w}$ and the probability of CCA failure for such a node would be $\alpha_1 = 1$.
- If one or more nodes finish their CSMA waits during the collision window, which happens with probability $p_2 = 1 - e^{-12(m-1)/E_w}$, the active period would sojourn over states *S0* and *S2* and last for $12 + 294 = 306$ symbols. The probability that an active node finishes its CSMA wait during such an active period is $q_2 = 1 - e^{-306/E_w}$. Since, the CCA would not fail during the collision window, the probability of CCA failure for such a node would be $\alpha_2 = 294/306$.

As before, the overall probability of CCA failure can be expressed as a function of m, the number of active nodes, and E_w, the average CSMA wait duration, as follows:

$$\alpha(m, E_w) = \sum_{i=1}^{2} p_i \frac{(m-1)q_i}{1 + (m-1)q_i} \alpha_i \tag{11}$$

As before, for a given value of m, α is a function of E_w, which in turn is a function of α Eq. (1). This cyclic relationship can be exploited to determine the unique value of α for a given m.

5.1.2. The probability of collision for a transmission

Out of $m - 1$ active nodes still in the middle of their CSMA waits when an active period starts, the probability that i ($0 \leqslant i \leqslant m - 1$) nodes finish their CSMA waits during the collision window is $p_{\text{coll}}(i) = \binom{m-1}{i}(1 - e^{-12/E_w})^i (e^{-12/E_w})^{m-1-i}$. Thus, $p_{\text{coll}}(i)$ is the probability that $1 + i$ transmissions take place during an active period, including the transmission that triggers the active period. As before, more than one transmissions during an active period imply collisions for all the transmissions and hence the probability of collision β for a transmission is given by Eq. (3) (Section 4.3).

5.1.3. The packet latency

With CCA duration of 16 symbols, the average durations (including the CCA duration) of up to 5 CSMA waits during a transmission attempt would be 86, 166, 326, 326 and 326 symbols respectively. Thus, the expected duration of a transmission attempt that ends in a channel access failure would be $d_{\text{CAF}} = 86 + 166 + 3 \times 326 = 1230$ symbols and the expected duration in symbols, if the transmission attempt does not end in a channel access failure, would be $d_{\text{noCAF}} = \frac{1-\alpha}{1-\alpha^5}(86 + 252\alpha + 578\alpha^2 + 904\alpha^3 + 1230\alpha^4)$. The rest of the calculations are same as discussed in Section 4.5.

Figure 8 shows the performance under CCA durations 8 symbols and 16 symbols as predicted by the stochastic model (using $n = 100$ and $T = 0.2s$). Clearly, increasing the CCA duration to 16 symbols causes significant decrease in the probability of collision for a transmission while its impact on the probability of CCA failure is negligible. Also, decrease in the probability of collision, resulting from increased CCA duration, translates into slightly lower packet loss probability as well as slightly lower packet latency. A comparison of simulation results with CCA durations 8 symbols and 16 symbols revealed similar trends.

5.2. Understanding the impact of CSMA wait duration on performance

In Section 4.1, we described how the expected CSMA wait duration E_w depends on the probability of CCA failure α and the CSMA algorithm used in beaconless IEEE 802.15.4 MAC operation. Now, we assume that E_w is an independent parameter and analyze its impact on different performance metrics. The objective of this exercise is to see if changing the CSMA wait duration, by either changing the *macMinBE*/*macMaxBE* values or making the CSMA wait duration independent of α, results in better performance under given traffic load conditions. The following analysis assumes that CCA duration is 16 symbols and α and β are calculated as described in Sections 5.1.1 and 5.1.2 respectively (except that E_w is now an independent parameter and hence Eqs (11) and (3) can directly be used to calculate α and β). The probability of packet loss (λ) still depends on α and β as per Eq. (4) while the packet latency (δ) calculations would change as described next.

5.2.1. The packet latency under independent E_w

IEEE 802.15.4 MAC layer performs up to 5 (by default) CSMA waits to find the channel idle during a transmission attempt. We assume that each CSMA wait duration is exponentially distributed with average E_w symbols. Thus, the expected duration of a transmission attempt that ends in a channel access failure (5 back-to-back CCA failures) is $d_{\text{CAF}} = 5 \times (E_w + 16)$ symbols and the expected duration in symbols, if the transmission attempt does not end in a channel access failure, is $d_{\text{noCAF}} = \frac{(1-\alpha)(E_w+16)}{1-\alpha^5}(1 + 2\alpha + 3\alpha^2 + 4\alpha^3 + 5\alpha^4)$. The rest of the calculations are same as in Section 4.5 and the packet latency (δ) is given by Eq. (5).

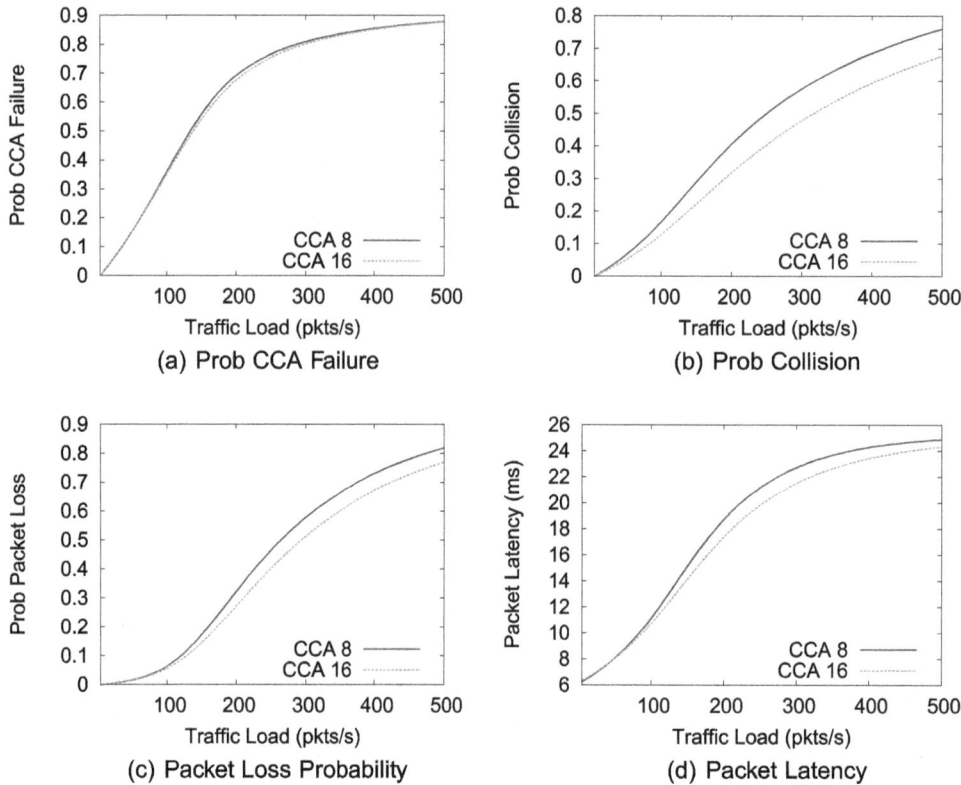

Fig. 8. CCA Duration 8 Symbols versus 16 Symbols.

5.2.2. Impact of CSMA wait duration on performance

Figure 9 shows the impact of E_w on different performance metrics calculated using the modified model. Similar trends were observed from the simulation results. The CSMA wait duration has a complex relationship with different performance metrics. The increase in E_w, in general, causes an increase in packet latency. However, as seen in *100 pkts/s* curve in Fig. 9d, the packet latency actually decreases slightly as E_w increases over a certain range. This can be attributed to sharp decrease in the probability of collision, and hence packet retransmissions, as E_w increases over this range. The increase in E_w reduces the probability that a CSMA wait ends during the 12 symbol collision window and hence is expected to reduce the probability of collision for a transmission. However, as seen in Fig. 9b, the probability of collision seems to stabilize as E_w becomes larger and larger. This can be attributed to sharp increase in the number of active nodes as the packet latency increases rapidly with increase in E_w beyond a certain value. The increase in the number of active nodes increases the probability that an active node finishes its CSMA wait inside an active period as well as the probability that an active node finishes its CSMA wait inside the collision window. Thus, increase in the number of active nodes, resulting from increase in E_w, increases both the probability of CCA failure as well as the probability of collision. However, increase in E_w also decreases the probability that an active node finishes its CSMA wait inside an active period and hence has a larger chance to trigger an active period itself, which should decrease the probability of CCA failure. These two factors tend to balance each other, however, as Fig. 9a indicates, the probability of CCA failure increases with increase in E_w beyond a certain value. Finally, Fig. 9c indicates that the impact of increase in E_w on the overall probability of packet loss depends on

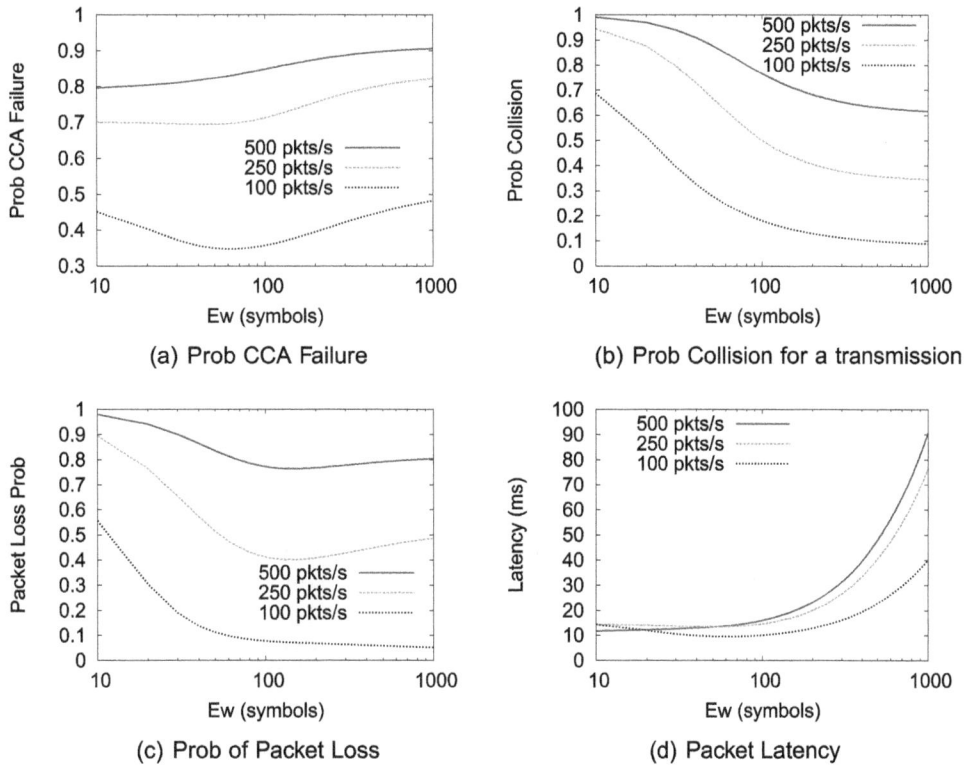

(a) Prob CCA Failure

(b) Prob Collision for a transmission

(c) Prob of Packet Loss

(d) Packet Latency

Fig. 9. Impact of E_w on Performance ($n = 100$).

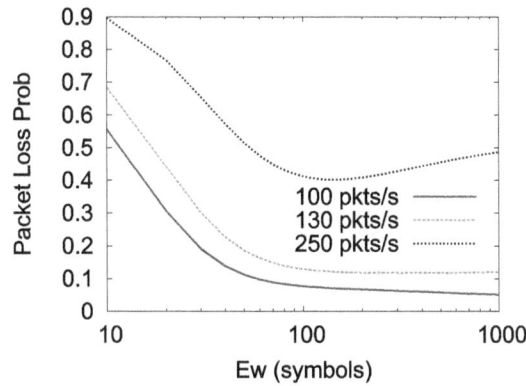

Fig. 10. Increase in E_w causes decrease in packet loss probability for traffic loads less than 130 pkts/s.

the traffic load. Under low and moderate traffic loads (less than 130 packets/s as shown in Fig. 10), increase in E_w decreases the probability of packet loss. Clearly, in this case, the decrease in probability of collision dominates the increase in probability of channel access failure. However, for high traffic loads, the increase in E_w increases the probability of packet loss once E_w crosses a certain value. This implies that the increase in probability of channel access failure dominates the decrease in probability of collision under high traffic loads.

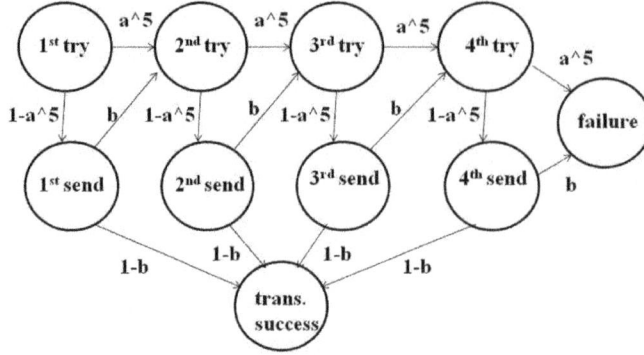

Fig. 11. Modified packet transmission process after eliminating channel access failures.

5.3. Eliminating channel access failures

IEEE 802.15.4 MAC layer makes at most *1+macMaxFrameRetries* (by default 4) attempts to transmit the packet and receive the acknowledgement. In a transmission attempt, the MAC layer performs up to *1+macMaxCSMABackoffs* (by default 5) CCAs to find the channel idle. Failure to find the channel idle even after fifth CCA causes a channel access failure and no further attempt is made to transmit the packet. In this section, we analyze the impact of eliminating the channel access failures, i.e., the MAC layer does not abandon the packet after 5 back-to-back CCA failures and simply proceeds with next transmission attempt. Figure 11 shows a state diagram for the modified packet transmission process. Let $p_l = \alpha^5 + (1 - \alpha^5)\beta$ be the probability that a transmission attempt fails to send the packet successfully. Clearly, the probability of packet loss, λ, under the modification, is p_l^4. We still assume that CCA duration is 16 symbols and CSMA waits are exponentially distributed with average E_w, which is a parameter independent of α and CSMA algorithm followed in IEEE 802.15.4 MAC. Thus, α and β are still calculated as described in Sections 5.1.1 and 5.1.2 respectively (except that E_w is an independent parameter).

The expected value of total time spent in CSMA waits and CCAs during a transmission attempt is $d_{\text{CSMA}} = (E_w + 16)(1 + \alpha + \alpha^2 + \alpha^3 + \alpha^4)$. A transmission attempt would result in a packet transmission with probability $1 - \alpha^5$, which would require $d_{\text{TA}} = 12$ symbols for *RX-to-TX* turnaround and $d_T = 266$ symbols to actually send the packet. The packet transmission would suffer a collision with probability β, in which case the sending node waits for $d_W = 54$ symbols for acknowledgement and then proceeds with next transmission attempt. Otherwise, the sending node would wait for the acknowledgement, which should be received in $d_A = 34$ symbols (12 symbols *RX-to-TX* turnaround for destination + 22 symbols to transmit 11 byte acknowledgement). Thus, the expected value of total time required to complete a transmission attempt is $d_{\text{trans}} = d_{\text{CSMA}} + (1 - \alpha^5)(d_{\text{TA}} + d_T + \beta d_W + (1 - \beta)d_A)$. Using p_l, the probability of failure in a transmission attempt, the expected packet latency in symbols can be expressed as $\delta(\alpha, \beta) = d_{\text{trans}}(1 + p_l + p_l^2 + p_l^3)$.

5.3.1. Impact on performance

Figure 12 shows the impact of eliminating channel access failure on different performance metrics as predicted by the model (using $n = 100$). Clearly, eliminating channel access failures causes both α and β to rise (Figs 12a and 12b), although increase in β is minor under moderate traffic load (100 packets/s). Eliminating channel access failures causes packet latency and hence the number of active

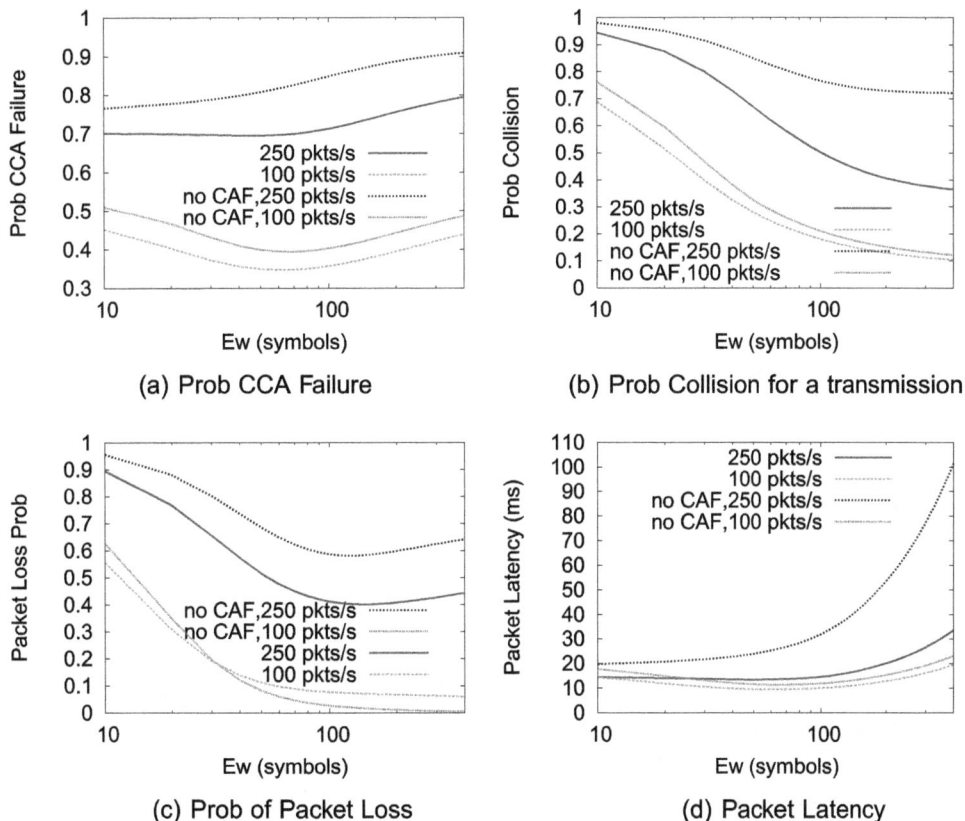

(a) Prob CCA Failure

(b) Prob Collision for a transmission

(c) Prob of Packet Loss

(d) Packet Latency

Fig. 12. Performance after eliminating channel access failures.

nodes to increase, which explains the increase in α and β. Minor increase in β value under moderate traffic loads can be explained as resulting from the corresponding minor increase in packet latency. Under high traffic load (250 packets/s), the increase in α and β more than offsets any beneficial impact of eliminating channel access failures and thus probability of packet loss is higher than before. However, under low and moderate traffic loads, eliminating channel access failures results in significant decrease in the packet loss probability for sufficiently high E_w values. Also, notice that eliminating channel access failures does not alter the impact of increasing E_w on the overall packet loss probability. As before, under high traffic loads (more than the observed threshold of *130 packets/s*), the packet loss probability increases with increase in E_w beyond a certain threshold and for lower traffic loads, the increase in E_w causes decrease in the packet loss probability. Note that eliminating channel access failures allows the packet loss rate under *100 packets/s* traffic load to decrease from about 6% to less than 1% (Fig. 12c) as E_w increases while causing only a minor increase in the packet latency (Fig. 12d).

Figure 13 shows the packet loss probability and packet latency values for E_w values 150, 310 and 630 symbols, corresponding to *macMinBE = macMaxBE = 4, 5* and *6* respectively, after eliminating the channel access failures and compares them against values with standard IEEE 802.15.4 MAC behavior with CCA duration 16 symbols. If we assume 5% to be the maximum acceptable packet loss probability then eliminating channel access failure with $E_w = 310$ symbols (corresponding to *macMinBE = macMaxBE = 5* increases the upper limit on traffic load on WPAN to 134 packets/s (with packet latency about 31.5ms) from 95 packets/s with standard IEEE 802.15.4 MAC behavior with *macMinBE = 3*,

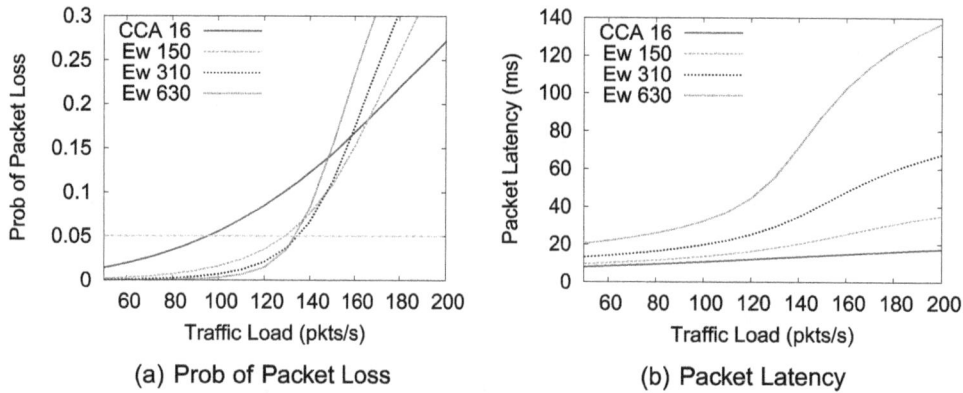

(a) Prob of Packet Loss

(b) Packet Latency

Fig. 13. Impact of E_w on Performance after eliminating channel access failures.

$macMaxBE = 5$ and CCA duration 16 symbols.

5.4. Reducing macMaxCSMABackoffs

In the previous section, we observed that eliminating channel access failures causes packet latency to increase, which in turn increases the number of active nodes and hence α and β. For high traffic loads, the increase in α and β values offsets beneficial impact of eliminating the channel access failures and the packet loss probability is higher than before. In this section, we examine the impact of reducing the packet latency by reducing the *macMaxCSMABackoffs* value. Reducing *macMaxCSMABackoffs* below its default value (4) would cause a channel access failure to be declared and packet to be abandoned for a smaller (than default 5) number of back-to-back CCA failures, which should tend to increase the packet loss probability. However, reduction in the packet latency, resulting from reduction in *macMaxCSMABackoffs* value, would reduce the number of active nodes and hence α and β values. Reduced α and β values would tend to reduce the packet loss probability. The stochastic model presented in Section 4 can be trivially modified to account for changes in the *macMaxCSMABackoffs* value.

Figures 14 and 15 show the impact of reducing *macMaxCSMABackoffs* on the probability of packet loss and the packet latency for different traffic loads as predicted by the model (using $n = 100$). As expected, reducing the *macMaxCSMABackoffs* value reduces the packet latency, however the impact on the probability of packet loss depends on the traffic load. For low traffic loads (*16.66, 100 packets/s*), the overall impact of reducing the *macMaxCSMABackoffs* is to increase the packet loss probability. However, for very high traffic loads (*500 packets/s*), there is a significant decrease in packet loss proobability as the *macMaxCSMABackoffs* value reduces from 4 to 1. For a traffic load of *250 packets/s*, reducing the *macMaxCSMABackoffs* value seems to help only for small E_w values.

We experimented with reducing packet latency by reducing the *macMaxFrameRetries* value as well. However, resulting increase in collision failures undid any beneficial impact of reduced latency for all traffic loads. We also experimented with dropping a certain percentage of packets at MAC layer without any attempt to send so as to reduce the loss probability for other packets. However, this approach did not prove as effective in reducing the packet loss probability for very high traffic loads as reducing the *macMaxCSMABackoffs* value.

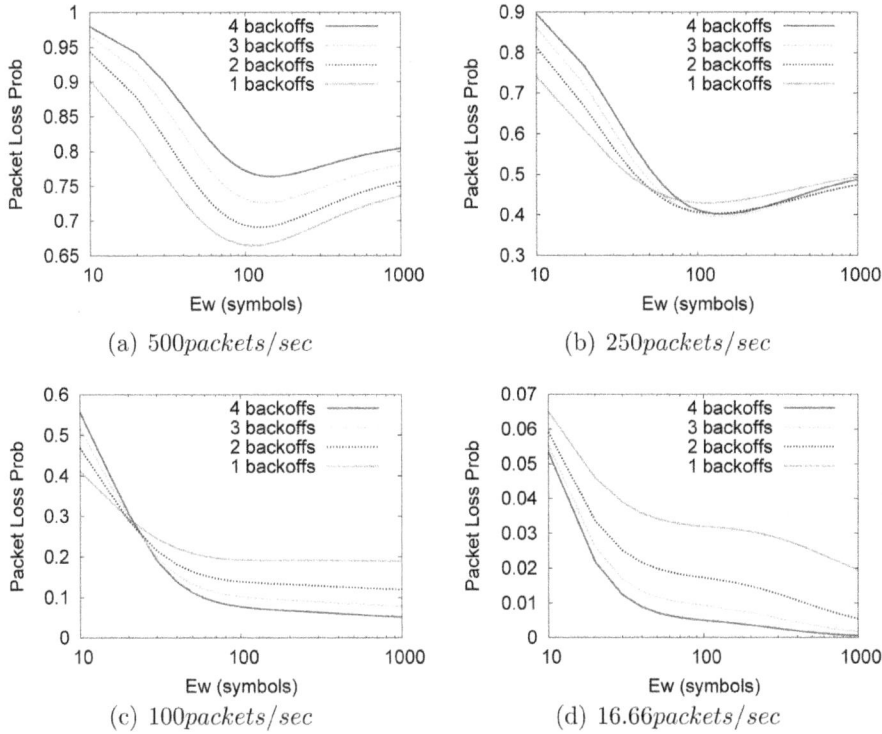

Fig. 14. Change in Packet Loss Probability with change in *macMaxCSMABackoffs*.

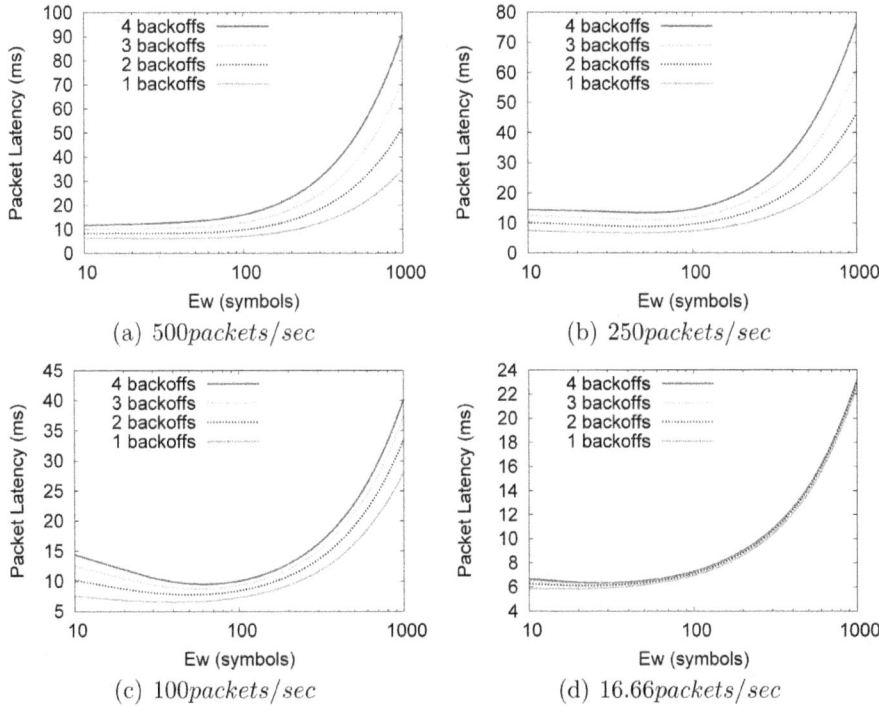

Fig. 15. Change in Packet Latency with change in *macMaxCSMABackoffs*.

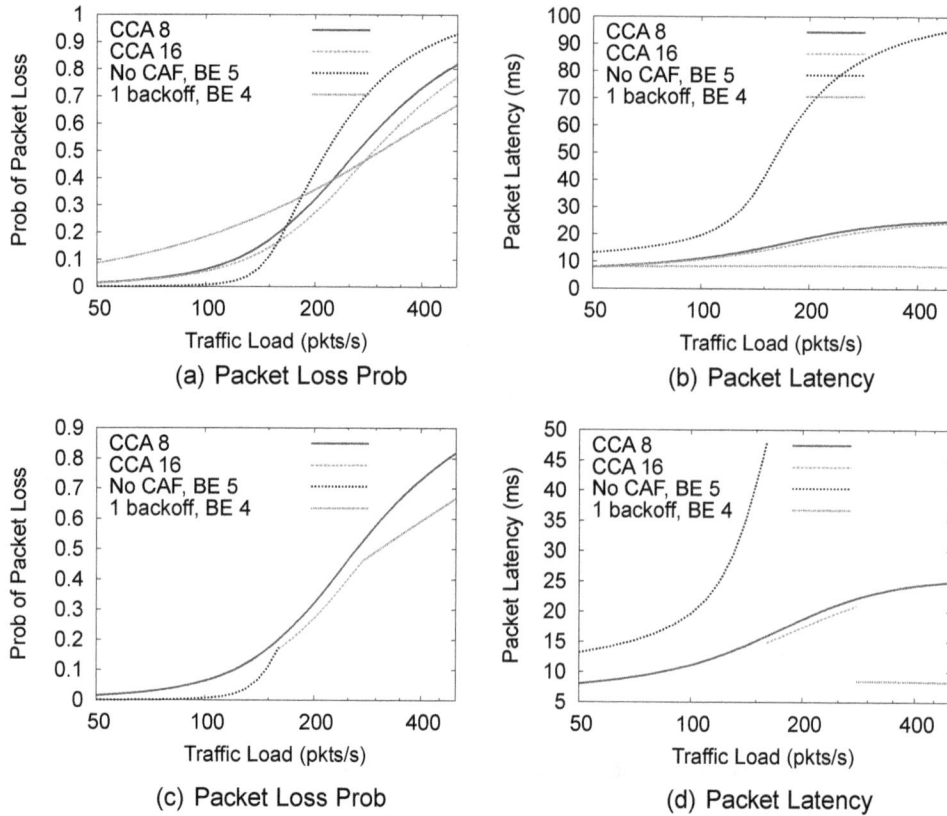

Fig. 16. Performance under Recommended Policies: Model Predictions.

5.5. Recommendations

The observations in Sections 5.1 through 5.4 imply that, in order to reduce the packet loss probability in beaconless IEEE 802.15.4 networks, different policies may have to be used under different traffic loads:

- Under low and moderate traffic loads, we suggest eliminating channel access failures and setting both *macMinBE* and *macMaxBE* to 5 (corresponding to $E_w = 310$ symbols).
- Under very high traffic loads, we suggest reducing *macMaxCSMABackoffs* to 1 and setting both *macMinBE* and *macMaxBE* to 4 (corresponding to $E_w = 150$ symbols).
- Under all traffic loads, we recommend increasing CCA duration from default 8 symbols to 16 symbols.

Figures 16 and 17 compare the performance under the recommended policies with the performance under standard IEEE 802.15.4 MAC operation with default parameter values, as predicted by the stochastic model and as observed in NS2 simulations respectively. Figures 16c and 17c show the traffic load range during which each policy should be used and the packet loss probability values under each policy during its applicable traffic load range. Figures 16d and 17d show the corresponding packet latency values. These figures also compare the performance under desirable policy for a certain traffic load with the performance under standard IEEE 802.15.4 MAC operation with default parameter values. Clearly,

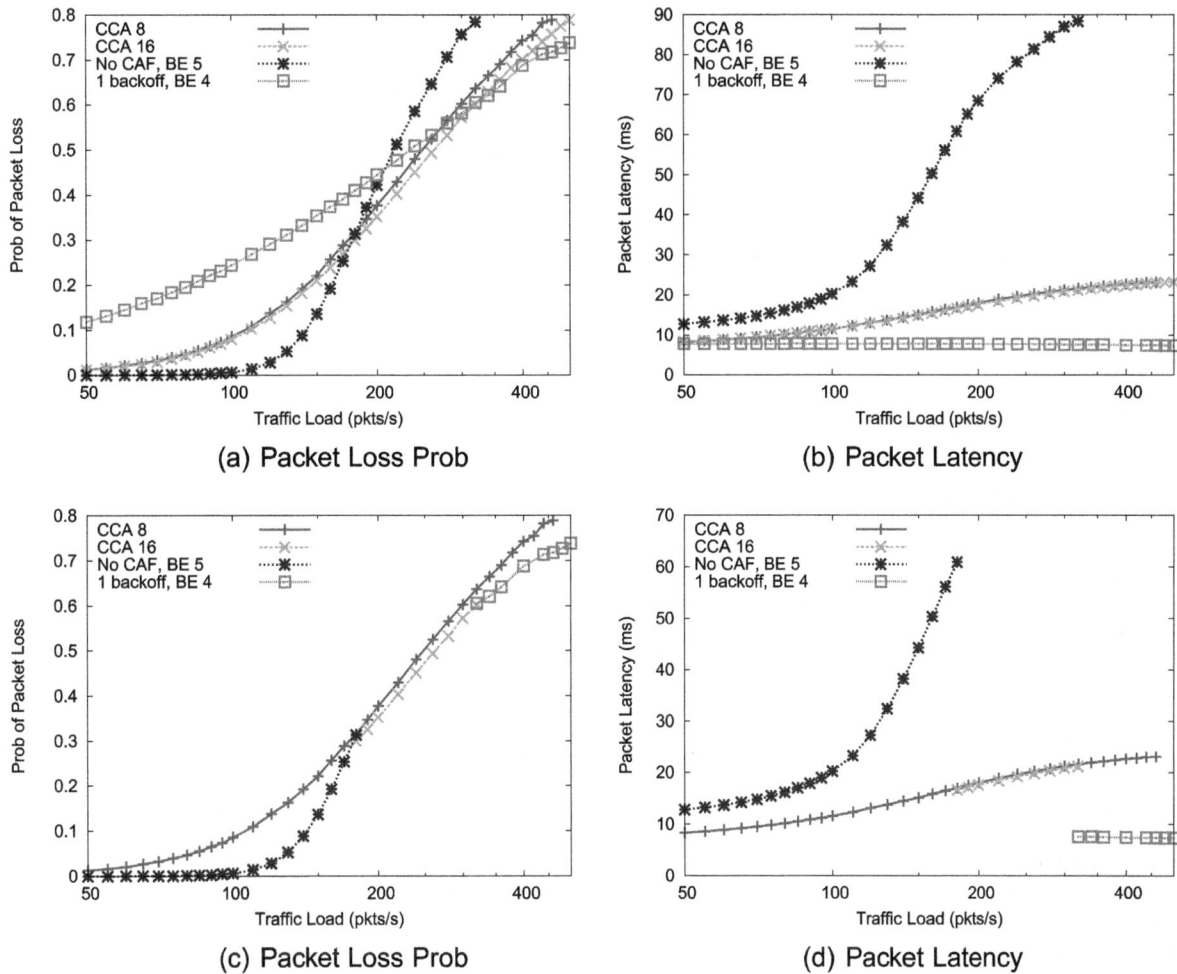

(a) Packet Loss Prob

(b) Packet Latency

(c) Packet Loss Prob

(d) Packet Latency

Fig. 17. Performance under Recommended Policies: Simulation Results.

we can achieve significant drop in the packet loss probability by using the appropriate policy for a given traffic load. For many WSN applications, the traffic load on the WSN may not be predictable and may vary with time over a large range, e.g. a WSN may observe sporadic intervals of very high traffic loads with moderate to low traffic for rest of the time. In such cases, we suggest that a node choose a policy based on the packet loss rate it has observed in recent past, which should be a good indicator of the current traffic load on the WSN [14]. Note that our analysis has assumed that no packet is lost due to signal attenuation/corruption at the physical layer. In situations where significant packet loss takes place due to signal attenuation/corruption at the physical layer, it may be necessary to apply a correction factor to the observed loss rate (to obtain the packet loss rate due to MAC layer effects) before choosing the appropriate MAC policy.

6. Conclusion

IEEE 802.15.4 enabled wireless sensor networks are finding increasing acceptance in various monitoring/control applications. In future, such networks may easily consist of several thousand nodes with each node possibly competing with hundreds of other nodes for access to the transmission channel, a situation that can severly impact the performance. In this paper, we presented a number of modifications to the beaconless IEEE 802.15.4 MAC operation with an objective of improving its performance. We evaluated the impact of these modifications via stochastic modeling and simulations. Our main observation was that the impact of these modifications is strongly dependent on the prevalent traffic load on the network and hence these modifications may be incorporated in IEEE 802.15.4 MAC operation as optional features that can be deployed as per the recommendations made in this paper.

References

[1]   Part 15.4: Wireless MAC and PHY layer specifications for low-rate wireless personal area networks. IEEE Std 802.15.4–2006, 2006.

[2]   A. Brandt, J. Buron and G. Porcu, Home Automation Routing Requirements in Low-Power and Lossy Networks. Request For Comments 5826, IETF, April 2010.

[3]   M. Dohler, T. Watteyne, T. Winter and D. Barthel, Routing Requirements for Urban Low-Power and Lossy Networks. Request For Comments 5548, IETF, May 2009.

[4]   A. Durresi and M. Denko. Advances in wireless networks, *Mobile Information Systems* **5**(2) (2009), 101–103.

[5]   A. Durresi, P. Zhang, M. Durresi and L. Barolli, Architecture for mobile heterogeneous multi domain networks, *Mobile Information Systems* **6**(1) (2010), 49–63.

[6]   J.Y. Goh and D. Taniar, Mobile data mining by location dependencies, In IDEAL, pages 225–231, 2004.

[7]   J. Goh and D. Taniar, Mining frequency pattern from mobile users. In KES, pages 795–801, 2004.

[8]   M. Goyal, Zigbee/IEEE 802.15.4 module for NS2 simulator, 2008.

[9]   M. Goyal, D. Rohm, H. Hosseini, K. Trivedi, A. Divjak and Y. Bashir, A stochastic model for beaconless IEEE 802.15.4 MAC operation. In IEEE SPECTS 2009: Proceedings of the IEEE International Symposium on Performance Evaluation of Computer and Telecommunication Systems, July 2009.

[10] J. Martocci, P. DeMil, N. Riou and W. Vermeylen. Building Automation Routing Requirements in Low-Power and Lossy Networks, *Request For Comments* 5867, IETF, June 2010.

[11] S. McCanne and S. Floyd. ns network simulator.

[12] V. Misic and J. Misic. Improving sensing accuracy in cognitive PANs through modulation of sending probability, *Mobile Information Systems* **5**(2) (2009), 177–193.

[13] K. Pister, P. Thubert, S. Dwars and T. Phinney, Industrial Routing Requirements in Low-Power and Lossy Networks. Request For Comments 5673, IETF, October 2009.

[14] D. Rohm, M. Goyal, H. Hosseini, A. Divjak and Y. Bashir, A simulation based analysis of the impact of IEEE 802.15.4 MAC parameters on the performance under different traffic loads, *Mobile Information Systems* **5**(1) (2009), 81–99.

[15] D. Taniar and J. Goh, On mining movement pattern from mobile users, *IJDSN* **3**(1) (2007), 69–86.

[16] A.B. Waluyo, R. Hsieh, D. Taniar, J. Rahayu and B. Srinivasan, Utilising push and pull mechanism in wireless e-health environment. In EEE, pages 271–274, 2004.

[17] A.B. Waluyo, B. Srinivasan and D. Taniar, Optimal broadcast channel for data dissemination in mobile database environment, In APPT, pages 655–664, 2003.

[18] A.B. Waluyo, B. Srinivasan and D. Taniar, A taxonomy of broadcast indexing schemes for multi channel data dissemination in mobile database. In AINA, pages 213–218, 2004.

[19] A.B. Waluyo, B. Srinivasan and D. Taniar, Research in mobile database query optimization and processing, *Mobile Information Systems* **1**(4) (2005), 225–252.

[20] A. Wheeler, Commercial applications of wireless sensor networks using zigbee, *IEEE Communications Magazine* **45**(4) (April 2007), 70–77.

[21] K. Xuan, G. Zhao, D. Taniar and B. Srinivasan. Continuous range search query processing in mobile navigation, In ICPADS, pages 361¤C368, 2008.

[22] Ilsun You and Takahiro Hara. Mobile and wireless networks, *Mobile Information Systems* **6**(1) (2010), 1–3.

[23] G. Zhao, K. Xuan, D. Taniar and B. Srinivasan, Incremental k-nearest-neighbor search on road networks, *Journal of Interconnection Networks* **9**(4) (2008), 455–470.

[24] Z. Alliance, Zigbee specification, December 2006.

Mukul Goyal is an Assistant Professor in Computer Science department at University of Wisconsin Milwaukee. He received a PhD in Computer and Information Science from Ohio State University in 2003. His research Interests lie in the field of computer networks and performance evaluation.

Weigao Xie is a PhD candidate in Computer Science department at University of Wisconsin Milwaukee. Her research interests like in the field of Wireless Sensor Networks and Internet routing protocols such as OSPF.

Hossein Hosseini is a Professor in Computer Science department at University of Wisconsin Milwaukee. He received a PhD from University of Iowa in 1982. His research interests lie in the fields of Computer Networks, High Performance Distributed and Parallel Systems, Computer Architecture, Performance Evaluation, Fault Tolerance, Reliability, Operating Systems, Real Time Systems.

3

A routing strategy for non-cooperation wireless multi-hop ad hoc networks

Dung T. Tran*, Trang T.M. Truong and Thanh G. Le

Faculty of Information Technology, University of Science, Ho Chi Minh City, Vietnam

Abstract. Choosing routes such that the network lifetime is maximized in a wireless network with limited energy resources is a major routing problem in wireless multi-hop ad hoc networks. In this paper, we study the problem where participants are rationally selfish and non-cooperative. By *selfish* we designate the users who are ready to tamper with their source-routing (*senders could choose intermediate nodes in the routing paths*) or next hop selection strategies in order to increase the total number of packets transmitted, but do not try to harm or drop packets of the other nodes. The problem therefore amounts to a non-cooperative game. In the works [2,6,19,23], the authors show that the game admits Nash equilibria [1]. Along this line, we first show that if the cost function is linear, this game has pure-strategy equilibrium flow even though participants have different demands. However, finding a Nash equilibrium for a normal game is computationally hard [9]. In this work, inspired by mixed-strategy equilibrium, we propose a simple local routing algorithm called MIxed Path Routing protocol (MiPR). Using analysis and simulations, we show that MiPR drives the system to an equilibrium state where selfish participants do not have incentive to deviate. Moreover, MiPR significantly improves the network lifetime as compared to original routing protocols.

Keywords: Local routing, Network lifetime, Nash equilibrium

1. Introduction

In recent years, the problem of lifetime maximization has been studied extensively. Since wireless devices have limited resources, e.g. battery, it is essential to develop energy efficient routing algorithms which optimize the overall energy utilization of the network. As done in [2,3], we define the lifetime of the network as the time before the first node on a forwarding path runs out of battery. From the literature, we know that the minimum energy path routing does not yield maximum lifetime of the network. The reason is that they may create congested areas which cause some nodes to quickly lose energy. Typically, the problem of routing for maximizing network lifetime is more important than the problem of finding a path with minimum energy consumption.

Many works have focused on finding energy-efficient routing for maximizing the network lifetime. However, they assume that all participants will follow the rules of the protocol. Since wireless ad hoc networks might have diverse participants, there may exist selfish participants who basically act for their own optimality. Commonly, the result of local optimization with conflicting interests does not lead to any type of global optimality. This is because the maximum amount of data a node can transmit depends not only on its chosen path but also the current traffic on that path. Senders without cooperation may choose the same set of preferred paths, leading to a paralysis of the system very soon.

---

*Corresponding author: Dung T. Tran, Faculty of Information Technology, University of Science, Ho Chi Minh City, Vietnam. E-mail: ttdung@fit.hcmus.edu.vn.

In this work, we assume that all the participants' objective is to maximize the amount of their transmitted data before their routing paths are disconnected. Since participants select the routing paths to maximize the amount of transmitted data independently, the problem thus becomes one of a *noncooperative game* [1]. Because the total amount of data a participant can transmit depends on its chosen path and the number of participants choosing this path, this game hence is a *congestion game* [1]. Thus, we take the advantage of game theory into the design of the routing protocol.

Let us first recall some basic concepts of game theory. A normal game is defined by a tuple (*players, strategies, utility functions*), in which each player has a set of strategies to choose, called *strategy space*; a utility (*payoff*) function of a player takes as input a *strategy profile* (a specification of strategies for every player) and yields a representation of utility as its output. A *Nash equilibrium* of a game is the point where there exists a *strategy profile* which fully specifies all actions in the game such that no player could gain more by unilaterally changing its strategy. In a game there may exist many Nash equilibria, so that the *social profit* (sum of players' utilities) of the game at each Nash equilibrium may get different values.

The works [2,6,19,23] show that the general version of routing game admits Nash equilibria. In this work, we assume that nodes transmit data in packets. This means that the network flow is in integer as the model of Rosenthal [6]. The first part of this paper focuses on perspectives of the non-cooperative Lifetime Game. In this part, we summary prior results and present our results. We show that if cost function is linear, the *pure-strategy equilibrium flow* exists even through nodes have different demands.

Finding a Nash equilibrium in a normal game is complete for the complexity class PPAD [9], even in the case of two players [15], this is evidence of the intractability of the problem. Orda et al. [19] and Lewin et al. [2] show that if all participants choose the best response, i.e. the current maximum residual energy path[1], the systems will eventually converge to a Nash equilibrium. However, in the practical wireless ad hoc networks, in oder to find the maximum residual path frequently, a node needs to have both the global map of the network and the updated energy of the edges. This requirement, however, is practically impossible to fulfill. Therefore, in the second part of this work we design a new protocol which satisfies two conditions:

1. It can drive the network into a Nash equilibrium point.
2. It can apply to the practical noncooperative ad hoc networks.

2. Model and problem formulation

We assume that nodes can adjust their transmission radio to save the power when sending packets to its neighbors. The network is modeled by a directed graph $G = (V, E)$, where $n = |V|$ is number of nodes and $m = |E|$ is the number of edges. There are k source-destination pairs $(s_1, d_1), \ldots, (s_k, d_k)$. Each source node i has r_i data packets (integer variable) and wants to maximize its data transmitted. Here, we consider wireless ad hoc networks where:

- Data flow is *splittable* (be able to transmit through many paths at a time), and be transmitted in packets with integer size.
- Each source node may have different amount of data.

---

[1]The maximum remaining energy path.

Above assumptions make this model like the one in [6].

Let define an increasing function $c_e(x) : \mathbb{N}^+ \to \mathbb{R}^+$ as the cost function to transmit x packets through link e, i.e. every node which uses this link experiences this cost. The cost, here, presents the amount of energy consumed when a packet is transmitted through a link. Then, the cost of a path p is as:

$$c_p = \sum_{e \in p} c_e(x_e) x_e \tag{1}$$

where x_e is the number of packets passed through edge e.

Let us define *Lifetime Game* as a tuple (G, c, r), where $c = \{c_1, \ldots, c_m\}$, and $r = \{r_1, \ldots, r_n\}$. *Equilibrium flow* is now defined as follows.

Definition 1 *Given an instance (G, c, r), and a set of source-destination pairs, an equilibrium flow is an assignment of flow to routing paths in which no source node could decrease its cost by unilaterally changing its routing paths. Specifically, let x be the feasible flow for instance (G, c, r), if p is a path in an equilibrium flow then for any other path p'.*

$$c_p(x) \leqslant c_{p'}(x)$$

Before presenting our main results, we summary prior results of this problem.

3. Prior results

In 1952, Wardrop [8] has modeled unregulated traffic as network flow with all paths between a given source-destination pair having equal cost function. The equilibrium routing, thus, is called *Wardrop equilibrium*. Since then, many properties and extensions to this traffic model have been studied [11–21]. We may classify this problem based on characteristics of the flow. There are two main kinds: data is splittable and non-splittable.

3.1. Data is non-splittable

In this case, source nodes can only transmit data through a single path at a time instant. All works of this kind consider the data flow in continuous variable, and the results are also applicable to the case with data flow as discrete variable. Fotakis et al. [22] proved that the affine cost function is the necessary condition for existence of pure NE. Panagopoulou and Spirakis [23] showed that pure NE always exists for instances with uniform exponential cost function ($c_e(x) = \exp(x)$). Harks et al. [24] generalized this to non-uniform exponential cost functions of the form $c_e(f) = a_e e^{\phi f} + b_e$ for some $a_e, b_e, \phi \in \mathbb{R}$, where a_e and b_e may depend on link e, while ϕ must be equal for every link. Finally, Harks and Klimm [25] proved the sufficient condition for the existence of pure NE. In their work, authors show that cost function must be either an affine function or is of type $c_e(f) = a_e e^{\phi f} + b_e$. The price of anarchy in this case is at most 2.618 (see [9, p. 476]).

3.2. Data is splittable

In this case, source nodes can transmit data through many paths at a time instant. Most work of this kind consider the data flow in continuous variable. The problem always admits mix-strategy equilibrium flow by Nash theorem [10]. If the cost function is differentiable and convex then pure-strategy equilibrium flow exists and is unique [9, p. 468]. Moreover, if the cost function is linear then the price of anarchy is at most 4/3 (see [9, p. 485]).

Rosenthal [6] considered this problem in integer flow. Author showed that pure-strategy equilibrium flow always exists if source nodes have *same* demand. If source nodes have different demand and link cost functions are general, the problem needs not admit pure-strategy equilibrium flow. In our work, we prove that if the cost function is linear, the problem has pure-strategy equilibrium flow even through source nodes have *different* demand. These equilibrium flow are not unique. The price of anarchy in this case is also at most 4/3.

The work of Levin-Eytan et al. [2] studies game-theoretic aspects of lifetime maximization problem. However, authors study the maximum lifetime problem in multicast and anycast routing. In this work, authors modeled a single multicast section as a static game, this thus is quite different from the model we discussed above.

4. Our results

4.1. Existence of pure-strategy nash equilibrium

Let a *strategy* of a player include:

1. A set of used paths from source to destination
2. Fraction of data will be transmitted on each path

In Lifetime Game, the set of paths between any pair of source-destination is finite, because there are finite number of nodes and finite number of links in the network. This means that Lifetime Game is a finite strategic game. Thus, it admits a mixed-strategy equilibrium flow (by Nash theorem [10]). However, the mix-strategy NE does *not* guarantee for an integer solution. Here, we show that this game also admits pure-strategy NE.

Theorem 1 *Given an instance of Lifetime Game where players have different demand and data flow is splittable. If the cost function is linearly increasing then Lifetime Game has pure-strategy equilibrium flow.*

Proof See detail in the appendix.

5. MiPR protocol

We now consider practical wireless ad hoc networks, where a single source node does not know either the global network topology, or other source nodes. Since the players now do not know the chosen paths of others, a node with the strategy of choosing the maximum residual energy path may even yield the worst case, e.g. the case in Fig. 1.

Assume every source node S_i, $i = \{1, \ldots, k\}$, wants to send data to destination T at the same time, and the capacities of edges (S_i, A_j) and $(B_i, T) >> (A_i, B_i)$ $\forall i, j$. For any source node S_i, the path (S_i, A_1, B_1, T) gives the lifetime value[2] k, all the other paths give the lifetime value $k/2$. Thus, all

---

[2]The total number of packets a path/link could transfer until one node in the path runs out of battery.

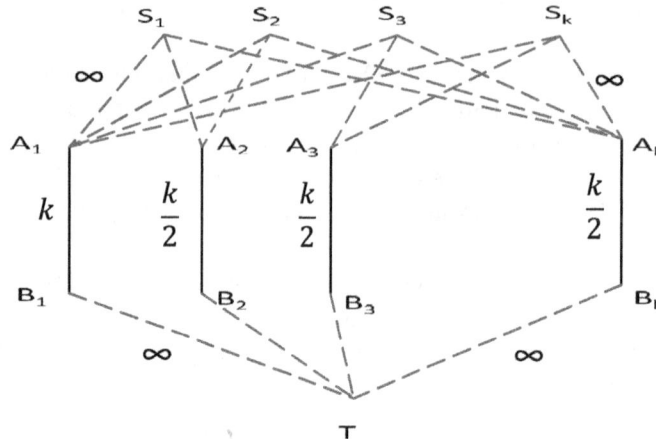

Fig. 1. Example of the worst case of non-cooperative game.

source nodes choose path (S_i, A_1, B_1, T) to send data, causing the network to transmit only 1 packet per node (because the capacity of edge (A_1, B_1) is k). The optimal solution for this case is to have each source S_i choose the path (S_i, A_i, B_i, T), so that each node could transmit at least $k/2$ packets. We note that, in this optimal solution, the utility of each node is not distributed fairly, since node S_1 can transmit up to k packets, while the other nodes can only transmit $k/2$ packets.

Typically, the central part of the network is likely to be the congested area because many transmission paths go through it. This not only causes nodes in the center to drain batteries faster, but also raises security issues. To overcome this problem, we make use of the concept of mixed strategies, i.e. source nodes will select each path in the set of available paths to destinations with some probability such that the network will converge to a mixed-strategy equilibrium flow. At this point, a single node cannot increase its lifetime by switching to another path. In other words, the network flow is balanced for all participants.

5.1. Problem of geographic routings

In the experiment of Mei and Stefa [3], the authors show that, in geographic routings,[3] the traffic load in the center area is always higher than that in the border area. This is the reason that nodes in the center area will be drained their energy soon. For example, nodes are uniformly distributed in a rectangle area. Source and destination nodes are both selected randomly. Then the traffic load in the network is as shown in Fig. 2.

5.2. Balanced energy routing protocol

In order to overcome the problem of geographic routings, we design a protocol, called *balanced energy routing protocol*. Nodes follow this protocol would not create any congested areas like in geographic routing.

The basic idea is that we transform the network area into a symmetric area which has no center part. For this purpose, we map the network area onto a sphere surface. First, we claim that the sphere

---

[3]Select the closest node to the destination to be the next hop.

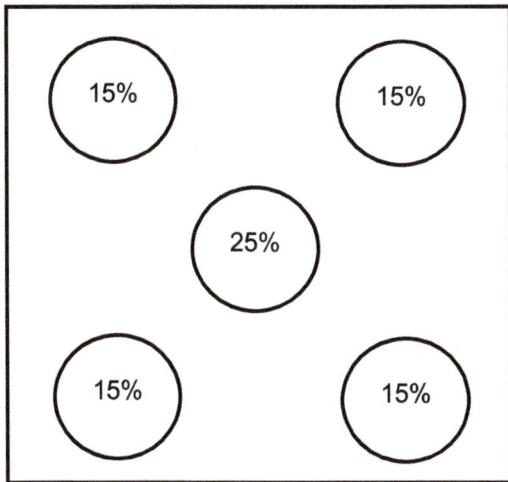

Fig. 2. Point $A(\phi, \theta)$ in sphere surface.

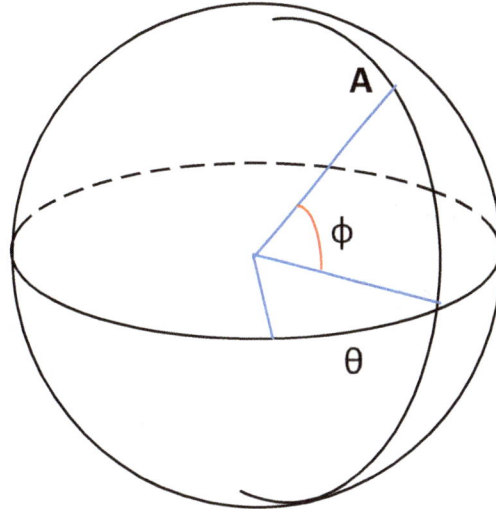

Fig. 3. Sample traffic load in geographic routing.

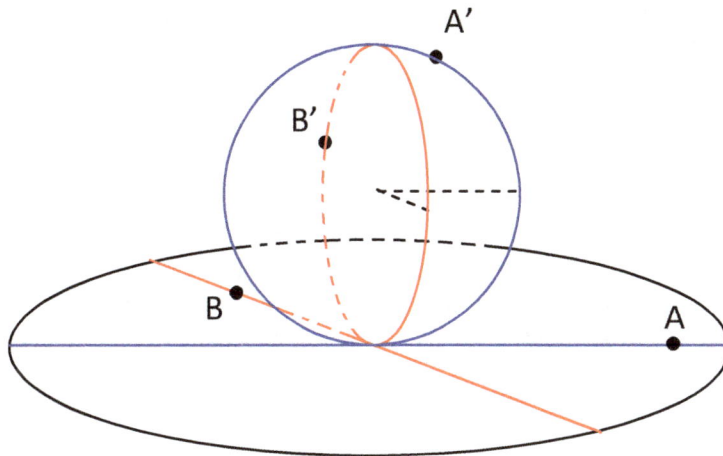

Fig. 4. Points A', B' are images of A, B on sphere surface respectively.

surface satisfies the condition that there is no center part in the network. Because the sphere surface is a symmetric region, if nodes are uniformly distributed on it, we easily see that every node on the surface can be on the shortest direction of a source-destination pair with the same probability.

The transformation process just maps nodes into new locations on the sphere surface; however, it keeps all edges in the original graph. This means that neighbors of node u in the new region are exactly those of u in the original region. A point on the sphere surface can be defined by its latitude and longitude.

For example, point A in Fig. 3, A's latitude is the angle ϕ, and its longitude is the angle θ. To transform the mapping of a network from a circle area to a sphere surface, let the circle be a set of all lines of the same length with the center of the circle as their midpoint. We then bend those lines upwards to create a sphere. We can visualize the mapping process as shown in Fig. 4.

Another issue is that nodes on the sphere surface have to be random, and uniformly distributed. The circle area can be divided into n small layers with equal area. Then we also divide the sphere surface into n equal slices along vertical line, as in Fig. 5. The mapping process will map each layer of the circle

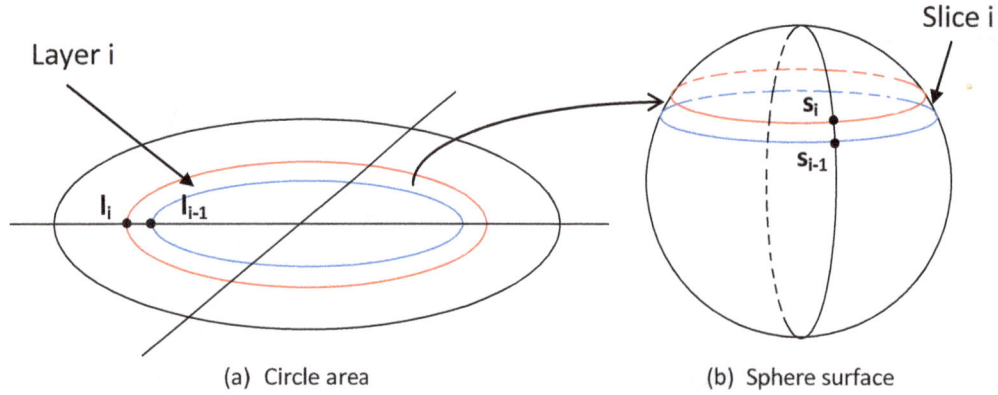

Fig. 5. Layer i^{th} of the circle region mapped to the corresponding slice i^{th} of the sphere surface.

region to its corresponding slice of the sphere surface. If node u belongs to the ith layer, then its image on the sphere belongs to the ith slice.

We now formally address the mapping process. Let a point $A(x, y)$ with coordinates (x, y) on the two-dimensional circle region with radius r. The center of the circle has coordinates (0,0). Let A' be the image of A on the sphere surface. We first find the formula to calculate the approximate longitude of A'. We can easily see that the longitude $\theta_{A'}$ of A' on the sphere surface is approximate the angle of A on the circle area, hence it can be calculated as follows:

$$\theta_{A'} = arctan\ (y, x) \tag{2}$$

The longitude is quite simple to find, but the latitude is more complicated. Assume the circle is divided into n equal layers which have the same center but different radius. Let layer i be the area between radii l_{i-1} and l_i, each layer be mapped to a corresponding slice i, which is the area between latitude s_{i-1} and s_i. Each area of layer i is $(\pi r^2)/n$. Given r and n, we can easily find the value of each l_i, taking $l_1 = 0$, as follows.

$$l_i = \sqrt{\frac{r^2}{n} + l_{i-1}^2} \tag{3}$$

We say a point $A(x, y) \in$ layer i^{th} if and only if $l_i \leqslant \sqrt{x^2 + y^2} < l_{i+1}$, i.e., the distance from A to the center is between l_i and l_{i+1}. We also divide the sphere into n slices, each slice have the same size in latitude π/n. Thus, slice i^{th} will be in range $[\frac{\pi}{n}(i-1) - \frac{\pi}{2}, \frac{\pi}{n}i - \frac{\pi}{2}]$.

Assume that $A \in$ layer i^{th}, latitude $\phi_{A'}$ can be found by the following formula:

$$\phi_{A'} = \left[\frac{\pi}{n}(i-1) - \frac{\pi}{2}\right] + \frac{\pi}{n}\frac{\sqrt{x^2 + y^2} - l_i}{l_{i+1} - li} \tag{4}$$

The distance between two points on the sphere surface with radius R is the shortest curve connecting them on the surface. We use Haversine's formula [4] for calculating the distance Δ between point $u(\phi_u, \theta_u)$ and $v(\phi_v, \theta_v)$ on the sphere surface:

$$\Delta = 2R.arcsin\sqrt{sin^2\left(\frac{\Delta_\phi}{2}\right) + cos\phi_u cos\phi_v sin^2(\frac{\Delta_\theta}{2})} \tag{5}$$

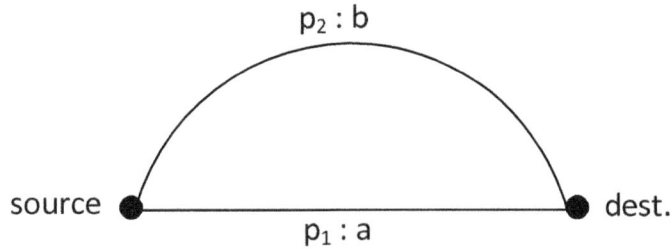

Fig. 6. Scenario of lifetime game.

On the sphere surface the logical distance between pairs of nodes changes, as all nodes near the border of the original area are now quite close together on the north area of the sphere. Therefore, a pair source-destination located on borders of opposite sides would prefer the routing path along the border rather than go through the central part. Thus, in the new domain, the shortest path between two nodes that are located near the border no longer goes through the congested center in the network. This path is clearly the shorter arc of the circle which contains the source and destination and is around the congested area.

We claim that if the nodes are uniformly distributed in the circle region, and if the size of each layer is small enough, then nodes are uniformly distributed on the sphere surface. We sketch the proof of our claim as follows. Since nodes are uniformly distributed, layers with the same size will contain an approximately equal number of nodes. We note that nodes are also uniformly distributed on each layer. Each layer is mapped to each slice on the sphere surface, hence, slices have approximately the same number of nodes. Slices and layers are similar in their shape (circular), and nodes in the same layer keep their relative distances. Nodes therefore are uniformly distributed in slices. Taking the limit, we obtain that nodes are uniformly distributed on the sphere surface.

We note that it is impossible to have a point-to-point mapping from a circle area to a sphere surface continuously. However, nodes are, in fact, distributed on the circle discretely. Therefore, we can have point-to-point mapping while approximately keeping the uniform distribution.

When the original region is a rectangle or any arbitrary shape in two-dimensional space, if we can scale them to an appropriate circle area in which nodes are randomly distributed, then we can apply this transformation to achieve a symmetric region.

5.3. MiPR protocol

The *balanced energy routing protocol* could not solve the problem of selfish nodes. If a node likes to reduce the delay by shortening its routing path, it might select the geographic routing without any penalty from the protocol. In this work, we assume that nodes don't have to strictly follow the protocol. Each single node may choose it routing paths for maximizing its utility. In this work, we propose MiPR protocol for this new issue.

We firstly model the network in the experiment above as a strategic game, called *Lifetime Game*. In this game, each source node is a player, and the set of paths from source to destination is a set of strategies. Consider two players, player 1 and player 2 are located at the same place on the border the network area, while their destinations are also located on the border, but at the other side of the network area. They share the same set of routing paths which could be divided into two groups: one group of paths created by geographic routing which will pass through the center network area, represented by path p_1 in Fig. 6;

Table 1
Lifetime game

		Player 2	
		p_1	p_2
Player 1	p_1	$a/2, a/2$	a, b
	p_2	b, a	$b/2, b/2$

the other group is created by *balanced energy routing protocol* which will go along the border of network area, represented by path p_2 in Fig. 6.

For a single transmission, path p_1, which goes through the center, has lifetime of a, and path p_2 has lifetime of b, $b < a$. Path p_2's lifetime is smaller because it is longer, with many nodes involved in the transmission. The total energy consumed then is larger. If two players choose path p_1 then p_1's lifetime is reduced to $a/2$. Or if they both choose path p_2, then p_2's lifetime is $b/2$. The matrix representing this game is shown in Table 1.

Lemma 1 If player 1 (player 2) chooses the path p_1 (p_2) with probability $q_1 = \frac{2a-b}{a+b}$ ($q_2 = \frac{2b-a}{a+b}$) then the other player could not improve its utility by changing its routing strategy.

The proof of Lemma 1 is shown in the following section.

5.3.1. Correctness of the protocol

Let q_1 and q_2 be the probability that player 1 and player 2 will choose path p_1 and p_2, respectively. The probability that player 1 and player 2 will transmit on path p_2 is $1 - q_1$, $1 - q_2$, respectively. Then the utility u_1 of player 1 is:

$$
\begin{aligned}
u_1 &= \frac{a}{2}q_1q_2 + aq_1(1 - q_2) + b(1 - q_1)q_2 + \frac{b}{2}(1 - q_1)(1 - q_2) \\
&= q_1\left(\frac{2a - b}{2} - q_2\frac{a + b}{2}\right) + \frac{b}{2}q_2 + \frac{b}{2}
\end{aligned}
\tag{6}
$$

Likewise, we have

$$
u_2 = q_2\left(\frac{2a - b}{2} - q_1\frac{a + b}{2}\right) + \frac{b}{2}q_1 + \frac{b}{2}
\tag{7}
$$

Consider Eq. (7), the utility u_2 does not depend on any probability q_2 if

$$
q_1 = \frac{2a - b}{a + b}
\tag{8}
$$

If $q_1 > \frac{2a-b}{a+b}$, the best response for player 2 is to set $q_2 = 0$. Otherwise, when $q_1 < \frac{2a-b}{a+b}$, the best response for player 2 is to set $q_2 = 1$. The best response of the two players can be graphically represented in Fig. 7. In the graphical representation, the best response function is the set of best response values when varying the other best response. Based on the concept of mutually best responses, we can identify the Nash equilibria as the crossing points of the best response functions in Fig. 7.

Therefore, if every node randomly chooses path p_1 and p_2 with probability $\frac{2a-b}{a+b}$, $\frac{2b-a}{a+b}$, respectively, then no single node can improve its lifetime by unilaterally deviating from this strategy.

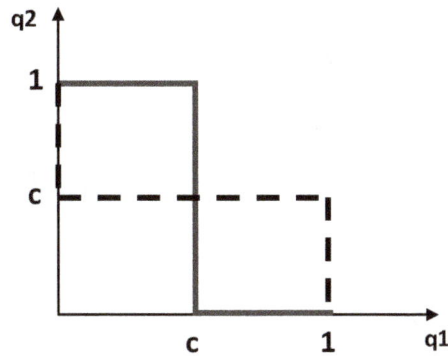

Fig. 7. The best response functions in Lifetime Game, q_1 as a function of q_2 and vice versa. The dash line represents the best response of player p_1, and the solid line is the best response of player p_2. Where $c = \frac{2a-b}{a+b}$.

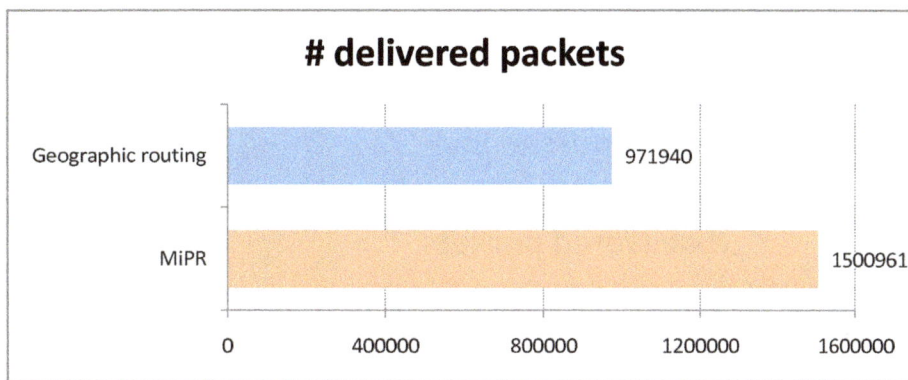

Fig. 8. Total number of packets delivered.

5.3.2. Protocol summary

In sum, MiPR follows two main steps:

1. Setting up path p_1, and path p_2.
2. Using mixed-strategy in Lifetime Game at Nash equilibrium to route data through these paths.

The next hop on path p_1 is the neighbor node which is closest to the destination in the two dimensional space. Similarly, the next hop on path p_2 is the neighbor node which is closest to the destination on the sphere surface. To find the probability of choosing a path, the sender node needs to know the value of a, b. Nodes could obtain the values of a, b by using request packets while setting up routing paths. The request packets will ask nodes about their energy capacity while finding routing paths. One could argue that nodes may act selfishly by lying in their response about their capacity, and this scheme therefore may fail. In this case, we could apply the truthful mechanism in [3] for finding the real value for the shortest paths p_1, p_2.

6. Simulation and evaluation

In order to assess the merit of our proposed protocol, we implement MiPR to compare it with other protocols in the same environment. We assume that the nodes know their position by being equipped

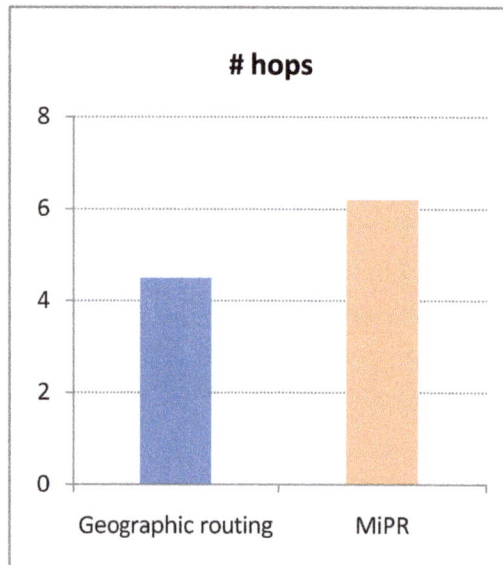

Fig. 9. Average hop-path length.

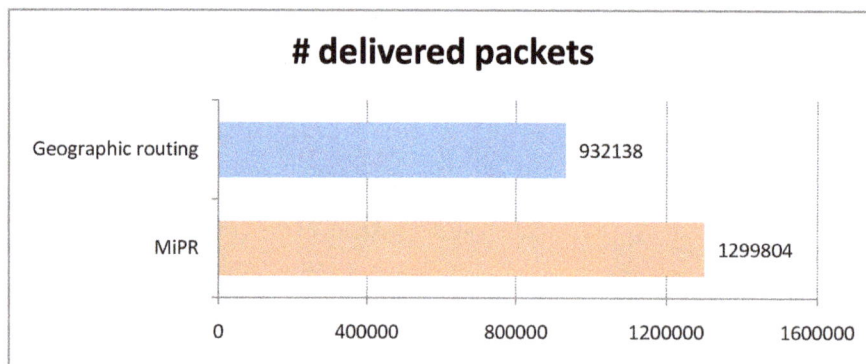

Fig. 10. Total number of packets delivered in the case of 300 nodes.

with a GPS (Global Positioning System) unit, or using some other location determination technique, such as the one in [5]. We consider the scenario of 500 nodes randomly and uniformly deployed on an area 1000×1000. Each node is equipped with a radio with the maximum transmission range of about 200. Nodes have the ability to adjust their transmission ranges to save energy. Thus, we assume nodes use the minimum energy required to transmit packets to the next hop. In *geographic routing*, we let the sender node choose the closest node to the destination as a next forwarder.

We initially assign each node equal battery energy. In each scenario, we ran over 1000 different topologies to get the average value. Figure 8 shows the simulation results of the protocols.

Simulation result shows that MiPR significantly improve network lifetime (1,500,951 vs 971,940 messages delivered), since it balances the traffic between paths, avoiding the creation of a congested area. There is, however, a different situation when we consider the average hop path length, see Fig. 9.

The average hop-length in MiPR is about 20% longer than that in Geographic Routing. This means that it takes more energy and a longer path to send a packet in MiPR protocol. On the other hand, for

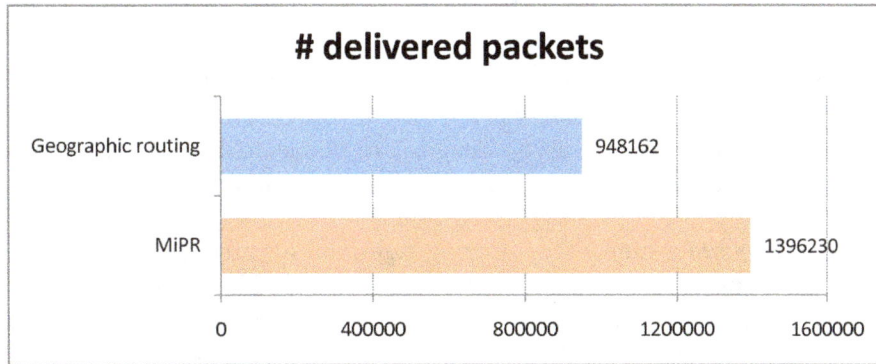

Fig. 11. Total number of packets delivered in the case of 400 nodes.

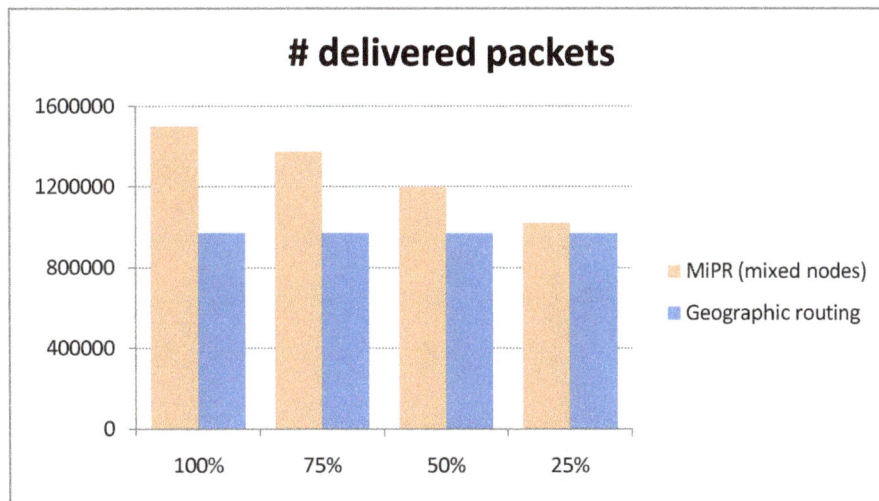

Fig. 12. Total number of packets delivered when the network contains both selfish and regular nodes.

fairness and network lifetime MiPR performed better. This is a common trade-off in lifetime routing problem.

We now keep the same settings while changing the number of nodes. Figures 10 and 11 show the results of the case of 300 and 400 nodes, respectively. These results are quite similar to that of the case of 500 nodes. However, since the graph is not as dense as the case of 500 nodes, each node on the transmission path may have to spend more energy for forwarding the packets to the next hop. Thus, the network lifetime is reduced.

We consider the case in which the networks contain two groups of nodes, a group of selfish nodes (which can choose the routes) and the group that has to follow the geographic routing. Figure 12 shows the results of four different cases in which 100%, 75%, 50% and 25% of the total participants are selfish. We again see that the MiPR gives better results than the original routing protocol.

In this protocol, we assume the network starts operating with equal initial battery energy at each node. When nodes have different capacity of batteries, we just need to make a minor modification in the protocol. When it chooses the next hop in each path, it considers all neighbors that are closer to the destination, then chooses the node with maximum remaining energy.

7. Conclusion

We have studied the problem of routing to maximize the network lifetime in wireless multi-hop ad hoc networks where selfish nodes are taken into account. We first provide insight into the network lifetime problem through game-theoretic perspectives. We used strategic game model to deal with this problem. With affine cost function, we show that the game possesses pure-strategy equilibrium flow even through the senders have different demands.

In the second part, we have proposed a simple local routing algorithm for practical wireless ad hoc networks. Although MiPR does not give the optimal lifetime for the network, it drives the network to some fair equilibrium point where participants do not have incentive to deviate from the protocol. Moreover, MiPR significantly improves the network lifetime, in comparison with the previous routing protocols. This routing protocol is also applicable for other routing objectives of the same situation such as minimizing the network latency or numbers of packet lost.

Acknowledgments

This research is supported by research funding from Vietnam National University in Ho Chi Minh City.

Appendix

Proof of Theorem 1
We recall some notations used in this proof,

- n: number of vertices
- m: number of edges
- x_{ik}: number of packets node i transmit through edge k
- $x_k = \sum_{i=1}^{n} x_{ik}$: total packets pass through edge k
- $c_k(x) : \mathbb{N}^+ \rightarrow \mathbb{R}^+$: is an affine function of edge k, where x is a number of packets pass through it.
- r_i: data amount at i

We use *potential function method* to prove this theorem.

Claim 1 Any solution to the following problem is a pure-strategy Nash equilibrium:

$$Minimize \sum_{k=1}^{m} \left(\sum_{t=0}^{x_k} c_k(t)t \right) \tag{9}$$

$$subject\ to\ x_k = \sum_{i=1}^{n} x_{ik}$$

for $k = 1, \ldots, m$; and $x_{ik} \in [0, r_i]$ for $i = 1, \ldots, n$; $k = 1, \ldots, m$.

Proof. Since the solution to Eq. (9) always exist. It suffices to establish that each is an equilibrium. Let (x_{ik}^*) solve Eq. (9). Assume it does not give rise to an equilibrium. Then for some node j taking some strategy $\mathcal{A}$ under (x_{ik}^*) can reduce its cost by switching to some strategy $\mathcal{B}$.

Here, we define a strategy $\mathcal{A}$ for a node i as a description of:

– Set of paths from i to its destination
– Number of packets are transmitted on each path

We prove claim 1 from the simple case to the general case as follows:

- *Case 1:* Node j moves one packet from path a to path b
- *Case 2:* Node j moves $\delta(>1)$ packets from path a to path b
- *Case 3:* Node j moves δ packets from path a to paths: $b_1, \ldots, b_l$ where each path b_i, $i = 1, \ldots, l$, is added δ_i packets which are taken from path a. We have $\delta = \sum_{i=1}^{l} \delta_i$.
- *Case 4 (general):* Node j changes from strategy $\mathcal{A}$ to strategy $\mathcal{B}$.

Case 1: Node j moves *one* packet from path a to path b
Node j made this move, i.e.

$$\sum_{k \in b \setminus a} c_k(x_k^* + 1)(x_k^* + 1) < \sum_{k \in a \setminus b} c_k(x_k^*)x_k^* \tag{10}$$

Consider the new values

$$(x_{ik}^o) = \begin{cases} x_{ik}^* + 1 & \text{if } i = j, k \in b \setminus a \\ x_{ik}^* - 1 & \text{if } i = j, k \in a \setminus b \\ x_{ik^*} & \text{otherwise} \end{cases}$$

(x_{ik}^o) is clearly feasible for Eq. (9). The objective function evaluated at (x_{ik}^o) is:

$$Objective(1) = \sum_{k \in b \setminus a} \sum_{t=0}^{x_k^*+1} c_k(t)t + \sum_{k \in a \setminus b} \sum_{t=0}^{x_k^*-1} c_k(t)t + \sum_{k \in (a \cap b) \cup (\bar{a} \cap \bar{b})} \sum_{t=0}^{x_k^*} c_k(t)t$$

$$= \sum_{k=1}^{m} \sum_{t=0}^{x_k^*} c_k(t)t + \sum_{k \in b \setminus a} c_k(x_k^* + 1)(x_k^* + 1) - \sum_{k \in a \setminus b} c_k(x_k^*)x_k^*$$

$$< \sum_{k=1}^{m} \sum_{t=0}^{x_k^*} c_k(t)t. \quad \text{A contradiction because of Eq. (10).}$$

Case 2: Node j moves δ packets from path a to b.
Consider the new values after this move.

$$(x_{ik}^0) = \begin{cases} x_{ik}^* + \delta & \text{if } i = j, k \in b \setminus a \\ x_{ik}^* - \delta & \text{if } i = j, k \in a \setminus b \\ x_{ik}^* & \text{otherwise} \end{cases}$$

(x_{ik}^0) is clearly feasible for Eq. (9). The objective function evaluated at (x_{ik}^0) is:

$$Objective(2) = \sum_{k \in b \setminus a} \sum_{t=0}^{x_k^*+\delta} c_k(t)t + \sum_{k \in a \setminus b} \sum_{t=0}^{x_k^*-\delta} c_k(t)t + \sum_{k \in (a \cap b) \cup (\bar{a} \cap \bar{b})} \sum_{t=0}^{x_k^*} c_k(t)t$$

$$= \sum_{k=1}^{m} \sum_{t=0}^{x_k^*} c_k(t)t + \sum_{k \in b \setminus a} \sum_{t=x_k^*+1}^{x_k^*+\delta} c_k(t)t - \sum_{k \in a \setminus b} \sum_{t=x_k^*-\delta+1}^{x_k^*} c_k(t)t$$

For a contradiction, we wish to show that,

$$\sum_{k \in b \setminus a} \sum_{t=x_k^*+1}^{x_k^*+\delta} c_k(t)t < \sum_{k \in a \setminus b} \sum_{t=x_k^*-\delta+1}^{x_k^*} c_k(t)t \tag{11}$$

We note that since cost function is an increasing function, if a node cannot decrease its cost by moving one packet from path a to path b. i.e.

$$\sum_{k \in b \setminus a} c_k(x_k^*+1)(x_k^*+1) \geqslant \sum_{k \in a \setminus b} c_k(x_k^*)x_k^* \tag{12}$$

then it cannot reduce cost by moving some δ (> 1) packets from path a to path b, because

$$\sum_{k \in b \setminus a} c_k(x_k^*+\delta)(x_k^*+\delta) \geqslant \sum_{k \in b \setminus a} c_k(x_k^*+1)(x_k^*+1) \geqslant \sum_{k \in a \setminus b} c_k(x_k^*)x_k^* \tag{13}$$

Otherwise, there are two cases: (i). every packet that node j moves from path a to b induces a lower cost for j; (ii). node j can decrease its cost by moving up to δ packets from path a to path b, but not β packets, for some $\beta > \delta$.

In the subcase (i), this means,

$$\sum_{k \in b \setminus a} c_k(x_k^*+1)(x_k^*+1) < \sum_{k \in a \setminus b} c_k(x_k^*)x_k^* \tag{14}$$

$$\vdots$$

$$\sum_{k \in b \setminus a} c_k(x_k^*+\delta)(x_k^*+\delta) < \sum_{k \in a \setminus b} c_k(x_k^*-\delta+1)(x_k^*-\delta+1) \tag{15}$$

$$\Rightarrow \sum_{k \in b \setminus a} \sum_{t=x_k^*+1}^{x_k^*+\delta} c_k(t)t < \sum_{k \in a \setminus b} \sum_{t=x_k^*-\delta+1}^{x_k^*} c_k(t)t \tag{16}$$

This implies Eq. (11) hold ((16) $\equiv$ (11)).

In the subcase (ii), because node j deceases its cost by moving δ packets from path a to path b, but not β packets, for some $\beta > \delta$, we have,

$$\sum_{k \in b \setminus a} (c_k(x_k^*+\delta)(x_k^*+\delta) - c_k(x_k^*)x_k^*) < \sum_{k \in a \setminus b} (c_k(x_k^*)x_k^* - c_k(x_k^*-\delta)(x_k^*-\delta)) \tag{17}$$

and,

$$\sum_{k \in b \setminus a} (c_k(x_k^*+\beta)(x_k^*+\beta) - c_k(x_k^*)x_k^*) \geqslant \sum_{k \in a \setminus b} (c_k(x_k^*)x_k^* - c_k(x_k^*-\beta)(x_k^*-\beta)) \tag{18}$$

Since $c_k(x)x$ is a linear function, $\sum c_k(x)x$ is also linear. Thus, Eqs (17) and (18) imply that $\sum_{k \in b\backslash a} c_k(x_k^*)x_k^*$ is linear and greater than $\sum_{k \in a\backslash b} c_k(x_k^*)x_k^*$ for some $x \geqslant x_k^* + \beta$. And $\sum_{k \in b\backslash a} c_k(x_k^*)x_k^*$ is linear and smaller than $\sum_{k \in a\backslash b} c_k(x_k^*)x_k^*$ for some $x \leqslant x_k^* + \delta$. $\quad$ (***)

Moreover, since node j transfers δ packets from path a to b, this means,

$$\sum_{k \in b\backslash a} c_k(x_k^* + \delta)(x_k^* + \delta) < \sum_{k \in a\backslash b} c_k(x_k^*)x_k^* \tag{19}$$

together with (***) implies,

$$\sum_{k \in b\backslash a} c_k(x_k^* + \delta - i)(x_k^* + \delta - i) < \sum_{k \in a\backslash b} c_k(x_k^* - i)(x_k^* - i), \forall i \in (0, \delta)$$

Then Eq. (11) holds. Therefore,

$$Objective(2) < \sum_{k=1}^{m} \sum_{t=0}^{x_k^*} c_k(t)t.$$

Case 3: Node j moves δ packets from path a to paths: $b_1, \ldots, b_l$ where each path b_i, $i = 1, \ldots, l$, is added δ_i packets which are taken from path a. We have $\delta = \sum_{i=1}^{l} \delta_i$.

If j can reduce its cost by moving δ_1 packets to path b_1 then it contradicts to proof of case 2. Otherwise, j does not make this move and it could reduce more cost. Likewise, for any other move δ_i to path b_i, node j cannot reduce its cost either. Therefore, this case is not true.

Case 4 (general): Node j changes from strategy $\mathcal{A}$ to strategy $\mathcal{B}$. In other words, j moves some packets from some of its current paths to some other paths.

First, consider path x_1 in j's current paths, assume that node j moves δ packets from path x_1 to paths $\{y_1, \ldots, y_k\}$ each path y_i with δ_i packets. Then this is case 3, i.e. j cannot reduce its cost. Likewise, j cannot reduce its cost on any of its other paths. In other words, node j cannot reduce its cost by unilaterally changing its strategy. $\square$

Claim 1 immediately implies Theorem 1. $\square$

References

[1] M.J. Osborne and A. Rubinstein, *A Course in Game Theory*. Cambridge, MA: The MIT Press, 1994.
[2] L. Lewin-Eytan, J. Naor and A. Orda, Maximum-lifetime routing: system optimization & game-theoretic perspectives. MobiHoc '07. ACM, New York.
[3] A. Mei and J. Stefa, Routing in outer space: fair traffic load in multi-hop wireless networks. MobiHoc '08. ACM, New York, 2008.
[4] R.W. Sinnott, Virtues of the Haversine, *Sky and Telescope* **68**(2) (1984).
[5] J. Hightower and G. Borriello, Location systems for ubiquitous computing, *IEEE Computer* **34** (August 2001), 57–66.
[6] R.W. Rosenthal, The network Equilibrium Problem in Integers. By John Wiley & Sons, Inc. 1973.
[7] R.W. Rosenthal, A class of games possessing pure-strategy Nash equilibria, *International Journal of Game Theory* (1973), 65–67.
[8] J.G. Wardrop, Some theoretical aspects of road traffic research. In proceedings of the Institute of Civil Engineers. Pt. II Vol 1, 1952, pp. 325–378.
[9] N. Nisan, T. Roughgarden, E. Tardos and V. Vazirani, *Algorithmic Game Theory*, Cambridge University Press, 2007.
[10] J. Nash, Noncooperative games, *Annals of Mathematics* **54** (1951), 286–295.

[11]   H.Z. Aashtiani and T.L. Magnanti, Equilibria on a congested transportation network. SIAM Journal on Algebraic and Discrete Methods, 1981, pp. 213–226.

[12]   S. Dafermos, Traffic equilibrium and variational inequalities, Transportation Science, 1980, pp. 42–54.

[13]   M. Florian, Nonlinear cost network models in transportation analysis, *Mathematical Programming Study* (1986), 167–196.

[14]   M.A. Hall, Properties of the equilibrium state in transportation networks, Transportation Science, 1978, pp. 208–216.

[15]   X. Chen and X. Deng, Settling the complexity of 2-player Nash-equilibrium, Electronic Colloquium on Computational Complexity, Fdns. Comp., USA, 2006.

[16]   A. Nagurney, Sustainable Transportation Networks. Edward Elgar, 2000.

[17]   Y. Nesterov, Stable flows in transportation networks. CORE Discussion Paper 9907, 1999.

[18]   Y. Nesterov and A. De Palma, Stable dynamics in transportation systems. CORE Discussion Paper 00/27, 2000.

[19]   A. Orda, R. Rom and N. Shimkin, Competitive routing in multi-user communication networks, IEEE/ACM Transactions on Networking, 1993, pp. 510–521.

[20]   Y. Sheffi, Urban Transportation Networks: Equilibrium Analysis with Mathematical Programming Methods. Prentice-Hall, 1985.

[21]   M.J. Smith, The existence, uniqueness and stability of traffic equilibria. Transportation Research, 1978, pp. 419–422.

[22]   S.K. Dimitris Fotakis and P. Spirakis, *Selfish unsplittable flows. Theoretical Computer Science*, 2005, pp. 226–239.

[23]   P.N. Panagopoulou and P.G. Spirakis, Algorithms for pure Nash equilibria in weighted congestion games, *Journal on Experimental Algorithmics* (2006).

[24]   M.K. Tobias Harks and R.H. Möhring, Characterizing the existence of potential functions in weighted congestion games. In Proc. of the 2nd Symposium on Algorithmic Game Theory (SAGT), 2009, pp. 97–108.

[25]   T. Harks and M. Klimm, Characterizing the existence of pure Nash equilibrium in weighted congestion games. Accepted to the 37th International Colloquium on Automata, Languages and Programming (ICALP), 2010.

Dung T. Tran received his Master degree in Computer Science at the SUNY at Buffalo in 2006 and received his PhD in Computer Science at the University of Texas at Dallas in 2010. His research interests focus on the line between computer networks and game theory.

Trang T.M. Truong received her Bachelor and Master degree in Computer Science at the University of Science in 2004 and 2008, respectively. Her research interests focus on the network routing protocols.

Thanh G. Le received his bachelor degree in Computer Science at the University of Science in 2009. He is working toward his Master degree in Computer Science.

4

A handover security mechanism employing the Diffie-Hellman key exchange approach for the IEEE802.16e wireless networks

Yi-Fu Ciou[a], Fang-Yie Leu[a,*], Yi-Li Huang[a] and Kangbin Yim[b]

[a]*Department of Computer Science, Tunghai University, Tunghai, Taiwan*
[b]*Soonchunhyang University, Asan, Republic of Korea*

Abstract. In this paper, we propose a handover authentication mechanism, called the handover key management and authentication scheme (HaKMA for short), which as a three-layer authentication architecture is a new version of our previous work, the Diffie-Hellman-PKDS-based authentication method (DiHam for short) improving its key generation flow and adding a handover authentication scheme to respectively speed up the handover process and increase the security level for mobile stations (MSs). AAA server supported authentication is also enhanced by invoking an improved extensible authentication protocol (EAP). According to the analyses of this study the HaKMA can effectively and efficiently provide user authentication and balance data security and system performance during handover.

Keywords: HaKMA, DiHam, PKM, WiMax, IEEE802.16, wireless security

1. Introduction

Recently, due to their popularity and the characteristics of convenience and high access speed wireless networks have been a part of our everyday life. Through wireless systems, people can surf web contents, send emails and watch video program outdoors anytime anywhere. To satisfy the requirements of high-speed mobile wireless networks the IEEE 802.16 Working Group in 2005 developed the IEEE 802.16e standard [1] known as the WiMax system which is an extended version of IEEE 802.16 by adding mobility management and a handover scheme so as to provide users with mobile broadband wireless services

To prevent malicious attacks, the IEEE802.16 standard employs a key management and authorization mechanism called privacy key management (PKM) to authenticate users and wireless facilities [9, 26]. However, several problems have been found [17] on this mechanism, such as lack of mutual authentication, and having authorization vulnerabilities and key management failures Also, the high complexity of its authentication mechanism and the presence of design errors [17] make the PKM fail to effectively protect the wireless system. To solve these problems, the IEEE Network Group proposed the PKMv2 in 2005 to fix the defects of the PKMv1 by adding mutual authentication and EAP support. But this enhancement also makes the PKMv2 more complicated and difficult to fix than the PKMv1 if

---

*Corresponding author: Fang-Yie Leu, Department of Computer Science, Tunghai University – No. 181, Section 3, Taichung Port Road, Taichung City 40704, Taiwan. E-mail: leufy@thu.edu.tw.

someday new shortcomings are found. On the other hand, Leu et al. [22] proposed a Diffie-Hellman-PKDS-based authentication method (DiHam for short) to improve some of the defects. However, the scheme does not guarantee full security since it only considers the initial network entry without providing handover and user authentication.

Therefore, in this paper we propose a handover authentication mechanism, called the handover key management and authentication system (HaKMA for short), which as a three layer authentication architecture is an extended version of the DiHam, one of our previous works by improving the key generation flow and adding a handover authentication scheme to respectively speed up the handover process and increase the security level for mobile stations (MSs). It also enhances the AAA server authentication by employing an improved version of the extensible authentication protocol (EAP). To meet different security levels of wireless communication, two levels of handover authentication are proposed. The analytical results show that the HaKMA is more secure than both the DiHam and the PKMv2.

The key contributions of this study are as follows.

(1) According to the security analyses and performance analyses of this study, the security level of the proposed the HaKMA is higher than those of the DiHam, and PKMv2. The authentication cost of the HaKMA is between the PKMv2 and DiHam.
(2) We apply a fast encryption and decryption scheme to protect wireless EAP messages, but keep the advantages of the PKMv2, such as mutual authentication and EAP support.
(3) The HaKMA provides fast and low cost handover authentication instead of skipping the handover authentication process.

The rest of this paper is organized as follows. Section 2 introduces the background and related work of this study. Sections 3 and Section 4, respectively describe the proposed handoff approaches for the DiHam and HaKMA Security analyses are presented in Section 5. Section 6 describes system simulation and gives discussion. Section 7 concludes this paper and outlines our future research.

2. Background and related work

2.1. The WiMax network architecture

Figure 1 shows a modern multi-layer wireless network configuration. The access service network gateway (ASN-GW) is connected to a network service provider (NSP) backbone network, and BSs are directly linked to their ASN-GWs. An ASN-GW can not only communicate with other ASN-GWs via the backbone network through R3 reference points, but also directly communicate with other ASN-GWs with direct links via R4 reference points [7,11]. An NSP may provide many ASN-GWs to serve users. An MS may currently link to a BS, or hand over between two BSs under the same ASN-GW, called the Intra-ASN-GW handover, or under different ASN-GWs, called the Inter-ASN-GW handover.

2.2. Privacy key management protocol

The PKM protocol first specified by the IEEE 802.16-2004 provides device authentication also known as facility authentication, and the PKMv2 proposed in the IEEE 802.16e-2005 is a new version of the PKM protocol which corrects design errors for security found in the PKMv1 [11] and supports user authentication.

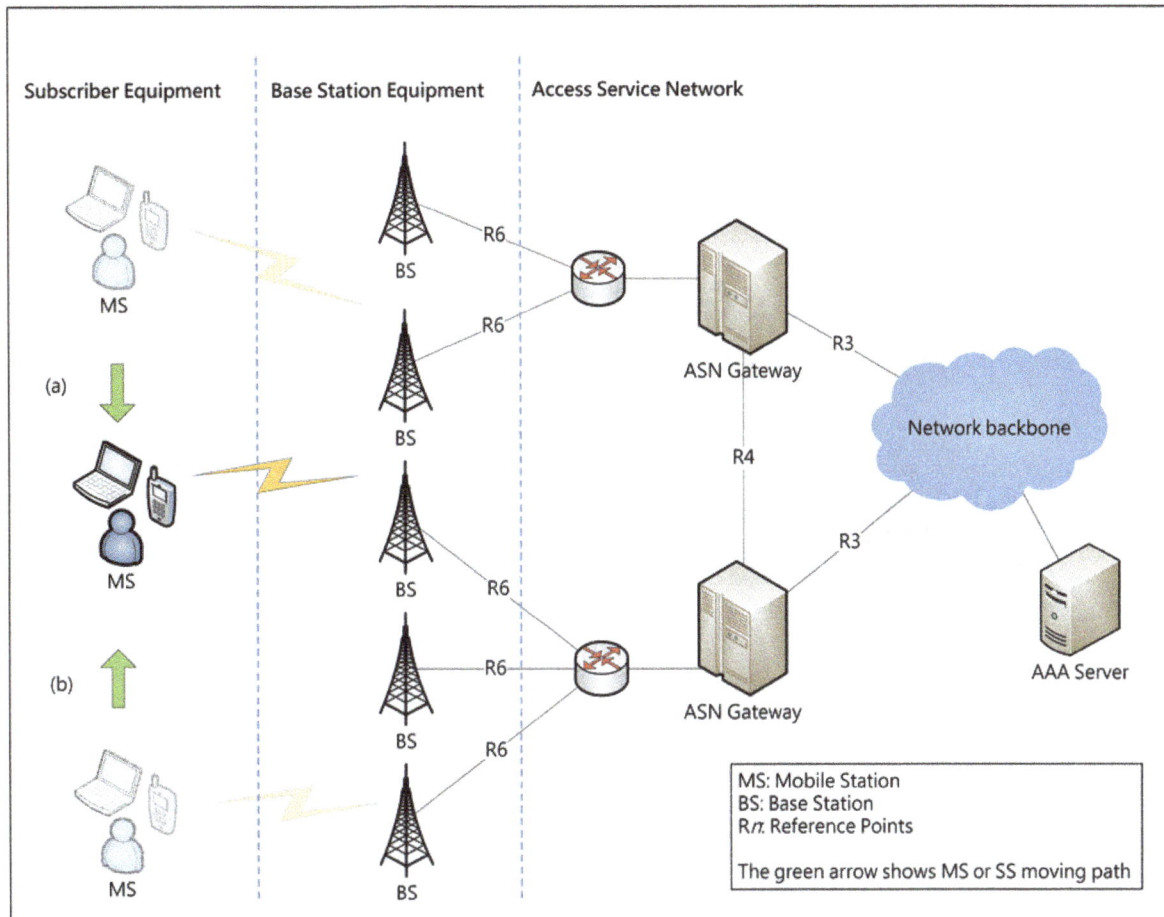

Fig. 1. The WiMax network configuration in which MS may perform an Inter-ASN-GW handover and an Intra-ASN-GW handover.

2.2.1. PKMv2

The PKMv2 uses X.509 digital certificates [15] together with either an RSA public-key encryption algorithm or a sequence of RSA device authentication to further authenticate communication facilities. After that, an EAP method is employed to authenticate users [18]. The encryption algorithms used by PKMv2 for key exchange between MS and the BS are more secure than those used by the PKMv1 [1].

The PKMv2 allowing mutual authentication or unilateral authentication, also supports periodic re-authentication/re-authorization and key renew. All PKMv2 key derivations are performed based on the Dot16KDF algorithm defined in IEEE802.16e standard [1].

Figure 2 illustrates the initial network entry procedure of the PKMv2 which consists of 9 steps [1,2, 20,21]:

(1) **Initiation of network entry**: In this step, MS performs the physical layer initialization process including initial ranging and network entry. After the wireless link between the BS and MS is established, Authenticator starts the EAP exchange step.

(2) **EAP exchange**: In this step, MS and the Authenticator (which could be in the BS or ASN-GW) negotiate with each other to choose a suitable EAP method through EAP Request/Identity and

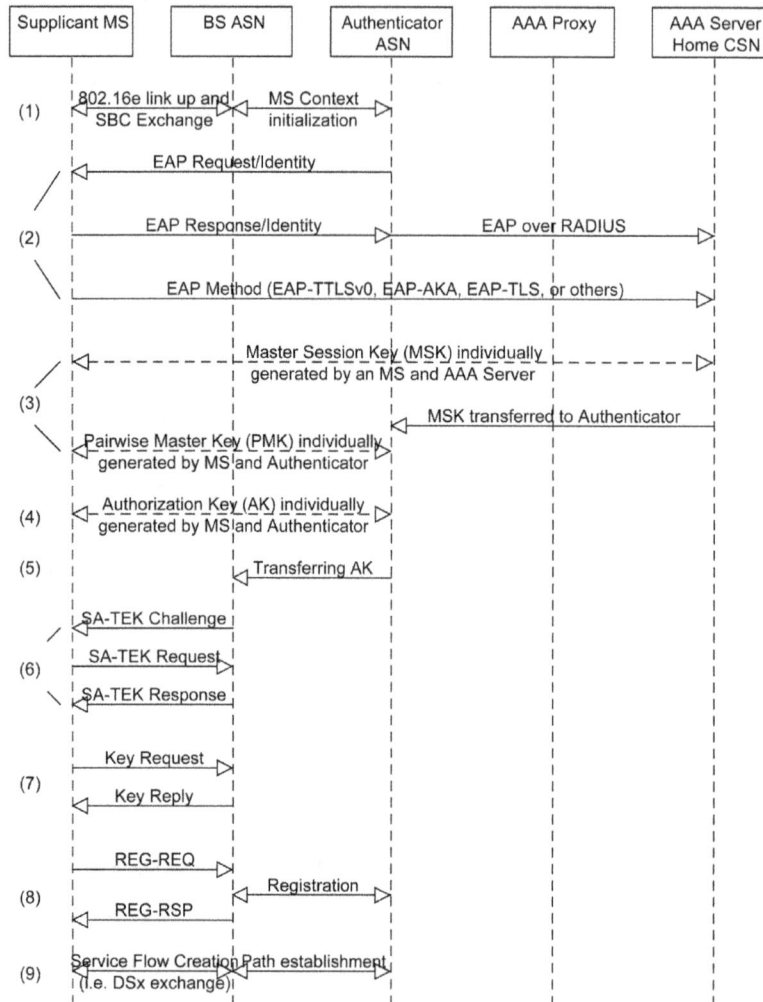

Fig. 2. The PKMv2 authentication procedure [2], where SBC stands for MS basic capability.

EAP Response/Identity messages. With the chosen method they mutually exchange certification messages so that the AAA server can authenticate the user.

(3) **MSK establishment**: In step 3, a master session key (MSK) is established between MS and the AAA server. The AAA server transfers the MSK to the Authenticator through a secure path [2,11] using the Radius protocol With the MSK, both the BS and Authenticator selfgenerate a pairwise master key (PMK).

(4) **Authorization Key (AK) generation**: Authenticator and MS individually generate the AK by using the PMK [1]. If the RSA-based authorization process is used, a pre-primary authentication key (pre-PAK) will be produced. The AK is derived from the pre-PAK. If RSA-based and EAP-based authorization is employed, a MSK and a pre-PAK will be generated. The AK is then calculated by using the two keys. If only EAP-based authorization is involved, only the MSK is generated, and the AK is derived from the MSK.

(5) **AK transfer**: Authenticator delivers the AK other keys (e.g., pre-AK and the MSK generated by itself) and parameters (e.g., MSK lifetime) to the serving BS. The BS caches them for the

following steps.

(6) **AK liveliness establishment and SA transfer**: This step consists of three phases. In the first the BS transmits a PKMv2_SA_TEK_Challenge message which includes a BS_Random challenge and identity of the AK to MS to establish Security Association (SA) between MS and BS In the second phase, MS responds with a PKMv2_SA_TEK_Request message which contains a random number *MS_Random* and other security attributes In the third phase, the BS transmits a PKMv2_SA_TEK_Response message containing the response of requested properties listed in the PKMv2_SA_TEK_Request message to MS to complete the SA.

(7) **TEK generation and transfer**: When MS would like to deliver data messages, it sends a Key-Request message to the BS. The BS then generates Traffic Encryption Keys (TEKs), encrypts them by using the Key Encryption Key (KEK) individually generated by MS and the BS in this step, and transfers the encrypted TEKs to the MS through a Key-Reply message. For each SA, the MS needs two TEKs to download streams and upload streams. This step repeats if multiple SAs are involved. Note that Leu et al. [22] challenged the TEK transfer since hackers could easily intercept the message and then decrypt the TEKs.

(8) **Network registration**: In this step, MS sends a REG_REQ message to the BS providing Access Service Network (ASN), and the BS responds with a REG_RSP message to negotiate network parameters such as Connection ID (CID), IP version, and so on. The REG process makes MS known to Authenticator and ASN-GW, and triggers the service or data transfer process.

(9) **Service flow creation**: In this step the BS uses DSA-REQ/RSP/ACK MAC management messages to create a new service flow and map an SA to that service flow thereby associating the corresponding TEKs with it [2].

2.2.2. PKMv2 handover procedure

Based on the IEEE 802.16 standard [1] and other related documents [2], the PKMv2 handover procedure shall be optimized within the same mobility domain. Sharing TEKs among BSs inside a trustable mobility domain is possible if the handover procedure can transfer TEK context information from BSs to neighbor BSs and these BSs are treated as entities of the same security levels.

In summary the PKMv2 protocol controls security key derivation, exchange and renew, and manages user authorization states to ensure the security of the user's data communication.

2.3. Diffie-Hellman PKDS-based authentication

The DiHam [22] was developed based on the PKMv1 by improving the key exchange flow and providing different data security levels. Basically, its key exchange process as shown in Fig. 3 consists of two phases the authentication phase and TEK exchange phase.

In the authentication phase, the AK is individually generated by the BS and MS after the delivery of the AuthenticationRequest message and AuthenticationReply message [22].

At first, MS generates three random numbers and the corresponding public keys, and sends an AuthenticationRequest message to the BS. The BS on receiving the message checks the correctness of the message generates three random numbers, the corresponding public keys, and common secret keys (CSK), and sends an AuthenticationReply message that contains a challenge field and information of the encrypted keys to the MS so that MS can authenticate the BS, and derive the CSKs. The formats of the authentication messages are shown in Fig. 4.

In the TEK exchange phase, three security levels of TEK generation processes are proposed to meet different user security requirements. This phase starts when MS sends a TEK-Exchange-Request message to the BS, and ends when the BS replies with a TEK-Exchange-Reply message.

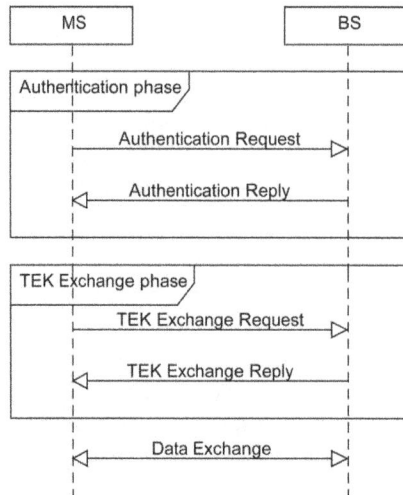

Fig. 3. The DiHam authentication process.

Authentication-Request (MS to BS):
    OP_Code|Cert(MS)|P_{RM1}|P_{RM2}|P_{RM3}
Authentication-Reply (BS to MS):
    OP_Code|Cert(MS)|P_{RB1}|P_{RB2}|P_{RB3}|Certfun(CSK1, CSK2)|EXOR(CSK2, pre_AK2)|
    EXOR(CSK3, pre_AK3)

Fig. 4. The DiHam authentication messages deployed in the authentication phase.

Level-1 TEK-Exchange-Request (MS to BS):
    OP_Code|Certfun(CSK1, pre_AK1)|*Security-capabilities*
Level-1 TEK-Exchange-Reply (BS to MS):
    OP_Code|Certfun(CSK2, pre_AK2)|EXOR(CSK2, old_TEK)|EXOR(CSK3, new_TEK)|
    TEK lifetime

Fig. 5. Formats of the exchanged messages for Level1 TEK.

2.3.1. Level1 TEK exchange

Level-1 TEK is designed for applications of low-security level such as web surfing. The BS is responsible for creating TEKs, encrypting TEKs with CSKs and transferring the TEKs to MS. Figure 5 shows the format of the exchanged messages

2.3.2. Level2 TEK exchange

In Level 2, MS generates a pre_TEK, and transfers the encrypted pre_TEK to the BS through a TEKExchangeRequest message. The BS then randomly chooses one of five generated TEKs and sends the TEK sequence number to MS telling MS which TEK is chosen. No keys actually used by MS and the BS are directly delivered through wireless channels. Only those deployed to generate keys are transferred. The security level of Level2 TEK is then higher than that of Level1 TEK. Level-2 TEK is suitable for voice phone calls and personal message communication. Figure 6 shows the formats of the two messages for Level-2 TEK.

Level-2 TEK-Exchange-Request (MS to BS):
    OP_Code|Certfun(CSK1, pre_AK1)|EXOR(CSK2, pre_TEK)|*Security-capabilities*
Level-2 TEK-Exchange-Reply (BS to MS):
    OP_Code|Certfun(CSK2, pre_AK2)|*TEK seq num*|*TEK lifetime*

Fig. 6. Formats of the exchanged messages for Level2 TEK.

Level-3 TEK-Exchange-Request (MS to BS):
    OP_Code|Certfun(CSK1, pre_AK1)|*Security-capabilities*
Level-3 TEK-Exchange-Reply (BS to MS):
    OP_Code|Certfun(CSK2, pre_AK2)|*TEK seq num*|*TEK lifetime*

Fig. 7. Formats of the exchanged messages for Level3 TEK.

2.3.3. Level3 TEK exchange

If a wireless communication needs a higher security level than that of Level2 TEK, Level3 TEK is then employed. In this level, the BS and MS individually generate 15 pre-TEKs and 75 TEKs. After that, the BS sends a TEKExchangeReply message that contains the sequence number of a chosen TEK for later data transfer to MS. Figure 7 shows the formats of the exchanged messages.

From the computation complexity viewpoint, a Level3 TEK exchange process consumes many more computation resources and has longer key generation delays than those of levels 1 and 2. But its security level is the highest among them.

However, the DiHam as stated above only considered initial network entry, without dealing with handover network re-entry. If it involves a handover procedure, the whole process needs to be fully performed on each handover, consequently causing serious service disruption time (SDT). Several fast handover schemes have been proposed. Mobility prediction proposed by Fülöp et al. [13] and the Client-based Mobility Frame System also introduced by Fülöp et al. [14] are two examples. Yang et al. [25] considered self-similarity in data traffic, handover, and frequency reuse to estimate the spectrum requirements of mobile networks so as to speed up communication and shorten SDT. However, the authors of these three papers did not deal with the security layer handover support. Otherwise, their re-entry delay would be long.

3. Handover involving the DiHam

A handover node, e.g., a node newly entering the system needs to perform initial ranging and authentication [1,21] before it can securely exchange data messages with a target BS. But this may interrupt the communication between MS and ASN-GW. To shorten the disruption time, a method to reduce the network re-entry delay is required. Reusing existing authentication keys and pre-calculating authentication keys are the possible methods.

If we temporarily ignore the difference between the three levels of TEKs proposed by the DiHam, and the DiHam is applied to the handover process, the target BS has to know MS's random numbers $RS_i, i = 1,2,3$, before it can generate TEK. However, if the serving BS can deliver Cert(MS) and MS's RS_i to the target BS (we call the message delivered the Inter-BS message) before MS enters the overlapped area of the target BS's and serving BS's communication areas the generation of new authentication parameters by the target BS and MS's handover network reentry to the target BS can be performed simultaneously. Therefore, the AuthenticationReply message can be sent to MS immediately by the target BS once

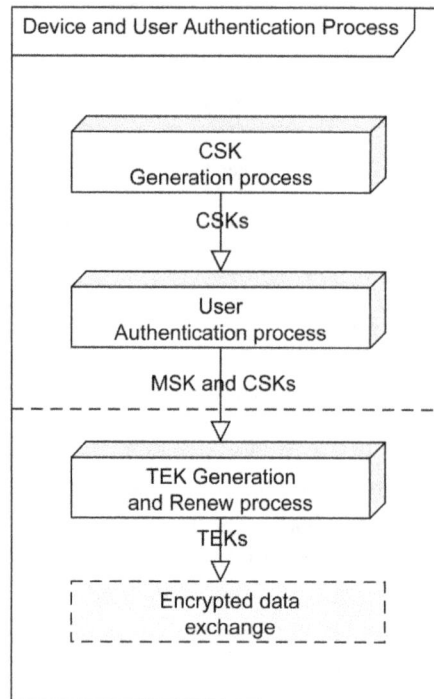

Fig. 8. Key hierarchy and sequence chart of the HaKMA authentication scheme. Processes above the dashed line are invoked by MS and Authenticator, and those beneath the dashed line are employed by MS and the BS.

the handover network reentry is completed. This change can effectively reduce handover delay, and MS's AuthenticationRequest message is no longer required since it is replaced by the Inter-BS message. Moreover, MS can also use MS's original private key RS_i and public key P_{RSi} $i = 1$–3 as well as the target BS's private keys RBs conveyed in the AuthenticationReply message to generate new *CSKs*, *pre_TEKs*, and then new *TEKs* to encrypt data messages.

The benefits of pre-calculating keys in the handover procedure are that the serving BS only needs to deliver MS's RS_i and Cert(MS) to the target BS and the SDT only relies on the PHYlayer handover delay [21].

According to the IEEE802.16 standard, the optimized handover can skip the security sublayer operation and reuses old keys such as TEKs [1,16], or provide handover support through mobile IPv6 with other proposed schemes [6,12,19,24]. In the DiHam if the serving BS can transfer all required security keys including CSKs Cert(MS), and RS_i to the target BS, the target BS can then reuse the TEKs currently employed by the serving BS without the requirements of re-calculating authentication keys and deriving current TEKs resulting in the fact that the AuthenticationReply TEK-Exchange and TEK-Reply messages can be further omitted. After entering the target BS's communication range and finishing its handover network re-entry MS can directly communicate with the target BS.

4. The proposed security system

The HaKMA, besides retaining the advantages of DiHam's Key Exchange and AK Generation processes [22] also involves an enhanced version of an EAP method to authenticate users. A handover support that improves the security level of the HaKMA wireless environment is added as well.

Table 1
Terms and functions used in this study

Term or function	Explanation
P	A strong prime number
g	The primitive root of P
RS_i and $RA_i, i = 1, 2$	Private keys generated by MS and Authenticator
P_{RSi} and $P_{RAi}, i = 1, 2$	Public keys generated by MS and Authenticator
$CSKi, i = 1,2$	Common secret keys
EXOR(x, y)	Exclusive OR function, i.e., $x \oplus y$
SEXOR(*key, data*)	Stream exclusive OR function, repeating *key* content to match the length of *data*, and performing exclusive OR bit by bit.
Certfun(a,b,...)	Modulus function, i.e., $g^{a+b+\cdots} \bmod P$
Encrypt(*PubKey, message*)	The standard RSA-OAEP-Encryption function for encrypting *message* into *ciphertext* with given public key *PubKey*
Decrypt(*PrivKey, ciphertext*)	The standard RSA-OAEP-Decryption function that decrypts *ciphertext* into *plaintext* with given private key *PrivKey*
ADR(a,b)	A binary adder, but ignoring the carry of the greatest significant bit

As stated above, the HaKMA architecture consists of three isolated processes (see Fig. 8): The CSK Generation process in which ASN-GW and MS mutually authenticate each other, the User Authentication process in which the AAA server authenticates users, and the TEK Generation and Renew process in which TEKs are produced to encrypt data messages. The three processes together are called the HaKMA authentication scheme. As a layered architecture any change in one of the three processes does not affect the functions of others, consequently making it easier for us to develop a new authentication process for IEEE802.16 wireless networks when some functions in one of the three processes need to be modified. The outputs of the three processes are sequentially CSKs, MSK and CSKs and TEKs. In this study, we move Authenticator from the BS to ASN-GW to simplify the HaKMA architecture and its handover process.

4.1. Initial process of network entry

With the HaKMA, when an error occurs, the BS sends an error message to MS and Authenticator. If MS has successfully completed one or two previous processes, but fails in an underlying process, the failed process is resumed from its beginning, instead of re-initiating the CSK Generation process. Of course, if the CSK Generation process fails, the HaKMA authentication scheme should be restarted. Table 1 lists terms and functions used in this study.

4.1.1. CSK generation process

The main objective of the CSK Generation process is to perform mutual authentication between Authenticator and MS, and produce two CSKs. Figure 9 shows the sequence chart. It first establishes a secure communication channel between MS and Authenticator, and completes the following steps for initial network entry. MS and Authenticator first mutually check each other's X.509 certificate, and perform a DiHam-like process to generate CSKs which are only known to MS and Authenticator and with which both sides encrypt those messages exchanged in the User Authentication process and TEK Generation and renew process. The BS recognizes a message received by accessing its OP_Code, and relays messages for MS and Authenticator without providing any authentication functions.

This process can be further divided into two phases: Authenticator-CSK phase and MS-CSK phase.

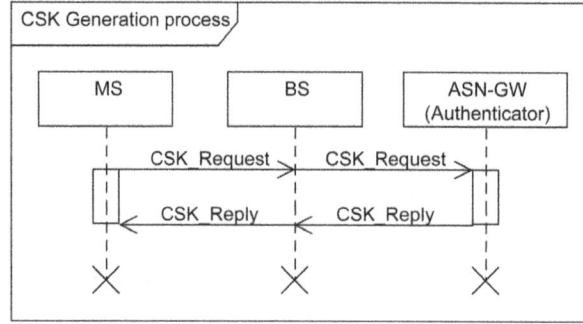

Fig. 9. Sequence chart of the CSK Generation process. BS only relays messages for MS and Authenticator.

4.1.1.1. Authenticator-CSK phase

In this phase, MS first sends a CSK_Request message the format of which is shown in Fig. 10, to Authenticator.

$$OP\_Code|\ NS_{MS}|\ Cert(MS)|\ P_{RM1}\ |\ P_{RM2}|Capabilites|SAID|HMAC(PubKey(MS))$$

Fig. 10. Format of a CSK_Request message sent by MS to Authenticator.

In this message, RM_1 and RM_2, two random numbers, are private keys generated by MS, and P_{RM1} and P_{RM2} are two public keys where

$$P_{RMi} = g^{RM_i} \bmod P, 1 \leqslant i \leqslant 2 \tag{1}$$

NS_{MS}, the nonce, is a timestamp indicating when this message is created, the capabilities field lists the security configurations acceptable by MS, a SAID field contains MS's primary SAID that is currently filled with the basic CID, and the hash-based message authentication code (HMAC) function produces a message signature by inputting all fields of the message as its plaintext and $PubKey(MS)$ as the encryption key.

Authenticator on receiving the message checks to see whether the message signature calculated by itself using $PubKey(MS)$ retrieved from Cert(MS) and the HMAC($PubKey(MS)$) sent by MS are equal or not. If not, implying the message has been altered, the Authenticator discards this message. If yes, it randomly selects two random numbers RA_1 and RA_2 as private keys to generate the corresponding public keys P_{RA1} and P_{RA2} where

$$P_{RAi} = g^{RA_i} \bmod P, 1 \leqslant i \leqslant 2, \tag{2}$$

It further produces two CSKs, i.e., *CSK1* and *CSK2*, where

$$CSKi = P_{RMi}{}^{RA_i} \bmod P, 1 \leqslant i \leqslant 2 \tag{3}$$

and calculates the certificate function Certfun($PubKey(MS)$, $CSK1,CSK2$).

After that, Authenticator sends a CSK_Reply message of which the format is shown in Fig. 11, to MS. To ensure that the message is securely delivered, a HMAC is also added.

$$OP\_Code|NS_{MS}|\ NS_{Authenticator}|\ Cert(Authenticator)|Encrypt(PubKey(MS), P_{RA1}|P_{RA2})|$$
$$Certfun\ (PubKey(MS),CSK1,CSK2)|HMAC(PubKey(Authenticator))$$

Fig. 11. Format of a CSK_Reply message sent by Authenticator to MS.

4.1.1.2. MS-CSK phase

MS on receiving the CSK-Reply message checks to see whether the message has been maliciously modified or not by comparing not only the HMAC value calculated with the value retrieved from the CSK_Reply message received, but also NS_{MS} retrieved from the message with previous nonce involved in CSK_Request message. They should be individually equal. Otherwise, the message is discarded. MS further records $NS_{Authenticator}$ for later authentication, and checks to see whether the Authenticator is trustable or not by comparing the authenticator's certificate Cert(*Authenticator*) with the certificate list provided by a trustable network provider and pre-installed in the MS. If yes, MS retrieves the public key P_{RA1} and P_{RA2} by performing RSA decryption function with its own private key, calculates CSKs, i.e., *CSK1* and *CSK2*, and the certificate function Certfun(*PubKey(MS), CSK1,CSK2*), and then compares the calculated Certfun() value with the one conveyed on CSK_Reply message sent by Authenticator where

$$CSKi = P_{RAi}^{RM_i} \bmod P, 1 \leqslant i \leqslant 2 \tag{4}$$

If the two Certfun() values are equal, the CSK Generation process terminates. MS starts the User Authentication process. Otherwise, MS discards the message and the calculated CSKs, and waits for a valid CSK-Reply message for a predefined time period.

If MS cannot receive a reply from the Authenticator until timeout, it assumes the CSK Generation process fails and then restarts the process by re-sending a CSK-Request message to Authenticator

4.1.2. User authentication process

In this process, MS and Authenticator first negotiate with each other to choose an EAP method. After that, Authenticator communicates with the AAA server to check to see whether the user is authorized to access requested services or not.

However, EAP was originally designed for wired networks [23]. When it is applied to wireless networks, hackers may intercept and decrypt sensitive information. To solve this problem, in this study, CSKs are invoked to encrypt messages exchanged between MS and Authenticator. In fact, some EAP methods do not provide security mechanisms (e.g., EAP legacy methods). Employing our encryption scheme can ensure the security of the originally insecure EAP communication between MS and the BS so that hackers cannot easily decrypt sensitive information, even though an insecure EAP authentication method is used.

Further, we employ EAP-AKA [5] as an EAP example, and suggest employing EAPauthentication based AK generation flow [1] to balance handover performance and the security level of the system under consideration. Basically, our modular design strategy results in the fact that any authentication methods designed for wireless authentication [5] can substitute EAP-AKA to perform user authentication. A general EAP authentication process can be found in [1].

Before this process starts, MS first generates a 160-bit random number *RAND*, and derives the EAP encryption key

$$\text{EAP Encryption Key} = \text{ADR(EXOR(}CSK1,RAND\text{), }CSK2\text{)} \tag{5}$$

which is generated individually by MS and Authenticator and used to encrypt or decrypt EAP messages by employing a streamed Exclusive OR method/function SEXOR(*EAP Encrypt Key, EAP-messages*).

Figure 12 illustrates our User Authentication process in which MS sends a PKMv2-EAP-Start message the format of which is shown in Fig. 13, to notify. Authenticator of the start of the process. Authenticator on receiving the message extracts *RAND* by using its private key, derives the EAP encryption key, and replies with an EAP-Request/Identity message containing a list of available EAP methods that

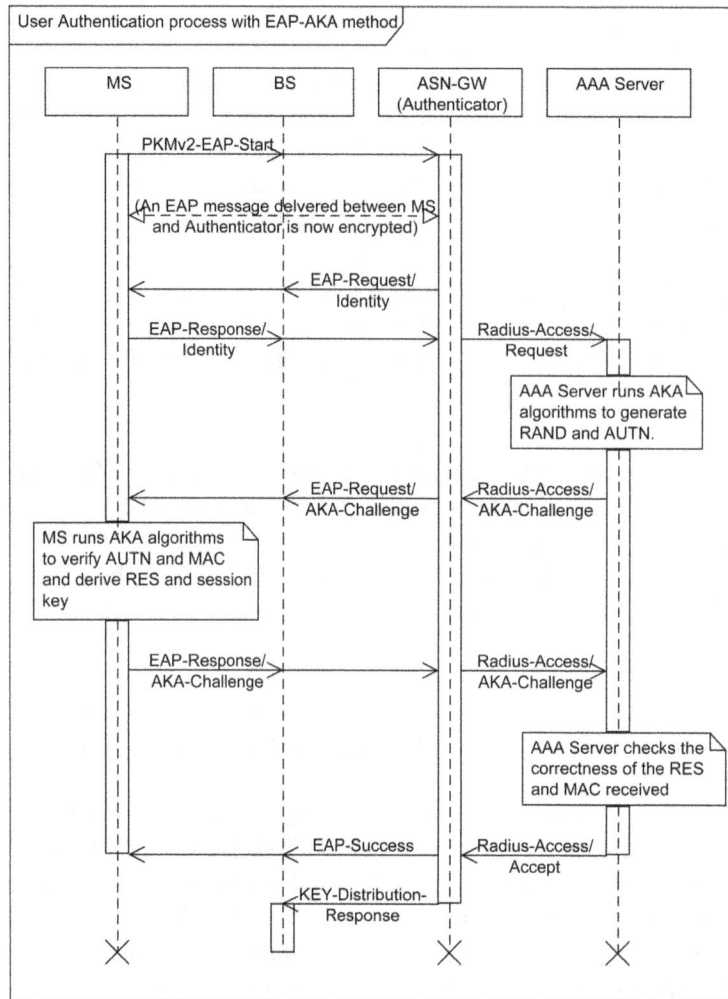

Fig. 12. The HaKMA's User Authentication process with the EAP-AKA method.

Authenticator supports to MS to start security negotiation. MS selects a suitable EAP method and sends an EAP-Response/Identity message to Authenticator. Authenticator then sends a Radius-Access request message which contains the Network Access Identifier (NAI) and other attributes conveyed in the EAP-Response/Identity message received to the AAA server via the Radius protocol. The AAA server invokes its AKA algorithms to generate a random number RAND and an AKA parameter AUTN, and then sends them to MS where AUTN is an authentication value produced by the AuC for authenticating the AAA server in the future and AuC is a mobile network element used to authenticate MS [5].

After that, the AAA server sends an AKA-Challenge message which contains AUTN to MS via Authenticator. MS on receiving the message runs its AKA algorithms to verify AUTN and MAC derives RES (an authentication result generated by MS) and the session key, and then sends an EAP-Response/AKA-Challenge message to the AAA server. After checking the correctness of the MAC and the RES received, the AAA server sends a Radius-Access/Accept message which contains an MSK 512 bits in length to Authenticator. Authenticator delivers an EAP-Success message to MS [5], indicating the user is authenticated.

OP_Code| NS$_{Authenticator}$| NS$_{MS}$| Encrypt($PubKey(Authenticator)$, $RAND$)|

HMAC($PubKey(Authenticator)$))

Fig. 13. Format of a PKMv2-EAP-Start message sent by MS to Authenticator.

OP_Code| BSID|SAID| CSK1| CSK2| MSK|BS-Random|TEK_Lifetime|TEK_Count| TEK1| TEK2

Fig. 14. Format of a KEY-Distribution-Response message sent by Authenticator to the BS.

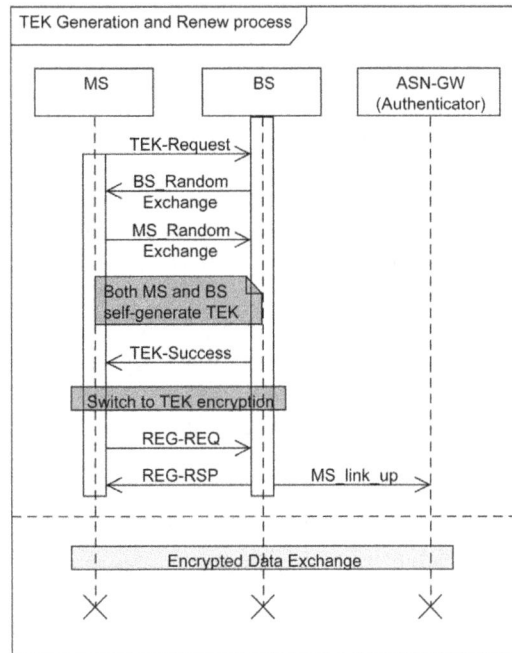

Fig. 15. Sequence chart of the TEK Generation and Renew process.

OP_Code|NS$_{MS}$|SAID|HMAC(ADR(CSK1,CSK2))

Fig. 16. Format of the TEK-Request message sent by MS to the BS.

4.1.3. Key distribution in the initial network entry

In the HaKMA, we design a key distribution message, prefixed by KEY-Distribution to deliver keys among the serving BS, target BS and Authenticator and from a HaKMA sub-process to the next sub-process. Authenticator sends a KEY-Distribution-Response message which contains the MSK and CSKs to the BS (see Fig. 12). Figure 14 shows the format of this message. Note that both TEK_Lifetime and TEK_Count are zero since TEKs have not been generated. Their usage will be described later.

BS on receiving this message extracts the MSK and CSKs from this message, and starts its TEK Generation and Renew process Currently, both the BS and MS possess the MSK and CSKs.

4.1.4. TEK generation and renew process

Like that in the DiHam, TEKs are individually generated by MS and the BS by employing required key parameters supplied by Authenticator. Figure 15 shows the process which starts when MS sends a TEK-Request message, the format of which is shown in Fig. 16 to the BS.

Both MS and the BS invoke the Dot16KDF algorithm which accesses the first 160 bits of MSK [1], to individually derive AK, where

$$AK = \text{Dot16KDF}(\text{``AK''},2,\text{TRUNCATE}(MSK,160),160) \tag{6}$$

The BS (MS) self-generates a random number called *BS_Random* (*MS_Random*) which is also 160-bits in length. *BS_Fingerprint* (*MS_Fingerprint*) is then generated by encrypting *BS_Random* (*MS_Random*) with CSKs and AK, and delivered to MS(BS) through wireless channels where

$$BS\_Fingerprint = \text{ADR}(\text{EXOR}(BS\_Random, CSK1), AK) \tag{7}$$

and

$$MS\_Fingerprint = \text{ADR}(\text{EXOR}(MS\_Random, CSK2), AK) \tag{8}$$

The purpose is to protect the random number from being intercepted by a hacker. After that, a BS_Random_Exchange message the format of which is shown in Fig. 17 and which carries *BS_Fingerprint*, OP_Code, and lifetime of TEKs, denoted by Key-lifetime, is sent by the BS to MS.

OP_Code|NS_{MS}|NS_{BS}|BS_Fingerprint|Key-lifetime|HMAC(ADR(CSK1,CSK2))

Fig. 17. Format of the BS_Random_Exchange message sent by the BS to MS.

MS on receiving the message decrypts the *BS_Random* by using one of the following formulas:

$$BS\_Random = \begin{cases} (BS\_Fingerprint - AK) \oplus CSK1, & \text{if } BS\_Fingerprin \geqslant AK \\ (BS\_Fingerprint + \bar{A}K + 1) \oplus CSK1, & \text{if } BS\_Fingerprint < AK \end{cases} \tag{9}$$

After that, MS sends an MS_Random_Exchange message the format of which is shown in Fig. 18 to the BS.

OP_Code|NS_{BS}|NS_{MS}|MS_Fingerprint|HMAC(ADR(CSK1,CSK2))

Fig. 18. Format of the MS_Random_Exchange message sent by MS to the BS. Note that Key-lifetime is not involved since it is determined by the BS.

Following that, MS generates TEKs where

$$TEK_i = \text{EXOR}(\text{ADR}(\text{EXOR}(MS\_Random,AK),BS\_Random), CSK_i), 1 \leqslant i \leqslant 2 \tag{10}$$

The BS on receiving the MS_Random_Exchange message retrieves the *MS_Random* and generates TEKs with the same formula where *MS_Random* is calculated by using one of the following formulas:

$$MS\_Random = \begin{cases} (MS\_Fingerprint - AK) \oplus CSK2, & \text{if } MS\_Fingerprint \geqslant AK \\ (MS\_Fingerprint + \bar{A}K + 1) \oplus CSK2, & \text{if } MS\_Fingerprint < AK \end{cases} \tag{11}$$

Now both sides possess the TEKs. A TEK-Success message shown in Fig. 19 is sent by the BS to inform MS of the success of the TEK generation. MS registers its terminal device with the BS by sending a REG-REQ message, and the BS replies with a REG-RSP message which are both defined in the IEEE802.16 standard to finish this process. The data exchange can now be started.

OP_Code|NS_{BS}|NS_{MS}|HMAC(ADR(TEK1,TEK2))

Fig. 19. Format of the TEK-Success message sent by the BS to MS.

4.2. Handover process of network Re-entry

The main objectives of a handover process include minimizing the handover delay by key reuse and pre-distribution in the employed security scheme. If a user is authenticated in the initial network entry as stated above, we assume that he/she is still authenticated after handover. This means we can reuse the MSK and CSKs generated in previous processes to avoid the re-authentication delay. For security reasons, we can also renew TEKs on each handover. That means each time when MS moves to a target BS, new TEKs are required to substitute for the two TEKs used by the serving BS, called previous TEKs (prev_TEKs for short to avoid confusing with the term pre_TEKs used in PKMv2).

In this study, two security levels of handover are proposed to meet different security requirements. With Level1 handover, before prev_TEKs expire, the target BS after MS's handover reuses the same TEKs to encrypt data messages so as to shorten communication disruption time. With Level2 on each handover, the target BS temporarily reuses prev_TEKs to communicate with MS, generates new TEKs, and encrypts data messages with the new TEKs.

4.2.1. Key distribution in the network Re-entry

To deliver key information between MS and Authenticator, the KEY-Distribution-HOInfo message shown in Fig. 20 is designed to provide MS with handover support. In this message, the TEK_Count indicates the number of generated TEK pairs, and the TEK_Lifetime shows TEKs' remaining life time in minutes.

We also provide a KEY-Distribution-Request message shown in Fig. 21 for the target BS to request MS's key information from Authenticator.

Before MS hands over to the target BS, the serving BS sends a KEY-Distribution-HOInfo message to its Authenticator. Authenticator stores the required keys in the MS's corresponding tuple in its authentication key table (AK Table for short), a table used to keep authentication keys including the MSK and CSKs for the Authenticator's subordinate MSs. The AK Table is indexed by SAID to identify which MS the keys being considered belong to. Figure 22 shows the fields of this table. Authenticator further checks to see whether or not the target BS that it should newly associate with is in its BS Table a table for recording the BSIDs of the Authenticator's subordinate BSs, including those of its own and those subordinated by all its successor Authenticators, implying Authenticators are organized as a hierarchy. The table has only one field BSID. If yes, a KEY-Distribution-Response message that carries CSKs, MSK, and prev_TEKs is then sent to the target BS. If not, Authenticator, e.g., X, needs to relay the KEY-Distribution-HOInfo message to another ASN-GW that subordinates the target BS.

OP_Code|Serving_BSID|Target_BSID| SAID|CSK1| CSK2| MSK|

TEK_Lifetime|TKE_Count|TEK1|TEK2

Fig. 20. Format of the KEY-Distribution-HOInfo message sent by a serving BS to Authenticator or Authenticator to another Authenticator.

OP_Code|BSID|SAID

Fig. 21. Format of the KEY-Distribution-Request message sent by the BS to Authenticator.

Basically, X can use the backbone routing scheme to deliver the message to target Authenticator Y, or check its own Neighbor BS Table, a table for recording the Authenticators that all the BSs directly neighbor to any one of X's subordinated BSs belong to, to identify the right ASN-GW U, and relay the

SAID	BSID	CSK1	CSK2	MSK	TEK_Lifetime	TEK_Count	TEKs

Fig. 22. The fields of an AK Table. TEKs field stores TEKs.

message to U where the scheme of the neighbor BS Table is shown in Fig. 23. Once U receives the message its Authenticator Y looks up its BS Table to see whether the BSID conveyed on the message is one of its subordinate BSs or not. If yes, Y stores the MS's keys in its AK Table and sends a KEY-Distribution-Response message to the target BS.

The AK table should be updated dynamically each time when MS performs a network entry or re-entry, MS is going to hand over, or an MS key's lifetime expires. When MS initially enters a network, the AK Table is updated on the completion of the User Authentication process. Authenticator saves MS's keys leaving the TEKs field empty. The field will be filled after Authenticator receives a KEY-Distribution-HOInfo message from its BS or another Authenticator (which will be described later) and stores them in its AK Table. Generally, when MS is going to hand over, Authenticator extracts MS's TEKs from the KEY-Distribution-HOInfo message received from the serving BS and stores them in the AK Table. The serving Authenticator on receiving this MS's MSHO_link_up message sent by the target Authenticator deletes this MS key record. Finally, if the TEK_Lifetime expires, the TEK Generation and Renew process should be reinitiated to reproduce TEKs. After that Authenticator replaces the TEKs with the new TEKs in its AK Table.

BSID	ASN-GW MAC Address

Fig. 23. The schema of a Neighbor-BS Table. This table is statically constructed, and only contains neighbor BSs that are subordinated by ASN-GWs other than the underlying Authenticator's ASN-GW.

4.2.2. TEK generation and renew process on handover

Both processes of the two handover security levels start when MS sends a HO_IND message to its serving BS (see Figs 24 and 25). The serving BS then sends a MSHO_link_down message to inform its ASN-GW to start transferring data messages received from MS's corresponding node (CN) to both the serving BS and the target BS, and delivers a KEY-Distribution-HOInfo message which contains MS security attributes such as CSKs, MSK and TEKs that the serving BS currently uses, to its Authenticator.

Authenticator stores the keys in its AK Table if the target BS is one of its subordinate BSs. Otherwise it sends the keys to another ASN-GW during MS handover. No matter which is the case, the target Authenticator delivers a KEY-Distribution-Response message to the target BS. The target BS on receiving the message retrieves security keys and saves them for future use. Now, data message transfer can be resumed before the TEK Generation and Renew process starts, i.e., the target BS can relay data messages to MS before a new random number exchange, i.e., exchanging new *MS_Random* and *BS_Random* is completed.

After the completion of the TEK Generation and Renew process, an MSHO_link_up message will be sent by target BS to its ASN-GW to terminate sending data messages to the serving BS, and the transmission of encrypted data messages can be continued.

4.2.3. Level-1 Intra-ASN-GW Handover: TEK reuse mode

Once MS chooses a Level-1 handover, the KEY-Distribution-Response message sent to the target BS by Authenticator includes TEKs used by the serving BS. The target BS then waits for MS to complete its network re-entry, and on receiving the TEK-Request message sent by MS as shown in Fig. 24 it

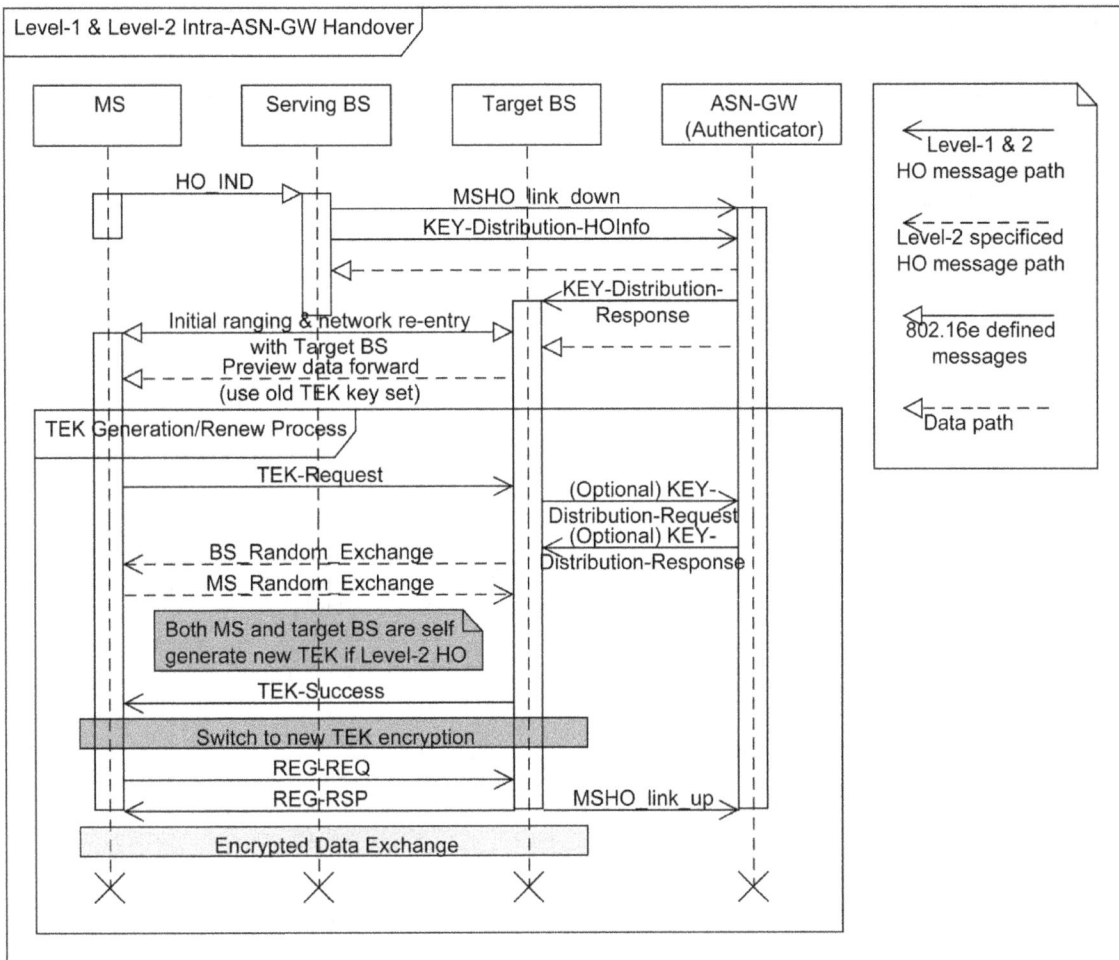

Fig. 24. The process of a Level-1 & Level-2 Intra-ASN-GW handover. The handover steps before the TEK Generation/Renew Process are the same in all Intra-ASN-GW handovers

delivers a TEK-Success message to MS, indicating the success of the TEK reuse mode. MS then sends a REG-RSP message to register itself with the BS. The BS replies with a REG-RSP message and sends an MSHO_link_up message to inform the ASN-GW of the termination of the handover service.

4.2.4. Level2 Intra-ASN-GW Handover: TEK regeneration mode

If MS selects a Level-2 handover the steps with which MS completes the network re-entry are mostly the same as those of a Level-1 Intra-ASN-GW handover. The following steps are a little different. The target BS on receiving a TEK-Request message from MS generates a new *BS-Random*, extracts CSKs and the MSK from the KEY-Distribution-Response message received from Authenticator uses the Dot16KDF algorithm to generate an AK, and then as shown in Fig. 24 sends a BS_Random_Exchange message containing a newly generated *BS-Fingerprint* (see Eq. (7)) to MS. MS then generates a new *MS_Random* and sends a MS_Random_Exchange message which contains a newly generated MS-Fingerprint (see Eq. (8)) to the target BS. The BS and MS individually generate new TEKs by using the new *BS-Random* (see Eq. (9)) and the *MS-Random* (see Eq. (11)). After that, MS and the target BS which is now MS's serving BS deliver data messages to each other by using the new TEKs. The following steps are the

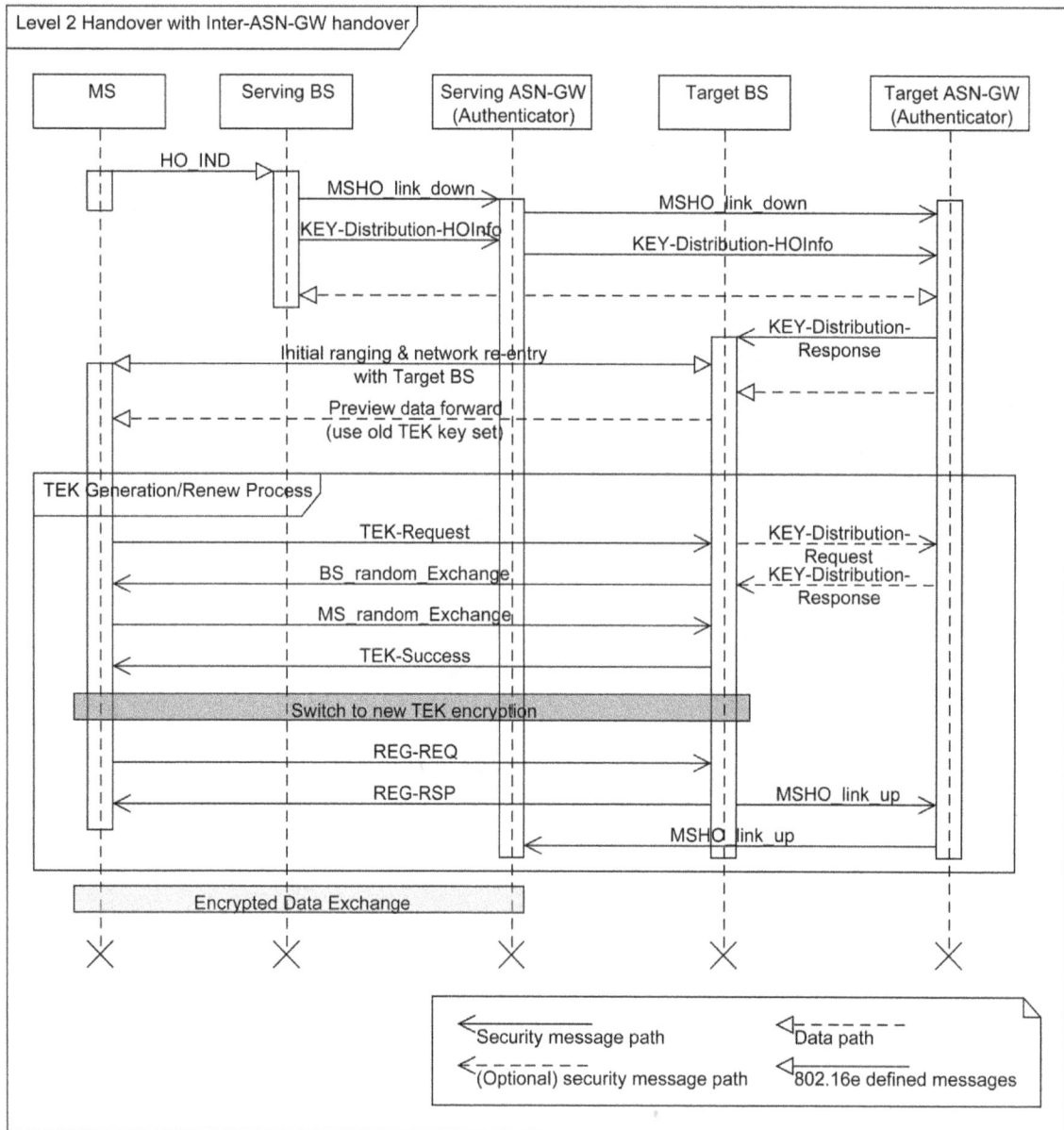

Fig. 25. The process of a Level-2 Inter-ASN-GW handover. The handover steps before the TEK Generation and Renew Process are the same as shown in Fig. 24

same as the corresponding steps of a Level-1 handover. Now the previous serving BSs can no longer communicate with MS since the prev_TEKs are out of date.

4.2.5. Inter ASN-GW handover

If MS hands over between two BSs which belong to different ASN-GWs, the serving Authenticator needs to transfer the KEY-Distribution-HOInfo message to the target Authenticator. As shown in Fig. 1, we assume any ASN-GW can communicate with other neighbor ASN-GWs via R4 reference points. Hence, the serving ASN-GW needs to know which ASN-GW the KEY-Distribution-HOInfo message should be forwarded to. To solve this problem, each ASN-GW, e.g., G as state above, maintains a BS Table to collect all BSIDs of its subordinate BSs and a NeighborBS Table to gather all the BSs subordinated by other ASN-GW but neighbor to one of G's subordinate BSs. Hence, from the Target_BSID conveyed in the KEY-Distribution-HOInfo message the serving Authenticator can determine which ASN-GW (Authenticator) is the one it should forward the message to. During the handover if the serving Authenticator could not find the corresponding ASN-GW of the target BSID in its BS and NeighborBS Tables, the target Authenticator would not provide security keys to the target BS, implying MS should re-enter the system, i.e., performing the initial process of network entry described above. An ASN-GW on receiving a KEY-Distribution-HOInfo message from another ASN-GW passes this message to its Authenticator.

Once the KEY-Distribution-HOInfo message arrives at the neighbor Authenticator, the Authenticator forces a Level-2 Handover to renew TEKs. Figure 25 shows the Inter-ASN-GW handover.

5. Security analyses

The objective of security analyses is to confirm that our study is secure enough to meet wireless security requirements presented in the IEEE 802.16 standard and related research [23]. The security under different attacks is also analyzed.

5.1. Message integrity and replay attack avoidance

Message integrity ensures that a message M has not been changed during its delivery. In this study, the receiving end on receiving M uses the HMAC function to detect data tampering retrieves the nonce conveyed on M and saves it. The HMAC code conveyed on M can act as a verification code for the message itself. If at least one parameter has been changed, including the nonce, the HMAC code varies M will be discarded If the HMAC code passes the verification, we further verify the nonce.

The first time a message M is sent, the receiving end R records the nonce contained in M. If R receives the same or similar message (with the same OP_Code) again, it confirms that this is not a replay attack by comparing the nonce previously saved and the one retrieved from M. If the nonce received is smaller than or equal to the one saved, then M is considered as an illegal one and will be discarded. All messages delivered in the CSK Generation process and TEK Generation and Renew process are detected by this method.

The DiHam scheme provides key integrity, rather than message integrity, by comparing the keys calculated by using the authentication function Certfun$(a, b, \ldots)$ and by using the data carrier function EXOR(x, y) individually with the corresponding value retrieved from the received message (see Figs 4–7). All messages exchanged in the authentication phase and TEK generation phase could be maliciously altered, but the receiving end cannot discover the change. The DiHam also lacks the involvement of the nonce. Hence, it cannot discover replay attacks issued by resending an intercepted Authentication-reply message.

The PKMv2 uses a cipher-based message authentication code (CMAC) or HMAC to authenticate authentication messages, and detects replay attacks by employing CMAC_KEY_COUNT after the success of EAP authentication or reauthentication [1,12]. However, due to involving no nonce, it cannot avoid replay attacks during the EAP authentication session.

5.2. Confidentiality

Confidentiality ensures the security of sensitive data such as encryption keys. Hence, hackers cannot directly acquire any unauthorized information from the intercepted messages. In our scheme, we analyze the confidentiality by checking to see whether exchanged information can be decrypted easily or not, and estimating the probability that a message being considered is cracked.

5.2.1. CSK confidentiality

The HaKMA uses the key exchange process of the DiHam to produce two CSKs. In this process, two public keys are exchanged between MS and Authenticator for each CSK and only the MS public keys P_{RM1} and P_{RM2} are transferred through wireless channels (see Fig. 10). The Authenticator public keys P_{RA1} and P_{RA2} are encrypted by using MS's certificate public key *PubKey(MS)* (see Fig. 11) which can only be decrypted by using MS's certificate private key *PrivKey(MS)*. Therefore, hackers who only know P and P_{RMi} cannot easily derive *CSK1* and *CSK2* where

$$CSK_i = x \bmod P = P_{RMi}{}^y \bmod P, 1 \leqslant i \leqslant 2 \tag{12}$$

in which $x = P_{RAi}{}^{RM_i}$ (see Eq. (4)) and $y = RA_i$ (see Eq. (3)) are known and need to be determined, thus

$$x = P_{RMi}{}^y, 1 \leqslant i \leqslant 2 \tag{13}$$

The possible combinations of x and y pair are infinite. Due to the difficulty of determining the real values for x and y, hackers can only generate CSKs by other methods, e.g., the brute-force method.

Further, the number of possible 160-bit CSK values is $2^{160} \approx 1.4615 \times 10^{48}$. The probability of successfully guessing the CSK on one trial is $1/2^{160}$ which is approximately zero However, two CSKs are used in the HaKMA The probability will be $1/2^{320}$. Therefore, we can conclude that the CSK confidentiality is high.

5.2.2. EAP encryption key confidentiality

In this study we use the EAP encryption key to encrypt and decrypt the messages exchanged between MS and Authenticator. Since this encryption key is static and may be illegally decrypted, we involve the random number *RAND*, which is encrypted by Authenticator's public key (see Eq. (5) and Fig. 13) during its delivery to generate encryption keys. Hackers cannot directly access *RAND*. Hence, it is hard to derive the EAP encryption key. Furthermore each EAP encryption key is used only by one session i.e., each different session uses a different EAP encryption key, making it more difficult for hackers to collect EAP messages and then accordingly decrypt the key. Thus, our scheme has high EAP encryption key confidentiality.

5.2.3. TEK confidentiality

Since TEKs are used to encrypt data messages, we need to keep them secure. TEKs are self-generated by MS and the BS. Two random numbers *BS_Random* and *MS_Random* are also involved in the key generation process. To prevent hackers from collecting random numbers so as to derive TEKs, the two random numbers are encrypted to the *BS_Fingerprint* and *MS_Fingerprint* Since our TEK generation scheme involves the ADR function [22] (see Eq. (10)) which ignores the carry to calculate TEKs from *BS_Random* and *MS_Random* which in turn are respectively derived from *BS_Fingerprint* (see Eq. (7)) and *MS_Fingerprint* (see Eq. (8)), hackers have to face the four different mathematical equations in Eqs (9) and (11). Since each formula's possible outputs are up to $2^{160} \approx 1.4615 \times 10^{48}$, and all four equations involve *AK* and *CSK1* or *AK* and *CSK2* as parameters which are unknown to hackers the number of possible parameter combinations for each equation is $2^{160 \times 3} = 2^{480} \approx 3.1217 \times 10^{144}$. Thus, we can conclude that the TEK confidentiality is high.

If hackers try to decrypt data messages, they must find the two correct TEKs for the uploading and downloading streams. If we assume that the time required to try a possible TEK is only one instruction, then it will take them about 1.4573×10^{29} years on a 159000 MIPS machine [4]. In other words, the HaKMA is a secure and safe system.

5.3. Mutual authentication

Mutual authentication between two nodes implies the two nodes authenticate each other before their communication starts. This can be securely achieved if the two nodes possess CSKs, or perform the public key infrastructure (PKI) public key exchange process. In this study, the mutual authentication focuses on device verification in the CSK Generation process. MS first sends its certification Cert(MS) through a CSK_Request message to Authenticator. Authenticator validates the correctness of a receiving certificate by running X.509 certificate signature algorithms, and/or it can also connect to CA to validate the effectiveness of the certificate through the NSP's backbone network. If the device certification is directly issued by the NSP or is already registered with the NSP, Authenticator can even validate the legitimacy of the device certification by contacting its authentication server [17].

Authenticator's certificate is contained in the CSK_Reply message (see Fig. 11). MS on receiving this message validates the correctness of the receiving certificate by running X.509 certificate signature algorithms too, and validates the legitimacy and effectiveness of the Authenticator's certificate by looking up the Authenticator blacklist and whitelist issued by the NSP.

The PKMv2 completes its mutual authentication by exchanging X.509 certificates. The authentication process is very similar to that described above. But the HaKMA enhances the process by involving MS's public key to encrypt two public keys P_{RA1} and P_{RA2}, implying that hackers can only intercept the two MS public keys, consequently increasing the difficulty of decrypting the two CSKs.

In the User Authentication process, the EAP message exchanged between MS and ASN-GW are encrypted by an EAP Encryption Key (see Eq. (5)) derived from the two CSKs. However, the two CSKs have never been transmitted through wireless channels, and they are only known to MS and Authenticator. Also, CSKs are produced at the end of the CSK Generation process. Only MS and Authenticator can start the User Authentication process with the valid EAP Encryption Key. Furthermore the TEK Generation and Renew process uses CSKs and MSK to generate TEKs (see Eqs (6)–(11)), implying the second and third processes of the HaKMA authentication scheme inherit mutual authentication from the first, i.e., the CSK Generation process. Hence, we can conclude that the HaKMA scheme provides mutual authentication in all three processes.

The DiHam process does not provide mutual authentication in the Authentication phase. It only uses CSKs generated in the Authentication phase to provide mutual authentication in the TEK Exchange phase.

5.4. User authentication

In the User Authentication process, an EAP method is involved. Since the PKMv2 also uses the EAP to perform user authentication, both the PKMv2 and HaKMA provide the same level of user authentication security But the HaKMA involves the EAP encryption key to encrypt EAP messages. Therefore, the HaKMA's user authentication security is higher than that of the PKMv2. Generally, the user identification security in the two schemes heavily relies on the selected EAP method [23]. The DiHam does not provide any user authentication. Its user identification is performed by recognizing MS's certification, or by using the certificate signed by the NSP [2].

5.5. Forward and backward secrecy on handover

Forward (backward) secrecy means the key K_n used in session n cannot be used in session $n + 1$ (session $n - 1$). In a Level-1 Intra-ASN-GW Handover, we reuse TEKs during and after the handover, implying a Level-1 Intra-ASN-GW Handover does not provide forward and backward secrecy. In a Level-2 Intra-ASN-GW Handover and Inter-ASN-GW Handover, we temporarily reuse TEKs to shorten the SDT, and generate new TEKs by involving the random numbers exchanged between MS and BS, i.e., the two handover processes provide forward and backward secrecy.

The PKMv2, due to considering performance optimization on Fast BS Switch (FBSS) and reusing all security attributes including TEKs, does not provide forward and backward secrecy.

Basically, the DiHam process has no forward and backward secrecy since it does not provide handover support. But if we apply the DiHam to the handover process, as described above, the BS and MS have to re-calculate TEKs for each handover, implying forward and backward secrecy. Note that if TEKs are reused after each handover, the DiHam's forward and backward secrecy will no longer exist.

5.6. Man-in-the-middle attack avoidance

A Maninthe-middle attack means hackers stay between valid MS and Authenticator to act as a legitimate Authenticator and MS. In the CSK Generation process and User Authentication process, MS and Authenticator exchange device certificate and determine whether the other side is legitimate or not But in the CSK Generation process, we use MS's and Authenticator's public keys i.e., *PubKey*(*Authenticator*) and *PubKey*(*MS*), to encrypt important keys such as P_{RA1} and P_{RA2} in the CSK_Reply message (see Fig. 11), and *RAND* in the PKMv2-EAP-Start message (see Fig. 13). The receiving end needs its own private key to decrypt those encrypted messages and keys. Now we assume that a hacker, H, is standing between a valid MS and Authenticator and wishes to steal EAP user passwords by eavesdropping EAP messages. Then H needs to act as an Authenticator so that it can acquire the valid CSK to continue the following User Authentication process since our EAP messages are all encrypted by using CSKs and other parameters like *RAND* (see Eq. (5) and Fig. 13). To complete the CSK Generation process besides relaying MS's and Authenticator's certificates H also needs to replace the Authenticator certificate with its own so that it can decrypt *RAND*. However, if H replaces the certificate with its own, this illegal certificate will not be recognized by MS and this session will be terminated On the other hand, if H continues using the real authenticator's certificate, it will not be able to decrypt the *RAND* carried on the next PKMv2-EAP-Start message sent by MS since *RAND* can only be decrypted by Authenticator's private key that H currently does not have, implying the User Authentication process is still secure because all EAP messages are encrypted by both the CSKs and *RAND*. As a result, our scheme can prevent man-in-the-middle attacks.

6. System experiments and discussion

In this study, several analyses and experiments were performed to evaluate the performance of the HaKMA and the compared schemes, including the PKMv2 and DiHam.

6.1. Performance analysis on key generation algorithms

Generally, in a Diffie-Hellman based authentication method, exponential operations dominate decisive performance differences [22]. In this study, two CSKs were individually generated by MS and Authen-

Table 2
Modular operations for different security schemes

Security scheme	Exponential operations			
	CSK	EAP	TEK	Total
DiHam with level-1 TEK	7	–	1	8
DiHam with level-2 TEK	7	–	6	13
DiHam with level-3 TEK	7	–	76	83
PKMv2 with EAP-AKA	–	2	–	2
HaKMA with EAP-AKA	5	2	–	7

ticator, and only Diffie-Hellman based public keys, are transmitted through wireless channels Deriving CSKs from exchanged public keys needs to solve discrete logarithm problems [8,10,22].

In the DiHam, Diffie-Hellman style keys are widely used, e.g., the generation of the CSKs, AKs and TEKs which provide a very secure method to protect the communication system, but the costs of key calculation are high. It has at least 7 exponential operations in the CSK Generation phase, and 1–76 exponential operations in different levels of the TEK Exchange phase. In the HaKMA, only 5 exponential operations are performed in the CSK Generation process, 1–2 exponential operations are involved in the User Authentication process depending on what EAP method is selected, and no exponential operations are involved in the TEK Generation and Renew process. The PKMv2 requires 0–2 exponential operations for the Diffie-Hellman style key exchange in its EAP Authentication process with a specific EAP method.

Other important algorithms employed in the HaKMA and PKMv2 are HMAC, CMAC and Dot16KDF. The HaKMA uses the HMAC algorithm six times in the CSK Generation process and TEK Generation and Renew process. The Dot16KDF algorithm is invoked only once in the TEK Generation and Renew process for generating AK. The PKMv2 uses this algorithm to derive AK, and the CMAC algorithm five times before the REG-REQ message is sent to the BS by MS. Note that the Dot16KDF algorithm invokes the CMAC or SHA-1 algorithm many times depending on the length of the key produced. But we ignore the difference since it is small and the output key lengths in both the PKMv2 and HaKMA are the same. Also, the costs of those algorithms performing fast operations such as exclusive OR operation and shift operation, are much smaller than those of exponential operations and can thus be ignored. Table 2 summarizes the costs of the evaluated schemes. We can see that the cost of the HaKMA operations is between those of the PKMv2 with the EAP-AKA method and the DiHam with level-1 TEK.

6.2. Costs and service disruption time

To evaluate the performance of the HaKMA, we calculate the processing cost for each message. The cost consists of two parts, message computation cost T, and message transmission cost T'. Thus, the processing cost $\mathbb{C}$ can be expressed as

$$\mathbb{C}_{CSK} = T_{CSK} + T'_{CSK} \tag{14}$$

$$\mathbb{C}_{MSK} = T_{MSK} + T'_{MSK} \tag{15}$$

$$\mathbb{C}_{TEK1} = T_{TEK1} + T'_{TEK1} \tag{16}$$

$$\mathbb{C}_{TEK2} = T_{TEK2} + T'_{TEK2} \tag{17}$$

The items used to evaluate the cost of the HaKMA and their descriptions are listed in Table 3.

Table 3
The items used to evaluate the cost of the HaKMA and their descriptions

Item	Description
$\mathbb{C}_{CSK}$, $\mathbb{C}_{MSK}$, and $\mathbb{C}_{TEKn}$	Costs of CSK generation process, User Authentication process, and Level-n TEK Generation and Renew process, respectively, where $n = 1$ or 2
$\mathbb{C}_{Intra_HOn}$ and $\mathbb{C}_{Inter_HO}$	Costs of the Intra-ASN-GW Handover and Inter-ASN-GW Handover, respectively, where $n = 1$ or 2
$\mathbb{C}_{Initial}$	Costs of the HaKMA initialization entry
T_{Exp}	Computation cost of an exponential operation
T_{Hmac}	Computation cost of a HMAC operation
T_{Enc}	Computation cost of a RSA-OAEP-Encryption operation
T_{Dec}	Computation cost of a RSA-OAEP-Decryption operation
$T_{Dot16KDF}$	Computation cost of a Dot16KDF algorithm

The costs of the HaKMA during the initial network entry and handovers can then be expressed as

$$\mathbb{C}_{Initial} = \mathbb{C}_{CSK} + \mathbb{C}_{MSK} + \mathbb{C}_{TEK2} \tag{18}$$

$$\mathbb{C}_{Intra\_HO1} = \mathbb{C}_{TEK1} + T'_{IntraHO} \tag{19}$$

$$\mathbb{C}_{Intra\_HO2} = \mathbb{C}_{TEK2} + T'_{IntraHO} \tag{20}$$

$$\mathbb{C}_{Inter\_HO} = \mathbb{C}_{TEK2} + T'_{InterHO} \tag{21}$$

6.2.1. Message computation costs

In the HaKMA, each message computation cost consists of the costs of message generation and receiving message verification. For example, the generation cost of a CSK_Request message on MS (see Fig. 10) is

$$2T_{Rand} + 2T_{Exp} + T_{Hmac} \tag{22}$$

The verification cost on Authenticator is T_{Hmac}. The total cost of the verification and generation of a CSK_Reply message (see Fig. 11) on Authenticator is

$$4T_{Exp} + 2T_{Rand} + T_{Enc} + T_{Exp} + 2T_{Hmac} \tag{23}$$

where $4T_{Exp}$ is the cost of invoking the Diffie-Hellman algorithm, $2T_{Rand}$ is the cost of producing two random numbers, and T_{Enc}, T_{Exp} and $2T_{Hmac}$ are the costs of invoking the encryption function, modulus function and HMAC function, respectively.

MS after receiving the CSK-Reply message spends $T_{Hmac} + T_{Dec}$ for verifying the message and another $3T_{Exp}$ for generating CSKs and invoking a modulus function. The total verification and key generation cost is

$$T_{Hmac} + T_{Dec} + 3T_{Exp} \tag{24}$$

The cumulative computation cost of the CSK Generation process is then (See Eqs (22), (23) and (24))

$$T_{CSK} = 4T_{Hmac} + 4T_{Rand} + 10T_{Exp} + T_{Enc} + T_{Dec} \tag{25}$$

The computation costs for other sub-processes can be calculated by a similar method. Table 4 lists the summaries, in which the CSK Generation process has higher cost than that calculated in Table 2 since what Table 4 lists are the cumulative costs. That means the operations are performed one by one instead of in parallel.

Table 4
Cumulative computation Costs of the sub-processes in the HaKMA

HaKMA sub-process	Computation Cost
CSK Generation Process	$T_{CSK} = 4T_{Hmac} + 4T_{Rand} + 10T_{Exp} + T_{Enc} + T_{Dec}$
User Authentication Process	$T_{MSK} = 2T_{Hmac} + 1T_{Rand} + T_{Enc} + T_{Dec}$
Level-1 TEK Generation & Renew Process	$T_{TEK1} = 4T_{Hmac}$
Level-2 TEK Generation & Renew Process	$T_{TEK2} = 8T_{Hmac} + 2T_{Rand} + 2T_{Dot16KDF}$

Table 5
The configurations used in the experiments

Variable	Configuration value
Bandwidth	Upward link: 1.5125 Mbps
	Downward link: 3.2425 Mbps
	Measured on:
	WiMax network / ISP: Vee Telecom Multimedia Corp.
Network topology delay	Between MS and BS (wireless connection): $d_w = 80.5$ ms
	Between BS and Authenticator, and between Authenticators (wired infrastructure connection):
	$d_l = 9$ ms
AAA server delay	$d_{AAA} = 923.220$ ms
	Measured in the situation: MTU = 1500 bytes, avg. message processing time = 90 ms, and a
	total of 5 messages are exchanged (including the network topology delay)

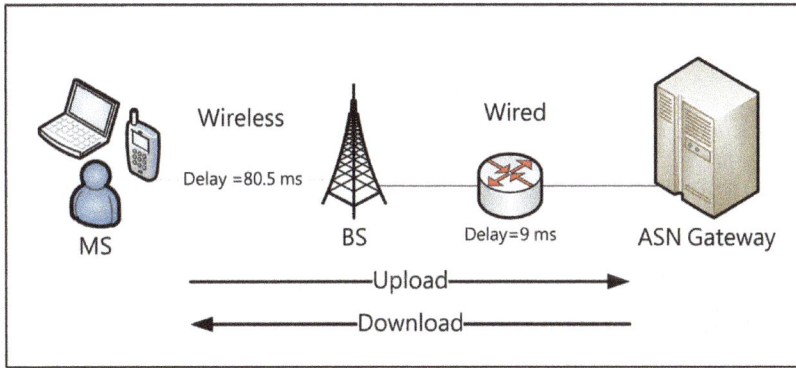

Fig. 26. The schematic diagram of the HaKMA performance evaluation.

6.2.2. Message transmission cost

The transmission cost of a message Msg in the HaKMA consists of delivery delay c and transmission delay d. Delivery delay c is the time required to deliver a message on a link where $c = Msg/bandwidth$, and transmission delay comprises the network topology delay including queuing delays on routers and the delays due to packet retransmission. In this study, we calculate the cumulative length of all messages generated, and compute the transmission cost under the network configuration shown in Fig. 26. Table 5 summarizes the measurements and specifications of the configuration, which are acquired on a real WiMax wireless network: Vee Telecom Multimedia Corporation, Taiwan [3].

Based on the configuration, if the message length is L in bytes, the upload delivery delay c_u in milliseconds from MS to Authenticator through the BS is

$$c_u(Msg) = \frac{8L}{1.5125 \times 2^{20}} \times 10^3 = \frac{L \times 10^3}{1.5125 \times 2^{17}} \tag{26}$$

Table 6
Maximum message lengths involved in the HaKMA

Message Msg	Length L (bytes)
CSK_Request	1104
CSK_Reply	1130
PKMv2-EAP-Start	66
KEY-Distribution-Response	182
TEK-Request	28
BS_Random_Exchange	48
MS_Random_Exchange	46
TEK-Success	26
KEY-Distribution-HOInfo	168
KEY-Distribution-Request	14

and the download delivery delay c_d in millisecond from Authenticator to MS is

$$c_d(Msg) = \frac{L \times 10^3}{3.2425 \times 2^{17}} \tag{27}$$

where all the values of Msg and the corresponding L are listed in Table 6.

Now, message transmission costs can be identified as:

$$T'_{CSK} = c_u(\text{CSK\_Request}) + c_d(\text{CSK\_Reply}) + 2(d_w + d_l) \tag{28}$$

$$T'_{MSK} = c_u(\text{PKMv2-EAP-Start}) + c_d(\text{KEY-Distribution-Response})$$
$$+d_w + 2d_l + d_{AAA} \tag{29}$$

$$T'_{TEK1} = c_u(\text{TEK-Request}) + c_d(\text{TEK-Success}) + 2d_w \tag{30}$$

$$T'_{TEK2} = c_u(\text{TEK-Request}) + c_d(\text{BS\_Random\_Exchange})$$
$$+c_u(MS\_Random\_Exchange) + c_d(TEK\text{-}Success) + 4d_w \tag{31}$$

Since the handover processes involve an extra KEY-Distribution-HOInfo message, the Intra-ASN-GW handover transmission and Inter-ASN-GW handover transmission costs for the message are

$$T'_{IntraHO} = c_u(\text{KEY-Distribution-HOInfo})$$
$$+c_d(\text{KEY-Distribution-Response}) + 2d_l \tag{32}$$

$$T'_{InterHO} = 2c_u(\text{KEY-Distribution-HOInfo})$$
$$+c_d(\text{KEY-Distribution-Response}) + 3d_l \tag{33}$$

6.2.3. System platform and experimental results

To evaluate the costs of the HaKMA to see whether it is feasible in practice or not, we implement various sub-processes and related algorithms used in the HaKMA. The specifications of the experimental system platform are listed in Table 7. The experimental results of all HaKMA subprocesses themselves and network entry/re-entry based on the configuration shown in Fig. 26 and those parameters listed in Table 5 are summarized in Tables 8 and 9, respectively.

Table 7
Specification of the experimental system platform

Component	Authenticator/BS equipment	MS equipment
H/W Platform	Intel × 86	ARM11
CPU	Intel Core i5 750 2.67 GHz	Samsung S3C6410 667 MHz
RAM	8GB	256MB
OS	Windows 7 Enterprise x64	Linux kernel 2.6.27 / Android 1.6

Table 8
Experimental results of the HaKMA sub-processes

HaKMA sub-process	Algorithm Experimental Result (ms)	System Experimental Result (ms)
CSK Generation Process, $\mathbb{C}_{CSK}$	2776.914	2964.142
User Authentication Process, $\mathbb{C}_{MSK}$	14.601	1037.083
Level-1 TEK Generation & Renew Process, $\mathbb{C}_{TEK1}$	0.160	161.362
Level-2 TEK Generation & Renew Process, $\mathbb{C}_{TEK2}$	4.075	326.622

6.2.4. Comparison of initial network entry and network Re-entry

For the MS, each of $\mathbb{C}_{Intra\_HO1}$, $\mathbb{C}_{Intra\_HO2}$, and $\mathbb{C}_{Inter}$ (i.e., network re-entry cost) is much shorter than $\mathbb{C}_{Initial}$ (i.e., initial network entry cost) because only the TEK Generation and Renew process is performed during the handover. Several keys originally generated in the CSK Generation process, e.g., CSKs, and the User Authentication process, e.g., the MSK, are now reused and as shown in Figs 20 and Fig. 14 delivered by key distribution messages to avoid introducing the authentication delay in the CSK Generation process, and the AAA Server delay in the User Authentication process. The experimental result shows that in the HaKMA the network re-entry cost ranges between 4.17% and 8.22% of the network initial entry cost. In other words, the HaKMA is feasible in a wireless system and has very low service disruption time.

7. Conclusions and future work

Wireless networks have been a part of our everyday life. Due to the mobility of end devices, we expect that wireless networks could someday substitute for wired broadband networks to serve users. However how to protect sensitive information delivered through wireless channels in a highly secure communication environment is an important issue in recent research.

In this paper, the HaKMA security scheme which provides fast and secure key generation process, mutual authentication and EAP based user authentication is proposed. The three-layer architecture simplifies key generation flows compared to those proposed in the DiHam and PKMv2. It further provides a fast and secure key renew process for handover. We also introduce two levels of handover processes to minimize SDT give connections between MS and BS forward and backward secrecy and analyze the HaKMA's security and performance. Table 10 summarizes the comparison on important issues. From this we can conclude that the HaKMA provides low-cost and effective handover, and its authentication approach is more secure than those of the DiHam and PKMv2.

In the future, we would like to enhance the HaKMA by developing its error handling capability. When the HaKMA receives an invalid message, it currently drops the message and waits for valid messages before timeout. If we wish to raise reliability for the HaKMA on error handling, the side that finds an error could send an error message to inform the other site of the occurrence of the error so that the compensative operations can be triggered immediately without wasting time to wait for

Table 9

Experimental results of the HaKMA for initial entry and handover network re-entry

Process	Cost (ms)
Initial Entry, $\mathbb{C}_{Initial}$	4327.847
Intra-HO/Level-1, $\mathbb{C}_{Intra\_HO1}$	180.638
Intra-HO/Level-2, $\mathbb{C}_{Intra\_HO2}$	345.898
Inter-HO/Level-2, $\mathbb{C}_{Inter}$	355.745

Table 10

Comparison of different security schemes with proposed scheme

Security Scheme	DiHam	PKMv2	HaKMA
Security enhancement and practical issues			
Messages Integrity	No	After EAP success	Yes
Confidentiality	Yes	Yes	Yes
Mutual Authentication	After Auth. phase	Yes	Yes, enhanced
User Authentication	No, rely on cert.	Yes, by EAP	Yes, by EAP
Efficient on key generation	Low	High	Medium
Handover related issues			
Handover support	No/Yes(see Sec. 3)	Skip	Yes
Forward and Backward Secrecy	−/Yes w/ renew TEK	No	Yes
Efficient under Handover	−/No	Yes when reuse	Yes
Low Cost under Handover	−/No	Yes	Yes
Fault-tolerant under Handover	−/No	No	Yes
Against threats			
Replay Attack	No	After EAP success	Yes
Man-in-the-middle Attack	No	Possible	Yes

valid messages In the handover support, we will design a flexible MS keys' routing scheme to deliver keys between/among Authenticators, and develop behavior and reliability models so that users can predict the HaKMA's behavior and reliability before using it. The handover authentication between two heterogeneous networks such as IEEE 802.11 or 3GPP LTE will also be developed. Those constitute our future research.

References

[1] IEEE Standard for Local and Metropolitan Area Networks Part 16: Air Interface for Fixed and Mobile Broadband Wireless Access Systems Amendment 2: Physical and Medium Access Control Layers for Combined Fixed and Mobile Operation in Licensed Bands and Corrigendum 1, *IEEE Std 802.16e-2005*, 2006.

[2] WiMAX Forum Network Architecture, *WMF-T32-003-R010v05*, 2009.

[3] Vee telecom network speed test, http://speed.vee.com.tw/, Accessed in 2011.

[4] Benchmark Results: SiSoftware Sandra 2011, http://www.tomshardware.com/reviews/core-i7-990x-extreme-edition-gulftown,2 874-6.html, Accessed in 2011.

[5] J. Arkko and H. Haverinen, Extensible Authentication Protocol Method for 3rd Generation Authentication and Key Agreement (EAP-AKA), *RFC 4187*, 2006.

[6] C.J. Bernardos, M. Gramaglia, L.M. Contreras, M. Calderon and I. Soto, Network-based Localized IP mobility Management: Proxy Mobile IPv6 and Current Trends in Standardization, *Journal of Wireless Mobile Networks, Ubiquitous Computing, and Dependable Applications* 1(2/3) (2010), 16–35.

[7] T.M. Bohnert, M. Castrucci, N. Ciulli, G. Landi, I. Marchetti, C. Nardini, B. Sousa, P. Neves and P. Simoes, QoS management and control for an all-IP WiMAX network architecture: Design, implementation and evaluation, *Mobile Information Systems* 4(4) (2008), 253–271.

[8] W. Diffie and M. Hellman, New directions in cryptography, *Ieee Transactions on Information Theory* 22(6) (1976), 644–654.

[9]   C. Eklund, R. Marks, K. Stanwood and S. Wang, IEEE standard 802.16: a technical overview of the WirelessMAN air interface for broadband wireless access, *IEEE Communications Magazine* **40**(6) (2002), 98–107.

[10] T. Elgamal, A Public Key Cryptosystem and a Signature Scheme Based on Discrete Logarithms, *Ieee Transactions on Information Theory* **31**(4) (1985), 469–472.

[11] M. Ergen, *Mobile broadband including WiMAX and LTE*, Springer Science+Business Media, LLC, Boston, MA, 2009.

[12] K. Etemad and M. Lai, *WIMAX technology and network evolution*, Wiley, Hoboken, N.J., 2010.

[13] P. Fülöp, S. Imre, S. Szabó and T. Szálka, Accurate mobility modeling and location prediction based on pattern analysis of handover series in mobile networks, *Mobile Information Systems* **5**(3) (2009), 255–289.

[14] P. Fülöp, B. Kovács and S. Imre, Mobility management algorithms for the Client-driven Mobility Frame System – mobility from a brand new point of view, *Mobile Information Systems* **5**(4) (2009), 313–337.

[15] R. Housley, W. Polk, W. Ford and D. Solo, Internet X.509 Public Key Infrastructure Certificate and Certificate Revocation List (CRL) Profile, *RFC 3280*, 2002.

[16] S.F. Hsu and Y.B. Lin, A Key Caching Mechanism for Reducing WiMAX Authentication Cost in Handoff, *IEEE Transactions on Vehicular Technology* **58**(8) (2009), 4507–4513.

[17] D. Johnston and J. Walker, Overview of IEEE 802.16 security, *IEEE Security & Privacy* **2**(3) (2004), 40–48.

[18] J. Jonsson and B. Kaliski, Public-Key Cryptography Standards (PKCS) #1: RSA Cryptography Specifications Version 2.1, *RFC 3447* (2003).

[19] H. Kim and J.-H. Lee, Diffie-Hellman key based authentication in proxy mobile IPv6, *Mobile Information Systems* **6**(1) (2010), 107–121.

[20] L.S. Lee and K. Wang, A Network Assisted Fast Handover Scheme for IEEE 802.16E Networks, in: *Proceedings of the IEEE International Symposium on Personal, Indoor and Mobile Radio Communications* (2007), 1–5.

[21] L.S. Lee and K. Wang, Design and Analysis of a Network-Assisted Fast Handover Scheme for IEEE 802.16e Networks, *IEEE Transactions on Vehicular Technology* **59**(2) (2010), 869–883.

[22] F.Y. Leu, Y.F. Huang and C.H. Chiu, Improving Security Levels of IEEE802.16e Authentication by Involving Diffie-Hellman PKDS, in: *Proceedings of the International Conference on Complex, Intelligent and Software Intensive Systems* (2010), 391–397.

[23] D.Q. Liu and M. Coslow, Extensible authentication protocols for IEEE standards 802.11 and 802.16, in: *Proceedings of the ACM International Conference on Mobile Technology, Applications, and Systems* (2008), 1–9.

[24] Z. Yan, H. Zhou and I. You, N-NEMO: A Comprehensive Network Mobility Solution in Proxy Mobile IPv6 Network, *Journal of Wireless Mobile Networks, Ubiquitous Computing, and Dependable Applications* **1**(2) (2010), 52–70.

[25] W.S. Yang, E.S. Yang, H.J. Kim and D.K. Kim, Estimation of spectrum requirements for mobile networks with self-similar traffic, handover, and frequency reuse, *Mobile Information Systems* **6**(4) (2010), 281–291.

[26] Y. Yang and R. Li, Toward Wimax Security, in: *Proceedings of the International Conference on Computational Intelligence and Software Engineering* (2009), 1–5.

Yi-Fu Ciou received the B.S degree in computer science engineering from the Tunghai University, Taiwan. He is currently a master student in Department of Computer Science at the Tunghai University, Taiwan. His research interests include wireless network standard, security, embedded systems, and cloud computing.

Fang-Yie Leu received his BS, master and Ph.D. degrees all from National Taiwan University of Science and Technology, Taiwan, in 1983, 1986 and 1991, respectively, and another master degree from Knowledge System Institute, USA, in 1990. His research interests include wireless communication, network security, Grid applications and Chinese natural language processing. He is currently a professor of Tunghai University, Taiwan, and director of database and network security laboratory of the University. He is also a member of IEEE Computer Society.

Yi-Li Huang received his master degrees from National Central University of Physics, Taiwan, in 1983. His research interests include security of network and wireless communication, solar active-tracking system, and grey theory. He is currently a senior instructor of Tunghai University, Taiwan, and director of information security and grey theory laboratory of the University.

Kangbin Yim received his B.S., M.S., and Ph.D. from Ajou University, Suwon, Korea in 1992, 1994 and 2001, respectively. He is currently associate professor as he has joined Dept. of Information Security Engineering, Soonchunhyang University since 2003. He has served as an executive board member for Korea Institute of Information Security and Cryptology, Korean Society for Internet Information and The Institute of Electronics Engineers of Korea. As he is currently an editorial board member of JoWUA and JISIS, he worked as the track chair of several workshops such as BWCCA and IMIS. His research interests include vulnerability assessment, code obfuscation, malware analysis, insider threats, access control, secure hardware, and systems security. Related to these topics, he has worked on more than forty projects and published more than ninety domestic and international research papers.

5

Cooperation as a service in VANET: Implementation and simulation results

Hajar Mousannif[a],[*], Ismail Khalil[b] and Stephan Olariu[c]

[a]*Cadi Ayyad University, Guéliz, Marrakech, Morocco*
[b]*Johannes Kepler University, Linz, Austria*
[c]*Old Dominion University, Norfolk, VA, USA*

Abstract. The past decade has witnessed the emergence of Vehicular Ad-hoc Networks (VANET), specializing from the well-known Mobile Ad Hoc Networks (MANET) to Vehicle-to-Vehicle (V2V) and Vehicle-to-Infrastructure (V2I) wireless communications. While the original motivation for Vehicular Networks was to promote traffic safety, recently it has become increasingly obvious that Vehicular Networks open new vistas for Internet access, providing weather or road condition, parking availability, distributed gaming, and advertisement. In previous papers [27,28], we introduced Cooperation as a Service (CaaS); a new service-oriented solution which enables improved and new services for the road users and an optimized use of the road network through vehicle's cooperation and vehicle-to-vehicle communications. The current paper is an extension of the first ones; it describes an improved version of CaaS and provides its full implementation details and simulation results. CaaS structures the network into clusters, and uses Content Based Routing (CBR) for intra-cluster communications and DTN (Delay – and disruption-Tolerant Network) routing for inter-cluster communications. To show the feasibility of our approach, we implemented and tested CaaS using Opnet modeler software package. Simulation results prove the correctness of our protocol and indicate that CaaS achieves higher performance as compared to an Epidemic approach.

Keywords: Intelligent transportation systems, VANET, publish/subscribe, cluster, CBR, DTN routing

1. Introduction

Over the last few decades, the need for better transportation systems has grown significantly. The number of vehicles on the road has approached critical mass [32], forcing government transportation departments across more and more countries to develop Intelligent Transportation Systems [4], which refer to broad range of diverse technologies, including information processing, sensors, communications, control, and electronics. Combining these technologies in innovative ways and integrating them into the transportation system will save lives, time and resources by simplifying data exchange between roadside infrastructure and vehicles. Some of this data is collected to support real-time traveler information and traffic control, whereas other data is collected and used off-line to help characterize typical travel patterns and project future traffic conditions [10].

Recently, the emphasis in the area of Intelligent Transportation Systems has turned to cooperative systems in which the vehicles communicate with each other and/or with the infrastructure. Such cooperative systems can greatly increase the quality and reliability of information available about the

---

*Corresponding author: Cadi Ayyad University, FSTG, B.P 549, Av. Abdelkarim Elkhattabi, Guéliz, Marrakech, Morocco.
E-mail: hajar.mousannif@gmail.com.

vehicles, their location and the road environment. They enable improved and new services for the road users, which, in turn, will lead to greater transport efficiency, by making better use of the capacity of the available infrastructure and by managing varying demands, and increased safety, by improving the quality and reliability of information and allowing the implementation of advanced safety applications.

These cooperative systems make a combined use of both Vehicular Sensor Networks (VSN) and Vehicular Ad-hoc Networks (VANET). Unlike a traditional wireless sensor network in which optimizing energy consumption is the main challenge [26], vehicles in a vehicular sensor network are typically not affected by limitations in power, computational capacities or memory. In fact, vehicles can be easily equipped with powerful processing and storage units, multiple wireless interfaces (e.g. Wifi, Bluetooth and 2G/3G), Event Data Recorders (EDRs) [12], and sensing devices of some complexity (e.g. GPS receivers, cameras, vibration sensors and acoustic sensors). Depending on its equipment, a car 'knows' about its speed (tachometer), the actual location (GPS), distance to the next car (distance control), destination of the actual trip (navigation system) and the actual weather conditions (rain and temperature sensor) [20].

Many car manufacturers have also been installing wireless connectivity equipment in their vehicles to enable communications with roadside base stations and also between vehicles, for the purposes of safety, driving assistance, and entertainment. The two primary distinct features of vehicle networks are that: 1°) Vehicles can be highly mobile, with speed up to 30 m/s and 2°) Their mobility patterns are more predictable than those of nodes in Mobile Ad hoc Networks (MANET), extensively studied in the literature (see, for example [38], and the references therein), due to the constraints imposed by roads, speed limits, and commuting habits. Therefore, these networks require specific tradeoffs and identify a novel research area, i.e., Vehicular Ad hoc Networks (VANET).

VANETs support two types of communication: Vehicle-to-Vehicle (V2V) and Vehicle-to-Infrastructure (V2I). While V2V deals with communication among vehicles themselves, V2I is concerned about transmitting information between a vehicle and the fixed infrastructure installed along the road, e.g. roadside base stations [11] that can be connected with each other or, depending on the deployment scenarios, can also be connected to the Internet. V2V and V2I communications are made possible via the DSRC/WAVE (Dedicated Short Range Communications/Wireless Access in a Vehicular environment) standard [19] which is a short to medium range communication technology operating in the 5.9 GHz. Readers can find a detailed overview of the DSRC standards in [19].

The potential applications of vehicular networks offer vast opportunity [14]. While the past decade has witnessed a proliferation of mainly vehicular safety applications [3], such as Electronic Brake Warning (EBW), Vehicle Stability Warning (VSW), application On-coming Traffic Warning (OTW) and Lane Change Warning (LCW), many other innovative applications can be achieved by combining high accuracy positioning, inter-vehicular communication technologies and the on-board array of sensors. Vehicle infotainment system (VIS) [17], as an example, has gained much attention recently due to its promising usage in a wide range of Internet-based services, ranging from location-aware services such those in [7], on-demand traveling information and traffic conditions, to rich media news and video distribution [13]. Vehicles can also act like mobile sensors monitoring parameters such as road and weather conditions, parking lots availability (like in [1]) or traffic density (like in [8,29]). Such information can be shared among vehicles in order to perform route optimization or adaptations of driving behavior. Internet access was one of the earliest applications proposed for VANET. The idea is to allow drivers to share their underutilized network resource with other drivers who may need to access the Internet while on the move. Free Internet access can also be combined with some form of advertisement distribution like in [21], where cars carry and distribute advertisement using mainly single-hop inter-vehicle communication. The same idea is found in [31], where the authors propose AdTorrent;

an extension of the physical billboards that allows drivers to download advertisements of interest using a location-sensitive distributed mechanism. Other applications such as RoadSpeak [39], which allows drivers to communicate on the road via voice chat messages, interactive online games [35], and video applications (video-phoning and teleconferencing) [36] might not be a driving force for VANET in the immediate future.

Motivated by finding solutions to problems such as the lost of worker productivity and fuel, and the high level of CO2 emissions due to traffic congestion, the increase of the number of fatalities directly attributable to traffic-related incidents and the huge cost related to Intelligent Transportation Systems, we described in previous papers [27,28], a new service-oriented solution for VANETs, referred to as Cooperation as a Service or CaaS, that extends the two novel types of Vehicular Cloud services: Network as a Service (NaaS) and Storage as a Service (SaaS) introduced in [33]. CaaS allows providing vehicles/drivers, which are willing to cooperate, with some sets of services using very minimal infrastructure, by taking advantage of Vehicle-to-Vehicle (V2V) communications. CaaS enables the integration of a set of improved services for the road users through a novel hybrid publish/subscribe mechanism we introduced. In [27], the proposed mechanism structures the network into clusters, and uses Content Based Routing (CBR) for intra-cluster communications and geographic routing for inter-cluster communications.

The main limitation of this previous version is the use of geographic routing for inter-cluster communications which assumes that nodes are capable of determining their own position either through GPS devices (such as Navigator Systems) mounted on the vehicles, or by deploying virtual coordinates [2]. Since our main objective is to provide all vehicles with services they subscribed to, regardless of whether they are equipped with a navigator device or not, it would be unfair to consider that all vehicles in the network need to be equipped with such devices to benefit from a certain service. In this paper, we overcome the limitations of the previous version by using DTN routing [18] for inter-cluster communications instead of geographic routing. Our choice will be justified throughout this paper.

The remainder of this paper is organized as follows: Section 2 reviews some related work in the area of data dissemination in VANETs and highlights our contribution. Section 3 describes the improved version of our cooperative solution for VANET. Implementation details and simulation results are provided in Section 4. Finally, Section 6 offers concluding remarks and directions for future work.

2. Data dissemination in VANETs

Before a VANET-based application can start to process and propagate data, local measurements need to be made. These local observations are application-dependant and are usually obtained through the car integrated sensors. Reading information from these sensors alone can indeed provide important information. As an example, reading the speedometer of a vehicle may allow conclusions to be drawn about the traffic conditions. But, integrating and combining information from different sources will make such conclusions even more accurate. Authors in [34] examine a system that aims at integrating measurements from multiple sources through sensor fusion techniques in order to provide useful information about the current road condition.

After obtaining local measurements, information has to be disseminated to interested parties. Nevertheless, due to the capacity constraints [15] in VANET, it is technically unfeasible to deliver detailed and regularly updated information to all participants in the network. The key idea is to combine information from a cooperative VANET using measurement summarizing and aggregation mechanisms which aim at reducing the generated amount of data. Further details about data fusion techniques can be found in [30]. It is also important to determine whether an event (e.g., a traffic congestion) or a resource (e.g., an available parking space) is relevant to a vehicle. Authors [6] in propose a general data management architecture for vehicular networks that allows to select, among the events received, the events that are relevant to the vehicle and so may be also relevant to the driver using geographic vectors and maps.

Depending on the application, information needs also to be shared among vehicles that are interested in it. These vehicles might need to adapt their behavior based on the received information. One way of distributing this information inside the network is to use flooding which simply consists of rebroadcasting the information by each node which receives it. Since this naïve approach may lead to severe congestions in the network, many approaches have been proposed to deal with this problem and mainly aim at influencing the forwarding behavior of vehicles by either adapting the time to forward, the geographic area where to forward [24] or by simply placing rules on whether a vehicle should forward or not. Readers can find details on the use of these flooding techniques in [37,41].

Despite the use of these flooding techniques and due to wireless signal dynamics, node mobility and vehicular networks density, especially in big cities, poor performance of flooding-based routing protocols has been noticed [43]. As an alternative, geographic routing has been chosen in many routing algorithms used for VANET. Geographic-based routing protocols exploit both local information and information about the surrounding road topology to route packets. In some scenarios, information about speed, direction or route plan can also be used. Details about the most used geographic-based routing protocols in VANET can be found in [22,23,45]. The most important assumption that almost every geographic routing protocol makes is that nodes are capable of determining their own position. This can be done either by equipping vehicles with GPS devices; a navigation system available on more and more vehicles nowadays can provide location information, or by deploying virtual coordinates, such as in [2], which consists in assigning some elected nodes in the network coordinates and letting the rest of the nodes obtain their virtual coordinates either through triangulation techniques or by averaging the coordinates of their neighbors.

In many application scenarios, another class of routing paradigms known as Content-Based Routing (CBR) is used to achieve better performances. In CBR, the sender simply injects the message in the network, which then determines how to route it according to the nodes' interests (or subscriptions). CBR is proposed as an efficient publish/subscribe approach in many Service-Oriented Architectures (SOAs) [5].

The fact that vehicular networks are highly mobile and sometimes sparse complicates finding an end-to-end connection to disseminate such data efficiently. DTN (Delay – and disruption-Tolerant Network) routing [18] is based on the principle of store-and-forward; that is, a message can be buffered at a node until an appropriate next hop appears. Then, the node forwards the message to the next such hop. Therefore, instead of waiting for a path to the destination, messages can be forwarded to intermediate nodes, which in turn would buffer these packets for some period of time and then forward them to other nodes. This process can be continued until some intermediate node eventually comes in contact with the destination node and delivers the message to it. Let us take, as an example, an alert application that propagates emergency messages when accidents occur. Just after the accident the application forwards a message that is rapidly propagated in order for the upcoming vehicles to brake and stop. There will be no need for the last vehicle which received the message to send it if there are no more neighbors to send packet to. The intelligent decision is to buffer the packet and forward it as soon as a new approaching vehicle is detected which will save another accident

In our work, we aim at integrating as many VANET-based services as possible and allowing the driver to select those he/she is interested in. Cooperation among vehicles is the key point in our framework.

This collaboration is illustrated through the novel publish/subscribe mechanism we propose for VANET; participants can act as publishers who generate information (either local or collaborative measurements) and subscribers are drivers who express their interests in a set of services and who are willing to cooperate to provide other subscribers with the information they are interested in.

Our work deals with three major challenges in VANET:

- we suppose a partially structured vehicular network;
- not all vehicles would be interested in collaborating, so they should not be affected;
- we deal with network fragmentation and the resultant lack of continuous end-to-end connectivity at any given instant.

The proposed solution will enable improved driving conditions and an optimized use of the road network. In fact, CaaS will allow:

- A decrease in the cost related to Intelligent Transportation Systems, since our solution performs well even when no infrastructure is available.
- A reduction in the severity of road congestions and CO_2 pollution, at the same time, time and fuel saving. Vehicles act like mobile sensors monitoring parameters such as road and weather conditions, or traffic density. Our solution allows such information to be shared among vehicles in order to perform route optimization or adaptations of driving behavior.
- Improving driver's safety and preventing rear-ending accidents. Our solution allows vehicles to alert the surrounding cars of its braking manoeuvres for example.
- A more enjoyable driving experience (Internet access, interactive online games...).

3. CaaS: Cooperation as a service

As stated earlier, our framework deploys a publish/subscribe interaction scheme. With publish/subscribe models; participants can act as subscribers who express their interest in an event, or a pattern of events, and publishers who submit information regarding those events to the system. Readers can find a detailed survey on this communication paradigm in [9]. Such a scheme is well adapted to the loosely coupled nature of distributed interaction in large-scale networks, VANET for instance, mainly because of its decoupling properties. In Fact, publish/subscribe-based schemes achieve at least two dimensions of decoupling: $1°$) Space decoupling: subscribers are interested in getting the information they want regardless of who published or how this information is published in the network and $2°$) Time decoupling: publishers and subscribers do not need to interact at the same time. Our designed publish/subscribe mechanism for VANET insures both of them.

In this section, we will be discussing our proposed publish/subscribe interaction scheme from the algorithmic, the functional, and the architectural perspectives. But before doing so, we will start first by arguing our choices regarding the underlying routing protocols we use in our approach.

3.1. Discussion

A traditional publish/subscribe system model relies on an event notification service (or broker) that stores and manages subscriptions, thus, acting as a mediator between publishers and subscribers. In VANET, we cannot expect any dedicated server (or service) that will play such a role. Nodes themselves should act as mediators as well as publishers and subscribers. This makes designing a scalable publish/subscribe scheme well suited to VANET environments extremely challenging.

Two main approaches have been proposed to support the publish/subscribe paradigm in MANET in general and in VANET in particular: a structured-based approach (like [44]) and a gossip-based approach (i.e. BubbleStorm [40]). The former requires the nodes to be organized into a sort of overlay structure and builds the publish/subscribe methods on top of it. The latter uses gossiping for information exchange which supposes that a query or publication will sufficiently populate a large portion of the network so that their paths intersect at some rendezvous nodes with high probability. Although, the first approach suffers from the disadvantage of introducing an additional overhead for structure construction and maintenance, it achieves better efficiency in comparison to the second which introduces additional computation and storage costs due to the intermittent connectivity issues in VANET. Since we want to guarantee a high level of service delivery to the participants in the network without affecting non-interested parties, we favor the structure-based approach over the gossip-based one.

In most structure-based approaches proposed for VANET, Content based data dissemination, where information is routed based on the content rather than the destination address, is proposed as an efficient publish/subscribe scheme. But again CBR requires a one-tree structure network topology [25], almost impossible to maintain when the size of the network increases. The high mobility of nodes in the network may make this tree maintenance issue even worse since a tree may become partitioned into a number of trees leading to serious issues finding a path to merge those trees.

To take advantage of the benefits of CBR and reduce its disadvantages, we decide to allow more than one tree structure in the network. Each tree will represent a cluster whose size and depth are appropriately chosen to allow a proper maintenance of the trees. In our structure, we use CBR for intra-cluster communications; subscriptions of all members of the cluster are forwarded to a cluster-head and updated regularly to deal with the continuous movement of the nodes. This is done using the same algorithm we already presented in [27,28] and which we will be reviewing in the next section.

For inter-cluster communication, many possible solutions can be considered depending on whether infrastructure exists or not. If roadside stations are available, clusters will be interconnected using the infrastructure as depicted in Fig. 1. However, since a widespread presence of roadside stations cannot be guaranteed at any time and place, two options can be considered: The first is to use a flooding-based approach (i.e. Document flooding (DF) [44]) to exchange subscription summarizations and publications among clusterheads in the network. The second is to take advantage of the performance efficiency achieved by deploying DTN routing [18] to disseminate subscriptions and publications between clusters. In our structure, we decide to use DTN routing for inter-cluster communication and let cluster-heads which have heard a publication buffer the packet and forward it as soon as a new approaching cluster-head vehicle is detected or a roadside station is in its vicinity.

In summary, our approach suggests a hybrid publish/subscribe scheme for VANET where CBR is used for intra-cluster communication and DTN routing for inter-cluster communication. Figure 2 summarizes our proposed network structure.

3.2. Tree construction

As we mentioned earlier, we allow more than one tree structure to exist in the network. Each tree represents a cluster where the root of the tree (i.e. clusterhead) should maintain an up-to-date subscription summary of all members in its cluster. The clusterhead should also be able to route publications to interested vehicles using CBR. Since a node might join or leave the cluster at any time, the tree should also be able to maintain itself and, depending on the proximity of nodes, merge itself with another tree. Our intra-cluster structure considers the following roles to set up a routing infrastructure:

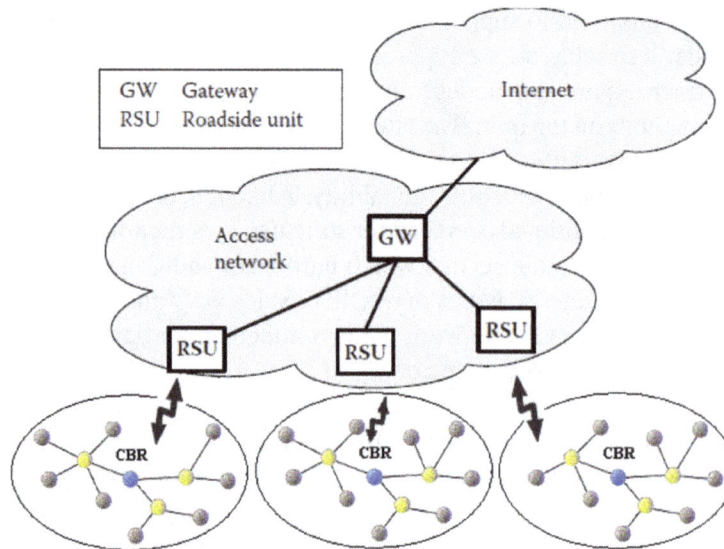

Fig. 1. Infrastructure-based vehicular network.

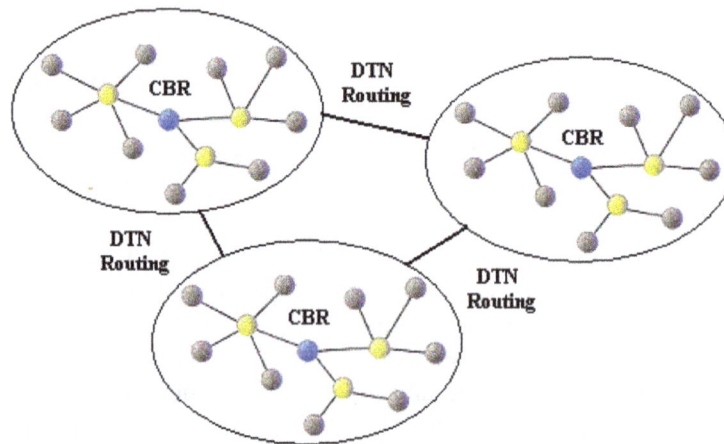

Fig. 2. Our proposed network structure for infrastructure-free vehicular network.

- Clusterhead: node responsible for summarizing subscriptions of the cluster members and forwarding them to other clusters. It is also responsible for delivering publications to interested nodes inside the cluster.
- Broker: node acting mainly as a relay. Each broker holds a subscription table used to determine how to disseminate subscriptions/publications along the tree.
- Subscriber: a node which expresses its interest in a service (or a set of services).
- Publisher: any entity in the network that publishes information about services in which vehicles might be interested.

To form a cluster, each node needs to know its neighbors' interests (i.e. subscriptions). Like other routing protocols for VANET, we consider that each node broadcasts a Hello message each Hello_Interval seconds to announce its existence to neighboring nodes. When a node is no longer receiving those

Table 1
Acronyms and definitions

NodeID$_i$	A unique identifier of node i, it could be a MAC address
ClusterID$_i$	The id of the root node of the tree to which node i belongs
Height$_i$	The maximum number of nodes separating a leaf from the root in the tree to which node i belongs. This value is the same for all members of a cluster, but could vary in time if a node joins or leaves the cluster
Max_Height	The maximum height allowed for a tree. This value is fixed to allow a proper maintenance of the trees
Level$_i$	The number of nodes separating node i from the root. The immediate neighbors of the root are on the first level
ParentID$_i$	The identifier of the neighbor parent of node i
Child_list$_i$	Identifiers of the neighboring childs of node i
CH_Flag$_i$	A flag indicating whether node i is a clusterhead or not.
Subscription Table	A list of pairs (NodeID, subscription list) maintained by each node

messages from a neighbor, it means that this neighbor disappeared or moved out of the transmission range of the current node. Hello messages will also serve to update cluster members and nodes' subscriptions. To achieve this, each Hello message should contain some necessary information to maintain the tree such as the node Id, the cluster Id, the height of the tree, the level of the node and the subscription table (or a part of it, in case a change in the nodes' subscriptions occurs) as described in Table 1. However, depending on whether the node is already a member of a cluster or not, the format of these messages might vary.

Initially, when a node enters the system, it tries to join an existing cluster by listening to the broadcasted Hello messages sent by neighbors and by considering as a parent the node whose signal is the strongest and whose level is strictly smaller than the maximum height allowed for the tree. A child cannot have two parents because the tree should not contain any cycles. If no such node exists, the node considers itself to be a clusterhead.

Each node broadcasts a hello message with the information listed above along with its interests if there are any. Upon receiving a hello message, a neighboring node will update its local information (ClusterID, level, Child_list... etc.). Moreover, each node should maintain heartbeats of its neighboring nodes (i.e. parent and children): After receiving a Hello message from a neighbor, a timer is set to a certain value. If this timer expires without receiving any hello messages from this neighbor, this neighbor will be considered unreachable and is removed from the list of neighbors maintained by the node. Particularly, if a node loses its parent, it should initialize its local parameters to the default ones (Fig. 3, initialization section) to either join another cluster or build its own tree. To avoid cycles, we also consider that only roots of the trees can issue the merging process. Figure 3 shows our proposed algorithm for processing hello messages.

3.3. Subscriptions and publications dissemination

Now that the trees are constructed and properly maintained thanks to the merging and dissociation processes explained earlier, subscriptions and publications can now be forwarded along the constructed links of the tree. Our idea is to exploit hello messages to communicate the node subscriptions to other nodes in the cluster and especially to the clusterhead. Each node (broker) will maintain a subscription table containing its interests, if any, along with the neighboring nodes' interests. Each node will send its subscription table only once; unless a change in this subscription table occurs (a node inserted a new interest or changed a subscription for example). If so, only the affected lines in the table will be sent in the following hello message.

Input: The Hello message received, it should contain at least the following fields: $NodeID_{sender}$, $ClusterID_{sender}$, $level_{sender}$, $Height_{sender}$, $ParentID_{sender}$, $Child\_list_{sender}$

Initialization:
//initialize local information
$ClusterID=NodeID_{current}$ //Initially, a node considers itself as the root of a tree
$Level=0$
$Height=0$
$ParentID=NodeID_{current}$ //Initially, a node is a parent of itself
$Child\_list=none$ //Initially, no children
$CH\_Flag=1$ //Initially, a node considers itself a clusterhead

Output:
1: **IF** $ClusterID_{sender}$ is different from the current ClusterID **THEN**
2:    **IF** (CH_Flag=1) && ($Height_{sender}+Height$ <Max_Height) **THEN**
3:        //Node should join the cluster of the sender (trees merging) and local information need to be updated
4:        $ClusterID=ClusterID_{sender}$
5:        $ParentID=NodeID_{sender}$
6:        $Level= level_{sender}+1$
7:        $Height=Height+Height_{sender}$
8:        $CH\_Flag=0$ //Current node is no longer a clusterhead since it joined a tree
9:    **END-IF**
10: **ELSE** //Sender belongs to the same cluster as that of the current node
11:    **IF** ($ParentID_{sender}= NodeID_{current}$) //The sender is a child of the current node **THEN**
12:        //Update child list
13:        $Child\_list=Child\_list+\{NodeID_{sender}\}$
14:    **END-IF**
14:    **IF** ($NodeID_{sender}=ParentID$) //The sender is the parent of the current node **THEN**
15:        execute instructions 4 , 6, and 7 // Local information update
16:    **END-IF**
17: **END-IF**

Fig. 3. Hello messages processing algorithm.

Publications will be routed along the path set up by subscriptions to interested subscribers only. This is the same approach as CBR. Moreover, each clusterhead will have to forward publications to other clusters either using infrastructure if available or DTN routing if no roadside stations are present.

4. Implementation details and simulation results

In order to assess the actual feasibility and prove the validity of our framework, we run exhaustive simulations and report on our proposed protocol performance over several scenarios, varying from a fully infrastructure-based scenario characterized by a large presence of roadside stations to a scenario in which no infrastructure is available as this represents the more challenging case for our protocol.

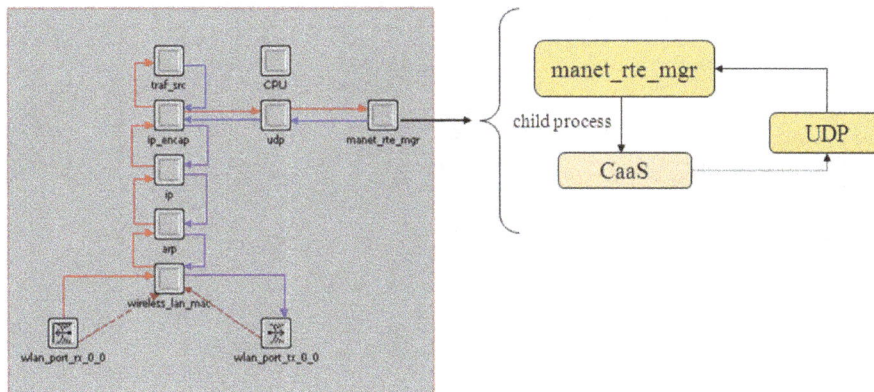

Fig. 4. CaaS model architecture in Opnet.

4.1. Simulation environment

As simulation platform, we used Opnet simulator, an event-driven, network simulation tool, which allows an easy implementation of all model elements. The hierarchical assembly in OPNET is done in three main layers: a) The node model which specifies the main blocks and parameters of a node and provides an interface to the network element, b) the process model which defines the states and the state transitions of the node model elements and abstracts the behavior of the network element. C/C++ code that governs each state of a process model can be rapidly customized. OPNET Kernel Procedure APIs exist to facilitate development and support common communications mechanisms, such as packets, queues, and traffic, and c) the network model which contains the set of nodes and defines links between them.

4.2. Implementation details

We implemented CaaS as a new custom proactive VANET routing protocol that interfaces with IP through UDP. Our protocol has the following features:

– Neighbor sensing mechanism.
– Fast topology change detection via periodic hello messages exchange.
– Neighboring nodes' subscription table maintenance and update.
– Maintenance of the cluster topology (route to members of the cluster and their subscriptions) by the clusterhead.
– DTN routing between a clusterhead and another clusterhead when no infrastructure is available.

Before explaining how these functionalities are evaluated, we need to briefly introduce the node model architecture of a VANET vehicle, running CaaS, in Opnet.

4.2.1. CaaS architecture model

The node model of a VANET vehicle running CaaS (or any other proactive VANET routing protocol) in Opnet is shown in Fig. 4. We created our protocol as a child process of manet_rte_mgr module which provides a common interface to multiple VANET routing protocols and is extensible for custom protocols. In Fact, manet_rte_mgr module is suitable for our situation because it is already interfaced to UDP, so no new "protocol number" is needed. IP simply thinks of the custom protocol as UDP on a port

```
                                          typedef enum
                                          {
                                             IpC_Dyn_Rte_Invalid        = -1,
                                             IpC_Dyn_Rte_Directly_Connected = 0,
    typedef enum                             IpC_Dyn_Rte_Ospf           = 1,
    {                                        IpC_Dyn_Rte_Rip            = 2,
       IpC_Rte_Default = -99,                IpC_Dyn_Rte_Igrp           = 3,
       IpC_Rte_Custom = -2,                  IpC_Dyn_Rte_Bgp            = 4,
       IpC_Rte_None = -1,                    IpC_Dyn_Rte_Eigrp          = 5,
       IpC_Rte_Rip = 0,                      IpC_Dyn_Rte_Isis           = 6,
       IpC_Rte_Igrp,                         IpC_Dyn_Rte_Static         = 7,
       IpC_Rte_Ospf,                         IpC_Dyn_Rte_Ext_Eigrp      = 8,
       IpC_Rte_Bgp,                          IpC_Dyn_Rte_IBgp           = 9,
       IpC_Rte_Eigrp,                        IpC_Dyn_Rte_Default        = 10,
       IpC_Rte_Isis,                         IpC_Dyn_Rte_Ripng          = 11,
       IpC_Rte_Misis,                        IpC_Dyn_Rte_Tora           = 12,
       IpC_Rte_Dsr,                          IpC_Dyn_Rte_Aodv           = 13,
       IpC_Rte_Tora,                         IpC_Dyn_Rte_Olsr           = 14,
       IpC_Rte_Aodv,                         IpC_Dyn_Rte_Mobile_IP      = 15,
       IpC_Rte_Olsr,                         IpC_Dyn_Rte_LDP            = 16,
       IpC_Rte_CaaS,                         IpC_Dyn_Rte_Custom         = 17,
       IpC_Rte_Grp,                          IpC_Dyn_Rte_Local          = 18,
       IpC_Rte_Ripng,                        IpC_Dyn_Rte_CaaS           = 19,
       IpC_Rte_Ospf3,                        IpC_Dyn_Rte_Number         = 20
       IpC_Rte_Static                     } IpT_Rte_Prot_Type;
    } IpT_Rte_Protocol;
```

Fig. 5. Set of routing protocols contributing entries to the common IP route table.

and once the port is set up, all we need is to send packets to UDP and have a receiver to get those packets from UDP. This module is either invoked by UDP through a stream interrupt (i.e., a packet for CaaS child process) or when CaaS child process directly sends packet to UDP on the connected port number. It is worth to mention that a process in Opnet is an instance of a process model and can dynamically create child processes and respond to interrupts.

In order to implement CaaS as a custom VANET routing protocol in Opnet, few steps need to be carefully performed. We need first to add CaaS to the list of dynamic routing protocol already implemented in Opnet and enumerated in the corresponding IP header files. Examples of these protocols include those listed in Fig. 5-a.

The CaaS process employs a routing table to keep track of valid routes to destination nodes. CaaS routing table is implemented as a hash table indexed by an IP address. CaaS routing table is populated and updated via Hello messages. OPNET implements packet forwarding within the IP module which uses a Common IP Routing Table. This routing table is updated and maintained by the routing protocol configured for the simulation study. Thus, CaaS and other routing protocols, in addition to maintaining their internal routing tables are also responsible for updating routing table at IP layer. Therefore, CaaS needs to be added to the list of those protocols that operate on IP Common Route Table as depicted in Fig. 5-b.

The second step is to create a new child process for CaaS and attach it to the parent manager process. This will be fully explained in the next section.

4.2.2. CaaS process model

Before proceeding with the implementation of any protocol in Opnet, a process model of this protocol needs to be created. Process Modeling Methodology (PMM) is a systematic approach to creating process models in OPNET Modeler and is considered to be the quickest and most efficient method of development thanks to the consistency of results it provides. Figure 6 shows our proposed process model for CaaS.

Our process model is composed of a forced (green) state and an unforced (red) state. The difference between the two is that the former is a non blocking state that deals with the initialization phase of nodes,

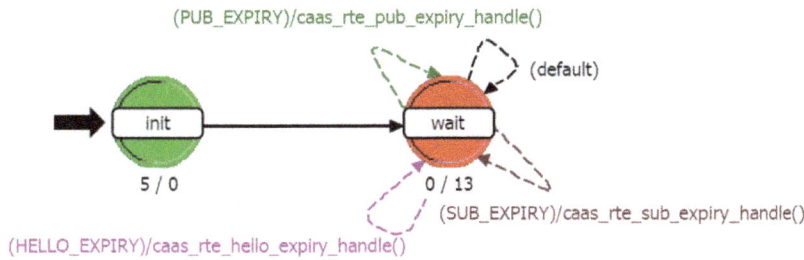

Fig. 6. CaaS process model.

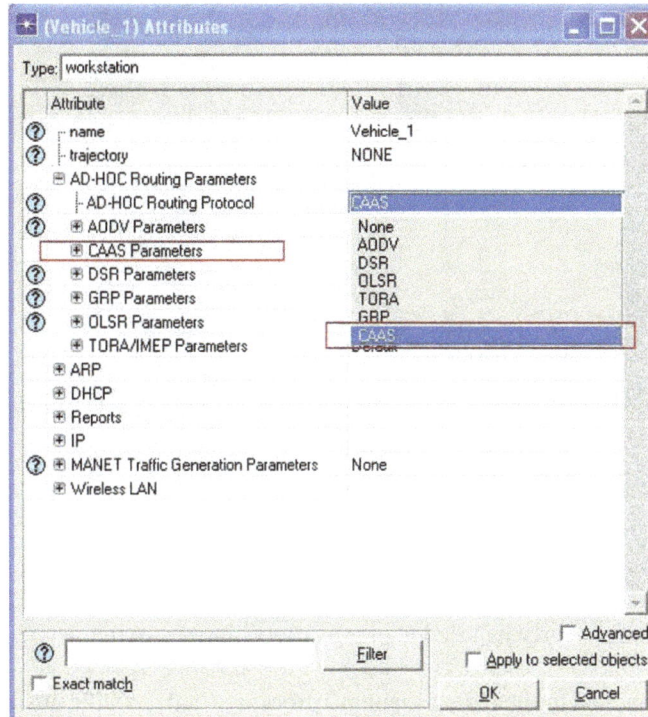

Fig. 7. A vehicle node attributes.

local and global parameters definition and attributes reading while the latter is a blocking state waiting for invocation.

The Simulation Kernel operates by maintaining an event list for the entire simulation. As the simulation executes, the Simulation Kernel manages a list of events to take place. As each event reaches the top (or head) of that list, it becomes an interrupt. Interrupts are often delivered to specific modules, and this occurrence is what activates the module's process model. All transitions in CaaS process model are from idle back to itself. Three interrupts are scheduled to occur: 1°) a publication is scheduled to be sent (PUB_EXPIRY condition), 2°) a subscription needs to be advertized (SUB_EXPIRY condition), and 3°) A simple hello message (mainly used for topology control) needs to be sent. A default transition is used because if there is a different type of interrupt, other than the three mentioned above, there must be a transition that the process model can follow. The default transition handles these different interrupt types. A transition with condition "default" is true if and only if no other conditions are true.

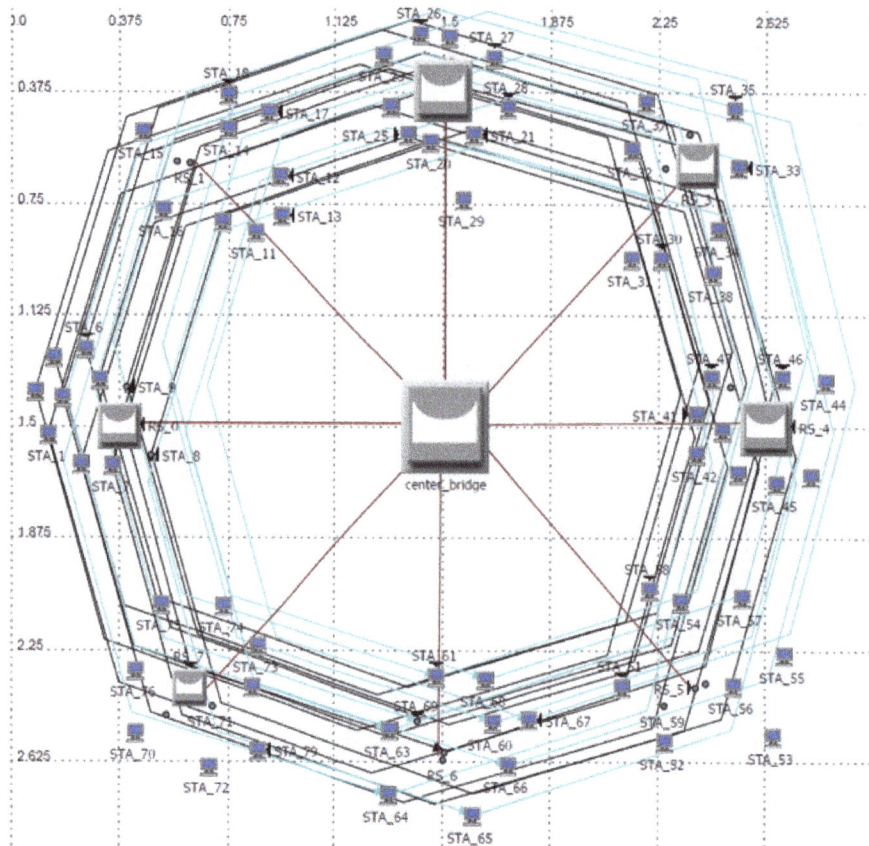

Fig. 8. CaaS network model.

Three different timers manage those three interrupts. In the waiting state, the arrival of one of the three packet types previously sent is properly handled using the algorithm we explained in details in the previous section.

The child process can later be attached to the parent manager process. CaaS specific attributes, such as the hello interval attribute, should also be added to manet_rte_mgr model attributes (Fig. 7) so they can be seen while configuring the protocol at a node. Attributes will be parsed by the child process model.

4.3. Simulation settings

As a first experiment, we focused on a fully infrastructure-based scenario in which a large presence of roadside stations is considered. Our goal is twofold: on one hand we want to prove the correctness of CaaS, and on the other hand we want to evaluate the dissemination of publications to interested parties inside the clusters (i.e., intra-cluster structure).

The simulation scenario we used is shown in Fig. 8. Our network model consists of a 3Km x 3Km area in which 80 vehicles and 8 roadside units running CaaS are present. Some vehicles move in clockwise direction, and others move in counter-clockwise direction, both at a speed of maximum 110 Km/h. Opnet allows defining custom and realistic mobility patterns for each vehicle to meet most complex road topology requirements. We use 802.11b wireless radio interface (data rate 11 Mbps and 2 mW transmission power). Other simulation parameters are presented in Fig. 9.

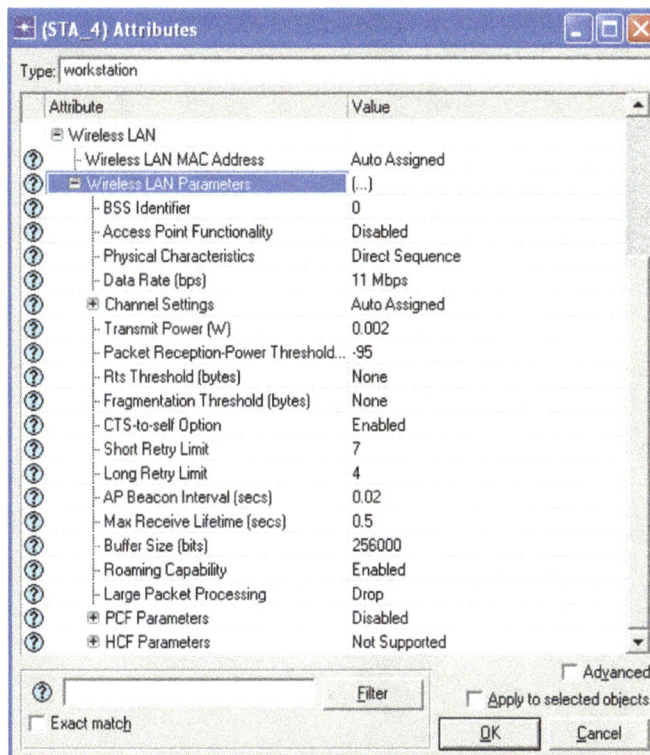

Fig. 9. WLAN simulation parameters.

Each roadside station acts like a router with two interfaces: a wireless interface running CaaS and automatically subscribing to receiving all kinds of publications) and an Ethernet interface connecting the station to the backbone network. The node model of the roadside base station is shown in Fig. 10.

We, later, progressively reduce the number of roadside units to assess the impact of the additional DTN routing in such scenario.

4.4. Simulation results

In all our experiments, we measure the publication delivery ratio, expressed as the number of subscribers that successfully received publications they subscribed to receiving, the average delay between publication advertizing and their reception, and the traffic overhead defined as the number of transmissions received per minute by each vehicle. We later compare CaaS against a classic epidemic approach [42]. Results are averaged over multiple runs.

Simulation results show that the delivery ratio is 100% in a fully infrastructure-based network as depicted in Fig. 11. All subscribers (a total of ten subscribers in our simulation) have successfully received at least one publication regarding the service they subscribed to, which demonstrates the correctness of CaaS intra-cluster structure. Nevertheless, despite this important result, assuming a widespread presence of roadside units is unrealistic in many scenarios. For this particular reason, we decided to reduce progressively the number of those roadside stations and observe the impact of this on the delivery ratio.

Remarkably, publication delivery is above 90% even if no roadside stations are available (as depicted in Fig. 12). This is mainly due to DTN routing deployment in which clusterheads buffer publications

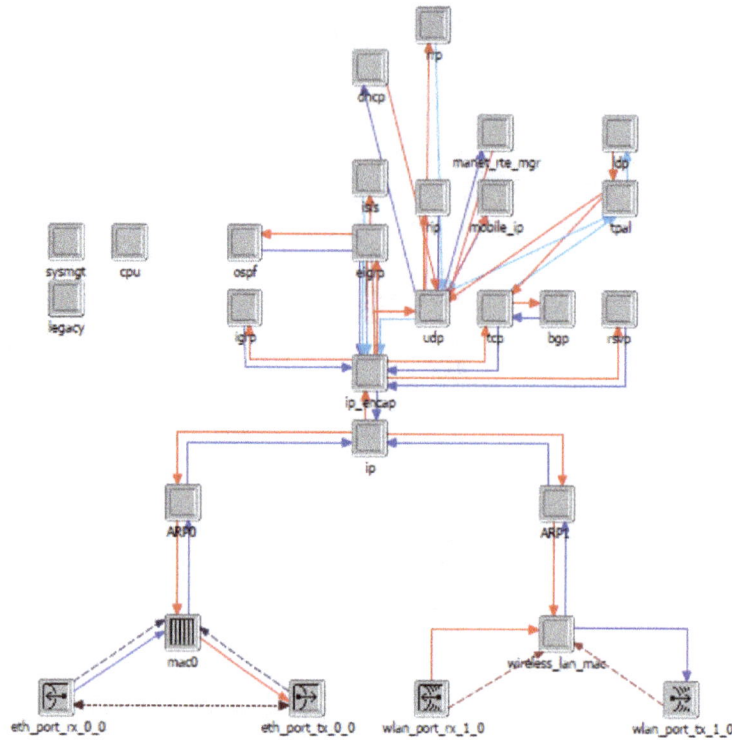

Fig. 10. Node model of the roadside base station.

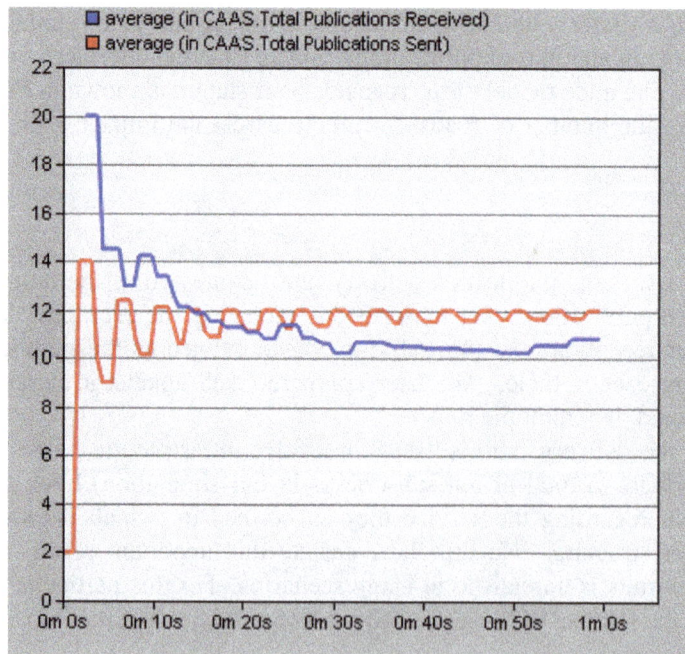

Fig. 11. Publication delivery in a fully infrastructure-based scenario.

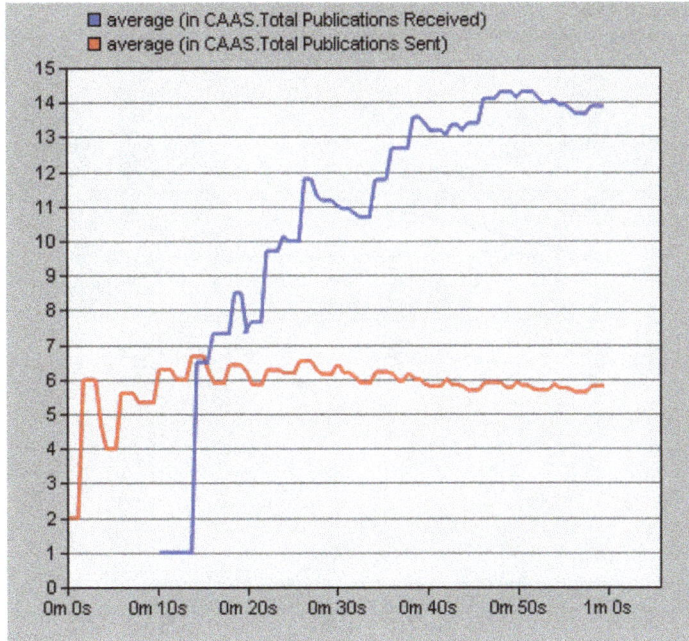

Fig. 12. Publication delivery in an infrastructure-free scenario.

and then forward them to other clusterheads when they are in their vicinity. When infrastructure is not available, transmitting publications to clusterheads that are going in the opposite direction represents the most favorable case for our protocol even if the connection between the two vehicles lasts only few seconds, as they act as information carriers that will rapidly alert vehicles going in the opposite direction about events affecting their route. The delay of publication reception in an infrastructure-free scenario is, however, higher than that of a fully infrastructure-based scenario. Figure 9-b shows that it takes at least 20 seconds to inform 90% of the subscribers about an event affecting their route while it was almost instantaneous in a fully infrastructure-based network. This is due to the multihop nature of transmission in pure VANET and intermittent connectivity issues, causing delays in packets buffering and appropriate neighbor finding.

To put our work in the context of related efforts, we also compared our protocol with an epidemic approach [42], in which all vehicles (subscribers and non-subscribers) buffer each message received and retransmit it to all neighbors which have not heard that message yet. In our experiment, we consider an infrastructure-free network since this is the most challenging situation for our protocol.

Simulation parameters are kept the same as in previous scenarios. We will not focus on comparing the delivery ratio since it is quite high for both protocols. Therefore, our main concern will be more about the network overhead induced by assuming that all communication in this experiment relies on vehicle-to-vehicle technology. Figure 13 shows that structuring the network into clusters significantly reduces the overall network overhead. This is explained by the fact that CaaS delivers publications only to interested vehicles inside the cluster, instead of all vehicles in the network. Moreover, since each clusterhead maintains a subscription table of all members of the cluster, it will not systematically forward all received publications to members of the cluster unless they are interested in receiving them, which confirm our claims of message overhead reduction.

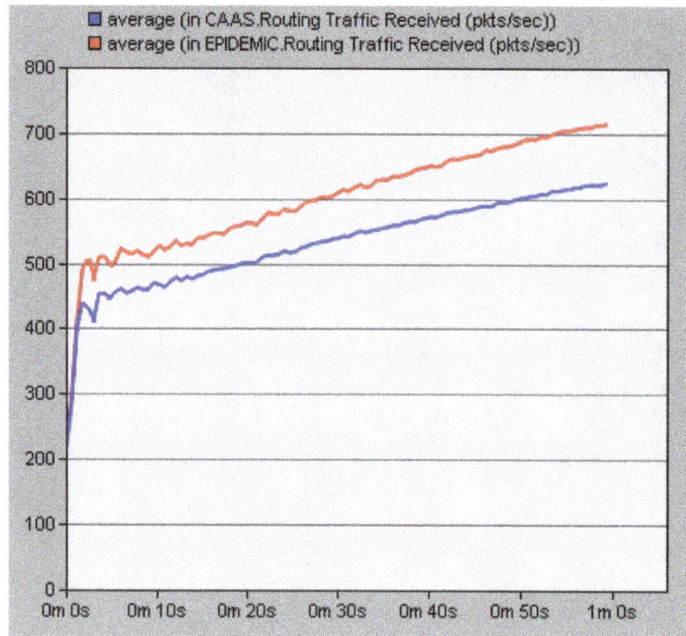

Fig. 13. Network overhead in CaaS and in epidemic routing.

5. Concluding remarks and future work

In this paper, We introduced a new solution for VANETs, referred to as Cooperation as a Service (CaaS), that allows providing vehicles/drivers, which are willing to cooperate, with some sets of services using very minimal infrastructure, by taking advantage of Vehicle-to-Vehicle (V2V) communications.

To tackle the infrastructure-less nature of these networks, organize our network into clusters that can be properly maintained and to use CBR for intra-cluster communications and DTN routing for inter-cluster ones.

We designed Cooperation as a Service (CaaS) to mainly provide drivers with services for free using vehicles' cooperation. There is one more theme that will contribute to shape CaaS in a more effective way: privacy (of location and motion patterns) and security (mainly, confidentiality and protection from DOS (Denial of Service) attacks). We did not dwell on the privacy/security topic in this paper because we chose to focus on conventional network layer aspects. However, two trends are clear. The need of a Certificate Authority (CA) will require efficient connection to Internet Servers. At the same time, to handle protection from bogus attacks in situations when they are disconnected from the Internet, or it is simply too time consuming to consult the Internet CA, the mobile users must organize in P2P communities and use majority rules and/or elect proxy mobile CAs to resolve security issues [16].

Acknowledgment

Professor Olariu was supported by NSF grants CNS 0721586.

References

[1] M. Caliskan, D. Graupner and M. Mauve, Decentralized discovery of free parking places, in *Proceedings of the 3rd ACM International Workshop on Vehicular Ad Hoc Networks*, (2006), pages 30–39.

[2] A. Carus, A. Urpi, S. Chessa and S. De, GPS-free coordinate assignment and routing in wireless sensor networks, in *Proc. 24th Annual Joint Conference of the IEEE Computer and Communications Societies* (INFOCOM'05), Vol. 1, (2005), pages 150–160.

[3] D. Caveney, in: *VANET: Vehicular Applications and Inter-Networking Technologies*, H. Hartenstein and K. Laberteaux, eds, A John Wiley and Sons, UK, 2010, pp. 21–48.

[4] T.G. Crainic, M. Gendreau and J. Potvin, Intelligent freight transportation systems: Assessment and the contribution of operations research, *Transportation Research Part C: Emerging Technologies* **17**(6) (2009), 541–557.

[5] G. Cugola and E. Di Nitto, On adopting Content-Based Routing in service-oriented architectures, *Information and Software Technology* **50**(1–2) (2008), 22–35.

[6] T. Delot, S. Ilarri, N. Cenerario and T. Hien, Event sharing in vehicular networks using geographic vectors and maps, *Mobile Information Systems* **7**(1) (2011), 21–44.

[7] M.D. Dikaiakos, A. Florides, T. Nadeem and L. Iftode, Location aware services over vehicular ad hoc networks using car-to-car communication, *IEEE Journal on Selected Areas in Communications* **25**(8) (2007), 1590–1602.

[8] S. Dornbush and A. Joshi, StreetSmart Traffic: Discovering and disseminating automobile congestion using VANET's, in *Proceedings of the 65th IEEE Vehicular Technology Conference*, (2007), pages 11–15.

[9] P.Th. Eugster, P.A. Felber, R. Guerraoui and A.M. Kermarrec, The many faces of publish/subscribe, *ACM Computers Survey* **35**(2) (2003), 114–131.

[10] M.D. Fontaine, in: *Vehicular Networks from theory to Practice*, S. Olariu and M.C. Weigle, eds, CRC Press, USA, 2010, pages 1-1/1-26.

[11] R. Frenkiel, B. Badrinath, J. Borras and R.D. Yates, The infostations challenge: balancing cost and ubiquity in delivering wireless data, *IEEE Personal Communications* **7**(2) (2002), 66–71.

[12] D.J. Gabauer and H.C. Gabler, Comparison of roadside crash injury metrics using event data recorders, *Accident Analysis & Prevention* **40**(2) (2008), 548–558.

[13] G. Gehlen and L. Pham, Mobile Web services for peer-to-peer applications, in *Proceedings of IEEE international conference on consumer communications and networking*, (2005), pages 427–433.

[14] M. Gerla and L. Kleinrock, Vehicular networks and the future of the mobile internet, *Computer Networks*, (2010), doi:10.1016/j.comnet.2010.10.015.

[15] P. Gupta and P.R. Kumar, The capacity of wireless networks, *IEEE Transactions on Information Theory* **46**(2) (2000), 388–404.

[16] X. Hong, D. Huang, M. Gerla and Z. Cao, SAT: building new trust architecture for vehicular networks, In *proceedings of ACM SIGCOMM Workshop on Mobility in the Evolving Internet Architecture* (MobiArch), 2008.

[17] R.C. Hsu and L.R. Chen, An integrated embedded system architecture for invehicle telematics and infotainment system, in *Proceedings of the IEEE International Symposium on Industrial Electronics* **4** (2005), pages 1409–1414.

[18] S. Jain, K. Fall and R. Patra, Routing in a Delay Tolerant Network. In SIGCOMM '04: Proceedings of the 2004 Conference on Applications, Technologies, Architectures, and Protocols for Computer Communications, ACM Press. ISBN 1-58113-862-8, (2004), pages 145–158.

[19] D. Jiang, V. Taliwal, A. Meier, W. Holfelder and R. Herrtwich, Design of 5.9 GHz DSRC-based vehicular safety communication, *IEEE Wireless Communications* **13**(5) (2006), 36–43.

[20] I. Khalil, R. Kronsteiner and G. Kotsis, A semantic solution for data integration in mixed sensor networks, *Computer Communications* **28**(13) (2005), 1564–1574.

[21] S.B. Lee, G. Pan, J.S. Park, M. Gerla and S. Lu, Secure incentives for commercial ad dissemination in vehicular networks, in *Proceedings of the 8th ACM International Symposium on Mobile Ad Hoc Networking and Computing*, (2007), pages 150–159.

[22] I. Leontiadis and C. Mascolo, GeOpps: Opportunistic Geographical Routing for Vehicular Networks, in *Proc. of the IEEE Workshop on Autonomic and Opportunistic Communications*, Helsinki, Finland, 2007.

[23] C. Lochert, M. Mauve, H. Fußler and H. Hartenstein, Geographic routing in city scenarios, *ACM SIGMOBILE Mobile Computing and Communications Review* **9**(1) (2005), 69–72.

[24] C. Maihofer, A survey on geocast routing protocols, *IEEE Communications Surveys and Tutorials* **6**(2) (2004) 32–42.

[25] L. Mottola, G. Cugola and G.P. Picco, A Self-repairing tree topology enabling content-based routing in mobile ad hoc networks, *IEEE Transactions on Mobile Computing* **7**(8) (2008), 946–960.

[26] H. Mousannif, H. Al Moatassime and S. Rakrak, An energy-efficient scheme for reporting events over WSNs, *International Journal of Pervasive Computing and Communications* **7**(1) (2011), 44–59.

[27] H. Mousannif, I. Khalil and H. Al Moatassime, Cooperation as a Service in VANETs, *Journal of Universal Computer Science* **17**(8) (2011), 1202–1218.

[28]   H. Mousannif, I. Khalil and S. Olariu, Cooperation in static and mobile sensor-based platforms for situation, activity and goal awareness. Proceedings of the 2011 international workshop on Situation activity & goal awareness at the 13th ACM International Conference on Ubiquitous Computing (UbiComp2011), (2011), pages 5–14.

[29]   T. Nadeem, S. Dashtinezhad, C. Liao and L. Iftode, TrafficView: traffic data dissemination using car-to-car communication, *ACM SIGMOBILE Mobile Computing and Communications Review* **8**(3) (2004), 6–19.

[30]   E.F. Nakamura, A.A.F. Loureiro and A.C. Frery, Information fusion for wireless sensor networks: Methods, models, and classifi cations, *ACM Computing Surveys* **39**(3) (2007), 9/1–9/55.

[31]   A. Nandan, S. Das, S. Tewari, M. Gerla and L. Klienrock, AdTorrent: delivering location cognizant advertisements to car networks, in *The Third International Conference on Wireless On Demand Network Systems and Services* (WONS 2006), Les Menuires, France, 2006.

[32]   National Highway Traffic Safety Administration, Traffic safety facts – preliminary 2009 report, http://www-nrd.nhtsa.dot.gov/Pubs/811255.pdf, March 2010.

[33]   S. Olariu, I. Khalil and M. Abuelela, Taking VANETs to the cloud, *International Journal of Pervesive Computing and Communication* **7**(1) (2011), 7–21.

[34]   K. Ozbay, H. Nassif and S. Goel, Propagation characteristics of dynamic information collected by in-vehicle sensors in a vehicular network, in *Proceedings of the IEEE Intelligent Vehicles Symposium*, (2007), pages 1089–1094.

[35]   C.E. Palazzi, M. Roccetti, G. Pau and M. Gerla, Online Games on Wheels: Fast Game Event Delivery in Vehicular Ad-hoc Networks, in *V2VCOM'07*, Istanbul, Turkey, 2007.

[36]   N. Qadri, M. Altaf, M. Fleury and M. Ghanbari, Robust video communication over an urban VANET, *Mobile Information Systems* **6**(3) (2010), 259–280.

[37]   M. Saito, J. Tsukamoto, T. Umedu and T. Higashino, Design and evaluation of intervehicle dissemination protocol for propagation of preceding traffic information, *IEEE Transactions on Intelligent Transportation Systems* **8**(3) (2007), 379–390.

[38]   S. Sesay, Z. Yang and J. He, A Survey on Mobile Ad Hoc Wireless Network, *Information Technology Journal* **3**(2) (2004), 168–175.

[39]   S. Smaldone, L. Han, P. Shankar and L. Iftode, RoadSpeak: Enabling Voice Chat on Roadways using Vehicular Social Networks, in *SocialNets'08*, Glasgow, Scotland, UK, 2008.

[40]   W.W. Terpstra, J. Kangasharju, C. Leng and A.P. Buchmann, BubbleStorm: resilient, probabilistic, and exhaustive peer-to-peer search, in *Proceedings of the 2007 Conference on Applications, Technologies, Architectures, and Protocols for Computer Communications*, ACM, New York, NY, USA, 2007, pages 49–60.

[41]   O. Tonguz, N. Wisitpongphan, F. Bai, P. Mudalige and V. Sadekar, Broadcasting in VANET, in *Proceedings of the 2007 Workshop on Mobile Networking for Vehicular Environments*, 2007, pages 7–12.

[42]   A. Vahdat and D. Becker, Epidemic routing for partially connected ad Hoc networks, Tech. Rep., Duke University, 2000.

[43]   K.J. Wong, B.S. Lee, B.C. Seet, G. Liu and L. Zhu, BUSNet: Model and Usage of Regular Traffic Patterns in Mobile Ad Hoc Networks for Inter-vehicular Communications, in *Proc. 10th International Conference on Telecommunications* (ICT '03), 2003, pages 102–108.

[44]   S. Yoo, J.H. Son and M.H. Kim, A scalable publish/subscribe system for large mobile ad hoc networks, *Journal of Systems and Software* **82** (2009), 1152–1162.

[45]   J. Zhao and G. Cao, VADD: Vehicle-Assisted Data Delivery in Vehicular Ad Hoc Networks, in *Proc. 25th IEEE International Conference on Computer Communications* (INFOCOM '06), 2006, pages 1–12.

Hajar Mousannif received her M.S. degree in Telecommunications and Computer Sciences from the National Institute of Posts and Telecommunications (INPT), Rabat (Morocco) in 2005. She joined, the same year, the Faculty of Sciences and Techniques Marrakech (Cadi Ayyad University, Morocco) as a telecom engineer where she was mainly responsible for managing and administrating the network. Since 2006, she is a teacher within the Department of Computer Sciences at the same Faculty. Her primary research interests include wireless sensor networks, wireless adhoc networks, and next generation internet technologies with main focus on routing protocols development and mobility management. In addition to her academic experience, she was in the program committee chair of many national conferences.

Ismail Khalil is a senior researcher and lecturer at the institute of telecooperation, Johanes Kepler University Linz, Austria since October 2002. He is the president of the international organization of Information Integration and Web-based Applications & Services @WAS. He holds a PhD in computer engineering and received his habilitation degree in applied computer science on his work on agents interactions in ubiquitous environments in May 2008.

He currently teaches, consults, and conducts research in Mobile Multimedia, Agent Technologies, and the Semantic Web and is also interested in the broader business, social, and policy implications associated with the emerging information technologies. Dr. Khalil has authored around 100 scientific publications, books, and book chapters. He serves as the editor-in-chief of the International Journal on Web Information Systems (IJWIS), International Journal on Pervasive Computing and Communication

(IJPCC), Journal of Mobile Multimedia (JMM), International Journal of Mobile Computing and Multimedia Communication (IJMcMc), Advances in Next Generation Mobile Multimedia book series, and Atlantis Ambient and Pervasive Intelligence book series. He is on the editorial board of several international journals. His work has been published and presented at various conferences and workshops. (http://www.iiwas.org/ismail/)

Professor Olariu has held many different roles and responsibilities as a member of numerous organizations and teams. Much of his experience has been with the design and implementation of robust protocols for wireless networks and in particular sensor networks and their applications. Professor Olariu is applying mathematical modeling and analytical frameworks to the resolution of problems ranging from securing communications, to predicting the behavior of complex systems, to evaluating performance of wireless networks. His research interests are in the area of complex systems enabled by large-scale deployments of sensors and vehicular networks and cloud computing.

Currently, Professor Olariu is the Associate Editor in Chief of IEEE Transactions on Parallel and Distributed Systems and an Associate Editor of IEEE Transactions on Computers.

6

Emerging Wireless and Mobile Technologies

Fang-Yie Leu[a], Ilsun You[b] and Feilong Tang[c]

[a]*Department of Computer Science, TungHai University, Tunghai, Taiwan*
[b]*School of Information Science, Korean Bible University, Nowon-gu, Seoul, South Korea*
[c]*Department of Computer Science and Engineering, Shanghai Jiao Tong University, Shanghai, China*

Recent advances in wireless and mobile technologies have led to a new paradigm of the high-tech society and people's daily life. Accordingly, wireless and mobile technologies have been gaining tremendous attentions from researchers all over the world in recent years. However, a lot of new challenges, which go much beyond conventional network systems, still need to be solved for advanced applications. To further improve the quality of the modern communication, new techniques need to be continuously explored and developed. This special issue looks for significant contributions and high quality research results on wireless and mobile technologies in theoretical and practical aspects, especially on QoS routing, distributed authentication mechanism, automatic security assessment, fast handover security mechanism, enhancing MISP with Fast Mobile IPv6 and applications in wireless mobile networks, as well as ad-hoc networks.

This special issue grew out of selected best papers from the Fifth International Conference on Broadband and Wireless Computing, Communication and Applications (BWCCA 2010), held in Fukuoka, Japan, and the 4th International Workshop on Intelligent, Mobile and Internet Services in Ubiquitous Computing (IMIS 2010), held in Krakow, Poland. This event was an effort to take up the challenges and to bring together an international community in the area.

The first paper [1], "QoS Routing in Ad-hoc Networks Using GA and Multi-Objective Optimization" from Admir Barolli, Evjola Spaho, Leonard Barolli, Fatos Xhafa and Makoto Takizawa, proposes a QoS routing in ad-hoc networks. Distinguishing from existing works related to routing in ad-hoc networks, this paper designs a QoS-guaranteed solution based on Genetic Algorithms (GAs) and multi-objective optimization. In particular, the authors implemented a search space reduction algorithm, which reduces the search space for GAMAN (GA-based routing algorithm for Mobile Ad-hoc Networks) to find a new route.

In the second paper [2], Inshil Doh, Jiyoung Lim and Kijoon Chae propose a "Distributed Authentication Mechanism for Secure Channel Establishment in Ubiquitous Medical Sensor Networks", in which bio-data is collected and delivered to the server system through mobile devices or gateways. This scheme adopts candidate devices, which act as the intermediary when the original device cannot function properly. In this way, bio-data of patients can be sensed and collected to monitor their body status uninterruptedly.

The next paper [3] entitled "Design of a secure RFID authentication scheme preceding market transactions" by Chin-Ling Chen, proposes a RFID mutual authentication protocol for market application systems. In order to achieve mutual authentication, the proposed scheme integrates fingerprint biometrics, cryptology and a hash function mechanism to ensure the security of transmitted messages. The proposed scheme can resist tag impersonation attack, replay attack, trace attack and forgery attack. Also, it maintains privacy protection and achieves mutual authentication, anonymity and forward secrecy.

In the fourth paper [4] entitled "Automatic Security Assessment for Next Generation Wireless Mobile Networks", Francesco Palmieri, Ugo Fiore and Aniello Castiglione design an active third party authentication, authorization and security assessment strategy. In this strategy, the infrastructure automatically detects incoming devices and analyzes whether they are secure or not. If a device is found to be insecure, it is immediately taken out from the network and denied further access until its vulnerabilities have been fixed. As the fundamental component of the security assessment strategy, the security assessment module takes advantage from a reliable knowledge base containing semantically-rich information about mobile nodes. Consequently, this scheme supports for the secure protection of wireless and mobile networks through automatic and real-time security/risk evaluation to some extend.

The fifth paper [5] with the title "A Handover Security Mechanism Employing Diffie-Hellman Key Exchange Approach for IEEE802.16E Wireless Networks" from Yi-Fu Cioua, Fang-Yie Leua, Yi-Li Huanga and Kangbin Yim, proposes a handover authentication mechanism, called handover key management and authentication scheme (HaKMA) to protect sensitive information delivered through wireless channels. The HaKMA provides a fast and secure key generation process for handover. The three-layer architecture in the HaKMA simplifies key generation flows compared to related work. Moreover, the authors also design two levels of handover mechanisms to minimize service disruption time (SDT), guaranteeing secure bidirectional connections between a mobile station and a base station. The analyses in this paper demonstrate that the HaKMA can effectively provide user authentication, and balance data security and system performance during handover with a low cost.

In the last paper [6] "Enhancing MISP with Fast Mobile IPv6 Security", Ilsun You, Jong-Hyouk Lee, Yoshiaki Hori and Kouichi Sakurai proposes a secure fast handover scheme that combines the advantages of *MISP* and *Fast Mobile IPv6* (*FMIPv6*). The MISP, a combination of MIS and MISAUTH protocols developed by Mobile Broadband Association, provides secure and fast connection for wireless access networks but suffers from denial-of-service attacks due to its weak session key. The proposed scheme in this paper improves the MISP through making full use of the fast handover approach of FMIPv6 and minimizing an involvement of the authentication server. The formal analyses in this paper show that the proposed scheme is robust against session key, off-line dictionary, DoS attacks while reducing handover latency compared with the existing schemes.

We would like to thank the authors of above papers published in this special issue, and regret that more papers could not be included. We appreciate all reviewers for their time and effort with reviewing assigned papers on time and providing invaluable comments and suggestions for authors for improving their papers. We also want to thank Professor David Taniar, Editors-in-Chief of *Mobile Information System*. His generous help and support have made this special issue a reality.

Hopefully, this special issue will bring forth advancements in science and technology as well as improve practices and applications of wireless and mobile technologies.

References

[1]   A. Barolli, E. Spaho, L. Barolli, F. Xhafa and M. Takizawa, QoS Routing in Ad-hoc Networks Using GA and Multi-Objective Optimization, *Mobile Information System* (*MIS*), 2011.

[2]   I. Doh, J. Lim and K. Chae, Distributed Authentication Mechanism for Secure Channel Establishment in Ubiquitous Medical Sensor Networks, *Mobile Information System* (*MIS*), 2011.

[3]   C.-L. Chen, Design of a secure RFID authentication scheme preceding market transactions, *Mobile Information System* (*MIS*), 2011.

[4]   F. Palmieri, U. Fiore and A. Castiglione, Automatic Security Assessment for Next Generation Wireless Mobile Networks, *Mobile Information System* (*MIS*), 2011.

[5]   Y.-F. Cioua, F.-Y. Leua, Y.-L. Huanga and K. Yim, A Handover Security Mechanism Employing Diffie-Hellman Key Exchange Approach for IEEE802.16e Wireless Networks, *Mobile Information System* (*MIS*), 2011.

[6]   I. You, J.-H. Lee, Y. Hori and K. Sakurai, Enhancing MISP with Fast Mobile IPv6 Security, *Mobile Information System* (*MIS*), 2011.

7

Detection of cross site scripting attack in wireless networks using n-Gram and SVM[1]

Jun-Ho Choi, Chang Choi, Byeong-Kyu Ko and Pan-Koo Kim[*]
Department of Computer Engineering, Chosun University, Gwangju, Korea

Abstract. Large parts of attacks targeting the web are aiming at the weak point of web application. Even though SQL injection, which is the form of XSS (Cross Site Scripting) attacks, is not a threat to the system to operate the web site, it is very critical to the places that deal with the important information because sensitive information can be obtained and falsified. In this paper, the method to detect themalicious SQL injection script code which is the typical XSS attack using n-Gram indexing and SVM (Support Vector Machine) is proposed. In order to test the proposed method, the test was conducted after classifying each data set as normal code and malicious code, and the malicious script code was detected by applying index term generated by n-Gram and data set generated by code dictionary to SVM classifier. As a result, when the malicious script code detection was conducted using n-Gram index term and SVM, the superior performance could be identified in detecting malicious script and the more improved results than existing methods could be seen in the malicious script code detection recall.

Keywords: Malicious code, SQL injection attack, n-Gram, SVM

1. Introduction

With the development of wireless network and internet, many parts of offline services have been converted into online services and currently most parts of online service are occupied by web services. Due to the merit of being available in anytime and anywhere, the importance of the web has been increased more and more everyday and the attacks aiming it has also been increased. Large parts of attacks targeting the web are aiming at the weak point of web application and SQL injection attack which is the form of XSS (Cross Site Scripting) attacks is not a threat to the system that uses or operates the web applications compared to other attacks, but it is very critical to the places that deal with the important information because sensitive information can be obtained and falsified. Various techniques have been studied in different fields to detect and prevent this critical SQL injection attack, including typically web framework, static and dynamic analysis and method using machine learning [7].

The web framework in a wireless network provides the filtering methods for input values but it is only filtering the special characters entered, so there are many obfuscate techniques using XSS. The static analysis analyzes the types of user input so it is more effective than simple filtering method, but it has the disadvantage that the attacks matching with the input type can't be detected. The dynamic analysis can find the weak point without modifying the web application but it can't find all weak points. The static and dynamic analysis method which complements the disadvantages of both static analysis and dynamic

---

[*]Corresponding author: Department of Computer Engineering, Chosun University, Gwangju, Korea. Pan-Koo Kim, E-mail: pkkim@chosun.ac.kr.
[1]This paper extends our previous work [16] published on Internationalconference of NBiS in 2011.

analysis is effective in detecting SQL injection attacks, but it has disadvantage that it is complicate in using because it is used by mixing both static analysis and dynamic analysis. The method using machine learning has the advantage of being able to detect unknown attacks but the misuse detection (False-Negative and False-Positive) can occur [17,18].

In this paper, it was determined whether the malicious script code could be detected or not using SVM (Support Vector Machine), which provides efficiency and accuracy in the process of a binary classification and to increase the accuracy of binary pattern classification, the index term applying n-Gram and code dictionary were generated. In the generated code dictionary the binary values obtained from matching with lexical grammar in the malicious script code were applied to the input node of SVM.

This paper is organized as follows: In Section 2, the related studies for the malicious script code classification was explained, in Section 3, XSS attack script pattern learning using SVM was described, and in Section 4, the study results using SVM were explained. Finally, in Section 5, the conclusion and future study direction were suggested.

2. Related work

2.1. SQL injection attack

A large number of applications that use database including the web application makes SQL queries using the user input values. For user login, the web application makes SQL queries for user account and password to check whether the user enters the effective account and password. At this time, the normal action can be interfered by sending fabricated username and password to change the normal SQL queries through SQL input attack technique [21].

In situations that the queries for database are generated dynamically, the input values of the user are treated as a meaningful statement in database, but the input values are treated as a simple string in the web application. Such attack method is called SQL injection attack.

Table 1
An example of the weak point of SQL input attack

$user_name = $_POST['username']
$passwd = $_POST['password']
Mysql_query("SELECT COUNT(*) FROM Users
       WHERE username = '$user_name' AND password = '$passwd' ");

Table 1 is an example of the weak point of SQL input attack that can be seen in the web application. In this example, $user_name and $Passwd are where the user enters. At this time, if the attacker enters the statement in $user_name as below, the SQL statement like Table 2 is created.

Table 2
A generated SQL statement by SQL input attack

SELECT COUNT(*) FROM Users
WHERE username = "or 1 = 1 –' AND password = ''

" 'or 1 = 1 –"

In this case, the WHERE clause will be always true for "OR $1 = 1$". That is, the attacker will obfuscate the user authentication using SQL injection attack. These vulnerabilities of SQL injection attack of the above Web application have been found in many websites in reality. If unsafe input information from the outside is used in the sensitive tasks such as execution of the script or SQL statement, the vulnerability of the Web application is exposed.

SQL injection attack converts SQL query syntax fixed into the existing web application into a new malicious SQL query syntax using malicious data values and then it requests and handles the data in database abnormally [12]. To prevent such SQL injection attacks, the web developers are basically using filtering method of input data value, but there are lots of obfuscate methods so it is difficult to prevent SQL injection only with simple filtering method. Therefore, it is necessary to have more advanced SQL injection attack detection and prevention method than simple filtering method.

2.2. SQL injection attack detection method

2.2.1. Method using static and dynamic analysis

The mixed method of static and dynamic analysis analyzes SQL query statically in the Web application and compares and analyzes the dynamic SQL query generated from the outside.

SQLCheck defines SQL injection attack and proposes *sound and complete algorithm* based on Context-free grammars and Complier parsing techniques [27]. There are advantages of SQLCheck in that there is no misuses detection (false-negative and false-positive) and the source code can be modified and applied to the web application directly. AMNESIA finds hotspots that executes SQL queries in the web application and then it creates every possible SQL queries [11]. The static SQL query generated and dynamic SQL query received from the user are classified and analyzed using JSA Library.

Buehrer uses a parse tree that is used a lot in lexical analysis so it has advantage that SQL query syntax can be analyzed correctly [1]. However, since SQL query syntax is different for each DBMS, it has the disadvantage of being dependent on DBMS. Wei proposed the method to detect the SQL injection attack by comparing and analyzing the static SQL query which is the stored procedure in the Web application and dynamic SQL query generated dynamically by making *control flow graph* [26]. Different from detecting and preventing method of SQL injection attack at the CGI-tier, this method can generaly detect and prevent the SQL injection attack at DBMS and stored procedure.

2.2.2. Method using SQL query profiling

Park et al. [15] detected the SQL injection attack by comparing and analyzing the dynamic SQL query generated dynamically, profiling the SQL query of the web application using "Pairwise sequence alignment of amino acid code formulated" method. This method can detect SQL injection attacks without modifying the web application, but it is inconvenient in that whenever the web application is changed, it should be profiling again.

2.2.3. Method using machine learning

Valeur proposed the Intrusion Detection System using machine learning [10]. This method learns the SQL query that generates from the web application, creates the detection model, and identifies the SQL injection attack by checking the SQL query generated in real time is the same as training model. This method can effectively detect and prevent the unknown attacks such as Zero-day attack. However, if insufficient training data set is used, the misuses detection (false-negative and false-positive) can occur.

WAVES finds the weak point in the web application through web crawler and it forms the attack code based on the pattern list and attack techniques [12]. The weak point of SQL injection attack is found using formed attack code.

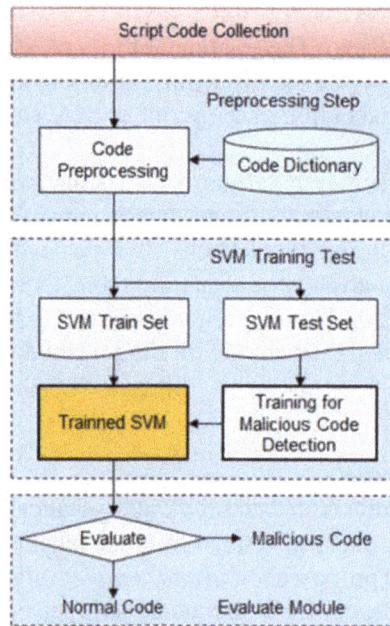

Fig. 1. A malicious script code detecting process.

3. XSS attack script pattern training using SVM

In case of existing malicious script code detection, the filtering is performed through pattern matching of specific words, but if the modified type of malicious script code is transmitted, this can't perform the filtering effectively and as the number of pattern is increased, it has a problem of increasing the time for filtering the malicious script code proportionally.

To solve such problems, in this paper the data set obtained through matching the malicious code pattern generated using n-Gram and code dictionary were applied to SVM classifier and the detection of malicious script was performed efficiently. N-Gram indexing and the malicious script filtering using SVM perform the following steps.

3.1. Entire configuration of mobile malicious script code detection method

To experiment the method proposed in this paper, after collecting the normal script code and malicious script code, the code dictionary and n-Gram index term were generated from collected code. Code dictionary was created manually by applying the frequency of words and the index term was created automatically by applying n-Gram for vocabulary in the script. In here, the training data set and test data set were formed by matching the generated index term and code dictionary.

The training data set generated from preprocessing step performs the training process of SVM classifier, and the test data set determines whether it is the normal script code or malicious script code by entering test pattern for malicious code. The entire configuration module for the malicious script code detecting using n-Gram index term and SVM is shown in Fig. 1.

3.2. Malicious script code pattern generation

The code dictionary is generated by extracting m number of high-frequency words, examining the frequency of lexical grammar in the code. In the generated code dictionary, the lexical grammar related

Table 3
An example of code dictionary

Code indexing	1	2	3	4	...	n
Code dictionary	SELECT	UPDATE	INSERT	Login	...	DELETE
Code indexing	201	202	203	204	...	m
Code dictionary	password	AND	username	account	...	Param1

Table 4
An example of code dictionary using n-Gram

Query	Unigrams	Bigrams	Trigrams
A	SELECT	SELECT_LastName	SELECT_LastName_FROM
	LastName	LastName_FROM	LastName_FROM_users
	FROM	FROM_users	FROM_users_WHERE
	users	users_WHERE	users_WHERE_UserID
	WHERE	WHERE_UserID	WHERE_UserID_1
	UserID	UserID_1	
	1		
B	SELECT	SELECT_FROM	SELECT_FROM_users
	FROM	FROM_users	FROM_users_WHERE
	users	users_WHERE	users_WHERE_login
	WHERE	WHERE_login	WHERE_login_'victor'
	login	login_'victor'	login_'victor'_AND
	'victor'	'victor'_AND	'victor'_AND_password
	AND	AND_password	AND_password_'123'
	password	password_'123'	
	'123'		

to malicious code is set. The index of code dictionary is composed of m number of words that match with the number of SVM input nodes and it is used for matching the generated index terms and words defined in code dictionary. Table 3 shows an example of generated code dictionary.

3.3. N-Gram index term generation

N-Gram is applied to the first collected codes to generate the training data set or test data set for collected codes, and the index term is generated by applying bigram to 4-gram using collected codes as shown in Table 4. The training data set or test data set is formed by matching index terms generated from this and code dictionary generated from Table 3.

– Query A: SELECT LastName FROM users WHERE UserID = 1
– Query B: SELECT * FROM users WHERE login = 'victor' AND password = '123'

3.4. SVM dataset generation

Figure 2 is the step that applies the index term generated by applying n-Gram and code dictionary. If the index term applying n-Gram and dictionary word are matched, it will have the value 1 (matching) as shown in Table 5 but if it can't find the matching words in code dictionary, it will have the value 0 (mismatching).

Table 5 shows the result obtained from matching the index term generated by applying n-Gram with code dictionary. After forming the training data set or test data set using matching result values as shown in Table 5, it is used as input vector of SVM. At this time the training step is performed using SVM classifier or the malicious script code is filtering using test data set.

Table 5
A matching result of code dictionary and n-Gramindexing

Indexing of code dictionary	1	2	3	4	5	......	200	......
Code set (1)	1	0	1	1	0	......	1	......
Code set (2)	1	1	0	1	0	......	0	......
Code set (n-1)					....			
Code set (n)	0	1	1	0	1	......	0	......

Table 6
A SVM training data set in experiment

Data SetCode type	Training set (Total 500)
Normal code	Normal code (250)
Malicious code	SQL injection related malicious code (250)

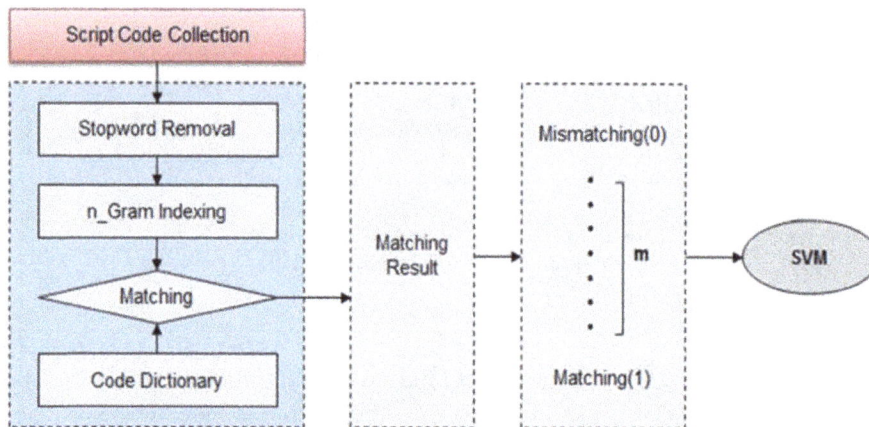

Fig. 2. A preprocessing for applying n-Gram indexing and SVM.

4. Experiments and results of XSS attack detection using SVM

4.1. Experiment environment of XSS attack detection

For XSS attack detection experiment, SQL injection attack code data among XSS attacks was collected, and with the collected sample, the training data set and test data set for SVM application were formed through preprocessing step explained in Section 3.

Each data set is composed of normal code and malicious code, and SQL injection related code from collected data set is classified as a malicious code and the rest is classified as normal code. Table 6 shows the SVM training data set for SQL injection malicious code filtering, and the number of training data set consists of 500. Among 500 training data, the normal code and malicious script code consists of 250, respectively.

The entire configuration for XSS attack detection proposed in this paper is shown in Fig. 3 and the experiment was performed using libSVM library [6] for spam mail filtering and the experiment results according to libSVM parameters were compared.

4.2. Results

The most efficient classification method of mobile malicious script code was studied and the performance for each classification method was measured. Total of 500 script codes were collected to perform

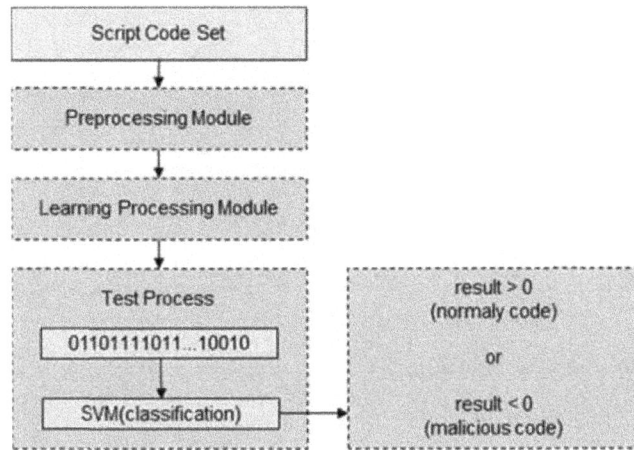

Fig. 3. A processing of malicious script code detecting using SVM.

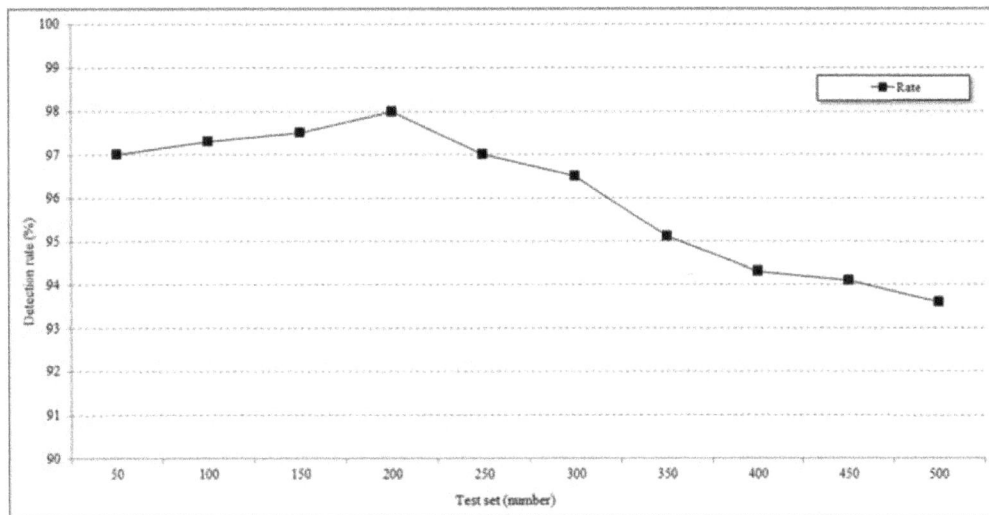

Fig. 4. A result of applying dot kernel.

the test, including 250 script codes were set as a training data set for SVM learning and 250 script codes were used as a test data set. In here, the normal script code and malicious script code were composed as 50% ratio, respectively.

The types of kernel used in test environment were Dot, Polynomial and Radial, and these three kernel functions were used to perform test. The performance of malicious script code detection according to kernel function is as follows.

4.2.1. Results of applying dot kernel

Dot kernel was used as a first test kernel function and m number of nodes which were matched with input node as the input node of SVM. The detection rate of malicious script code according to test data set increase is shown in Fig. 4.

The important characteristics according to the result are that as the number of test set was over 300, the normal malicious script code detection ratio decreased, while the false positive ratio and false negative

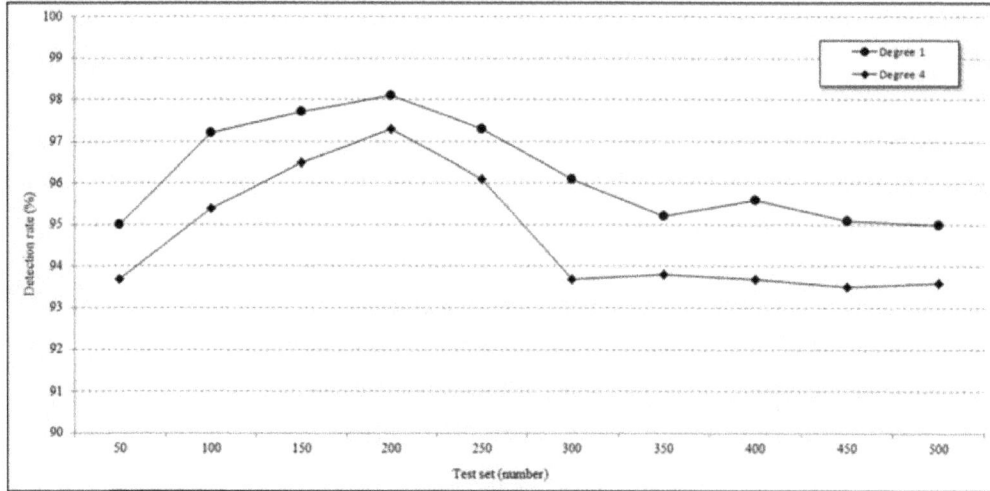

Fig. 5. A result of applying polynomial kernel.

ratio were increased. It is because there are many cases that the grammar of SQL syntax and diversity of variables in some script code are defined differently from the trained result through SVM.

4.2.2. Results of applying polynomial kernel

Polynomial kernel includes the parameter values and the test can be performed by adjusting the parameter values as shown in Eq. (1). The parameter has integer value and the test was performed by adjusting the degree value of parameter.

$$k(x,y) = (x * y + 1)^d \tag{1}$$

In the test using Polynomial kernel, it showed the similar false-positive ratio and false-negative ratio to Dot kernel method, but the malicious script code detection ratio showed more superior test results and the entire test result using Polynomial kernel is shown in Fig. 5. As shown in the test results of Fig. 5, when the degree value of parameter was set to 1, it had slightly better results than set to 4, and when the degree value was set to more than 4, the result value was impossible to derive due to the increase of false-positive and false negative ratios.

4.2.3. Results of applying radial kernel

As a last applied kernel function, Radial was used and the kernel function is defined as shown in Eq. (2). The parameter has real number value and the test was performed by adjusting the gamma value of parameter.

$$K(x,y) = \exp\left(-\frac{|x-y|^2}{\delta^2}\right) \tag{2}$$

In the test using Radial kernel, it showed the similar false-positive and false-negative ratio to the test result using Dot or Polynomial kernel method, but it showed the most excellent performance in the result of entire malicious script code detection.

The most important characteristics in the method using Radial kernel, the false-negative ratio was the best when the gamma value was set to 0.2 but it had a problem of increasing false-positive ratio relatively.

Table 7
Experimental resultsbyapplying each kernel function and arameter

Kernel definition	Feature	Parameter	Test set (500)			
			FP	FN	Detection recall	Detection precision
Dot Kernel	127	N/A	1.3	3.5	95.8	96.8
Polynomial	127	degree1	1.3	3.2	93.1	96.8
Kernel		degree 4	1.2	3.6	94.8	97.0
Radial	127	gamma0.2	2.1	1.1	96.4	97.7
Kernel		gamma0.6	1.9	1.1	96.4	97.5

– FP: False positive, FN: False negative.
– Detection precision = malicious code/classified malicious code.
– Detectionrecall = malicious code/total malicious code.

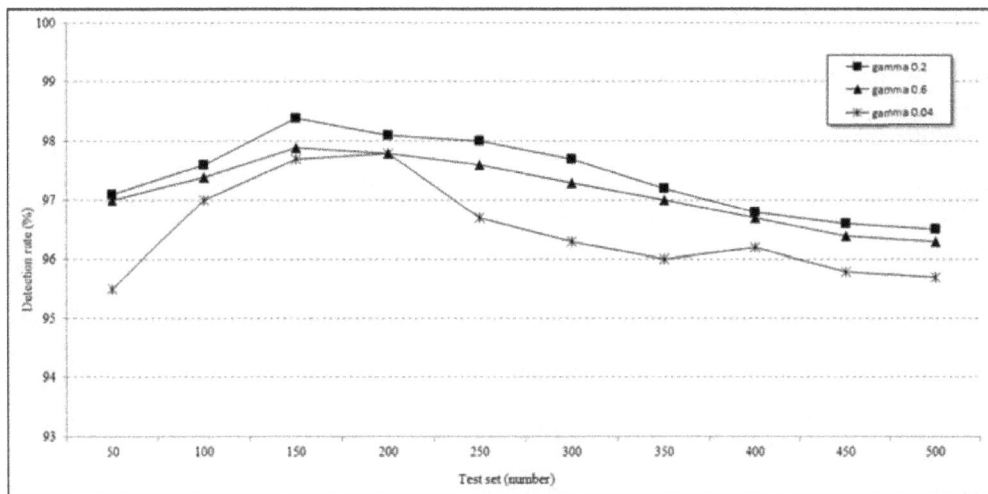

Fig. 6. A result of applying radial kernel.

In addition, when the gamma value was set to 0.01, the malicious script code detection ratio had the similar result to the malicious script code detection ratio using Dot or Polynomial kernel method, but it showed slightly less performance by comparing with different gamma values. The test results according to each kernel function and parameter are shown in Table 7.

The performance evaluation of malicious script code detection is expressed as detection precision and detection recall. As a result, the method applying Radial kernel as shown in Table 7 showed more excellent performance than Dot or Polynomial kernel in the test results of malicious script codedetection, especially when gamma was set to 0.2, the most excellent performance of malicious script code detection was shown.

4.3. Performance comparison of malicious script code detection

In order to compare the performance between the method proposed in this paper and existing detection methods of malicious script, Naïve Bayesian approach and keyword pattern were used as test filters for malicious script code detection. The method proposed in this paper collected 500 script codes and by having 65% and 35% composition of normal script and malicious script code ratio, the malicious script code detection test was performed.

Table 8
A result of malicious script code detection test

Filter used	Parameter	False Positive (%)	False negative (%)	Detection recall (%)	Detection precision (%)
SVM	gamma0.2	1.7	0.9	96.3.9	97.2
Keyword patterns	None	None	None	60.2	94.8
Naïve Bayesian	None	4.4	6.9	94.4	92.6

The malicious script code detection precision and malicious script code detection recall were used as the performance evaluation method for proposed method and the existing methods, and the method of applying n-Gram index term and SVM showed better performance than the method using keyword pattern in the malicious script detection precision and it had similar results to the Naïve Bayesian method. In addition, the malicious script code detection recall showed more excellent filtering performance than methods using Naïve Bayesian and keyword pattern and even though there is a slight difference according to the applied kernel function to the malicious script code vusing proposed method, it took average of about 1.5 seconds.

When the malicious script code detection was conducted using n-Gram index term and SVM as above, the superior performance could be identified in detecting malicious script code and the more improved results than existing methods could be seen in the malicious script code detection recall.

5. Conclusion

In this paper, the method to detect the SQL injection malicious script code which is the typical XSS attack using n-Gram indexing and SVM (Support Vector Machine) was proposed. In order to test the proposed method, the test was conducted after classifying each data set as normal code and malicious code, and the malicious script code was detected by applying index term generated by n-Gram and data set generated by code dictionary to SVM classifier. For kernel functions used in SVM classifier, Dot, Polynomial and Radial methods were used and the test results for each kernel method were explained in Table 7.

When the radial method was used as a kernel function of SVM classifier, it showed the most excellent performance, and comparing the existing studies such as Naive Bayesian methods showed the improved performance in the malicious script code detection. However, the codes containing a lot of SQL syntax has the problem of decreasing detection performance as the ratio of vectors that is out of the trained results values by SVM classifier increases. The solution for this can be resolved through pattern definition of SQL syntax in the preprocessing step. In this study, the malicious script code detection was performed limited to SQL injection attack which is the representative malicious script. However, it is necessary to detect various types of script-based malicious codes in the future.

Acknowledgments

This research was financially supported by the Ministry of Education, Science Technology (MEST) and National Research Foundation of Korea(NRF) through the Human Resource Training Project for Regional Innovation.

References

[1] G. Buehere, B.W. Weide and P.A. Sivilotti, Using Parse Validation to Prevent SQL Injection Attacks, InProceedings of the 5th international Workshop on software Engineering and Middleware, 2005, pp. 105–133.

[2] B.-Y. Zhang, J.-P. Yin, J.-B. Hao, D.-X. Zhang and S.-L. Wang, Using SupportVector Machine to Detect Unknown Computer Viruses, *International Journal ofComputational Intelligence Research* **2**(1) (2006), 100–104.

[3] B. Zhang, J. Yin and J. Hao, Intelligent Detection Computer Viruses Based onMultiple Classifiers, *Ubiquitous Intelligence and Computing* **4611** (2007), 1181–1190.

[4] C. Fung, Collaborative Intrusion Detection Networks and Insider Attacks, *Journal of Wireless Mobile Networks, Ubiquitous Computing, and Dependable Applications(JoWUA)* **2**(1) (2011), 63–74.

[5] C.-L. Chen, Design of a secure RFID authentication scheme preceding market transaction, *Journal ofMobile Information Systems* **7**(3) (2011), 201–216.

[6] C.C. Chang and C.J. Lin, LIBSVM: a library for support vector machines. Software available at http://www.csie. ntu.edu.tw/~cjlin/libsvm, 2001.

[7] Cisco Systems, What Is the Difference: Viruses, Worms, Trojans, and Bots, 2010.

[8] D. Moon, B. Park, Y. Chung and J.-W. Park, Recovery of flash memories for reliable mobile storages, *Journal of Mobile Information Systems* **6**(2) (2010), 177–191.

[9] F. Palmieri, U. Fiore and A. Castiglione, Automatic security assessment for next generation wireless mobile networks, *Journal of Mobile Information Systems* **7**(3) (2011), 217–239.

[10] F. Valeur, D. Mutz and G. Vigna, A Learning-Based Approach to the Detection of SQL Attacks, Proceedings of the Conference on Detection of Intrusions and Malware and Vulnerability Assessment, 2005, pp. 123–140.

[11] W.G. Halfond and A. Orso, AMNESIA: Analysis and Monitoring for Neutralizing SQL-Injection Attacks, Proceedings of the 20th IEEE/ACM international Conference on Automated Software Engineering, 2005, pp. 174–183.

[12] Y. Huang, S. Huang, T. Lin and C. Tasi, Web application security assessment by fault injection and behavior monitoring, Proceedings of the 12th international Conference on World Wide Web, 2003, pp. 148–159.

[13] I. You, Y. Hori and K. Sakurai, Enhancing SVO Logic for Mobile IPv6 Security Protocols, Journal of Wireless Mobilie Networks, *Ubiquitous Computing and Dependable Applications* **2**(3) (2011), 26–52.

[14] A. Jacob and M. Gokhale, Language classification using n-grams accelerated by fpga-based bloom, 2007.

[15] J. Park and B. Noh, SQL injection Attack Detection: Profiling of Web Application Parameter Using the Sequence Pairwise Alignment, *Information Security Applications LNCS* **4298** (2007), 74–82.

[16] J. Choi, H. Kim, C. Choi and P. Kim, Efficient Malicious Code Detection Using N-Gram Analysis and SVM, *International Conference of Network-Based Information Systems (NBiS)* (2011), 618–621.

[17] J. Kolter and M. Maloof, Learning to Detect and Classify Malicious Executables inthe Wild, *Journal of Machine Learning Research* **7** (2006), 2721–2744.

[18] M.G. Schultz, E. Eskin, E. Zadok and S.J. Stolfo, Data Mining Methods for Detectionof New Malicious Executables, Proc. of the 2001 IEEE Symposium on Security andPrivacy, CA: IEEE Computer Society Press, Oakland, 2001, pp. 38–49.

[19] M.P. Jayamsakthi Shanmugam, Cross site scripting-latestdevelopments and solutions: A survey, *International Journal of OpenProblems in Computer Science and Mathematics (IJOPCM)* **1**(2) (2008), 8–28.

[20] P. Nobles, S. Ali and H. Chivers, Improved Estimation of Trilateration Distances for Indoor Wireless Intrusion Detection, Journal of Wireless Mobile Networks, *Ubiquitous Computing, and Dependable Applications (JoWUA)* **2**(1) (2011), 93–102.

[21] W. Robertson, G. Vigna, C. Kruegel and R. Kemmerer, Using generalization andcharacterization techniques in the anomaly-based detection of web attacks. In: NDSS '06: Proc. 13th ISOC Symposium on Network and Distributed Systems Security, 2006.

[22] S. Caballé, F. Xhafa and L. Barolli, Using mobile devices to support online collaborative learning, *Journal of Mobile Information Systems* **6**(1) (2010), 27–47.

[23] M. Shafiq, S. Khayam and M. Farooq, Embedded Malware Detection Using Markov n-Grams, *Lecture Notes in Computer Science* **5137** (2008), 88–107.

[24] T.A. Zia and A.Y. Zomaya, A Lightweight Security Framework for Wireless Sensor Networks, Journal of Wireless Mobilie Networks, *Ubiquitous Computing and Dependable Applications* **2**(3) (2011), 53–73.

[25] V. Hamine and P. Helman, A Theoretical and Experimental Evaluation ofAugmented Bayesian Classifiers, American Association for Artificial Intelligence, Retrieved March 20, 2006.

[26] K. Wei, M. Muthuprasanna and S. Kothari, preventing SQL injections attacks in stored procedures, Software Engineering, 2006, pp. 18–21.

[27] Z. Su and G. Wassermann, The Essence of command Injection Attacks in Web Applications, In Conference Record of the 33er ACM SIGPLAN- SIGACT Symposium on Principles of Programming Languages, 2006, pp. 372–382.

Jun-Ho Choi received a doctoral degree in the Department of Computer Engineering at Chosun University of Korea in 2004. Currently, he is working as a lecturer at the same university. His research interests include multimedia processing, semantic information processing, ontology and semantic web.

Chang Choi is a Ph.D. Candidate in the Department of Computer Engineering at Chosun University of Korea. Currently, he is working as a lecturer at the same university and toward the Ph. D degree. His research interests include semantic information processing, semantic web and multimedia data processing.

Byeong-Kyu Ko is a student for the doctoral degree in the Department of Computer Engineering at Chosun University of Korea. He is received a master degree at the same university in 2012. His research interests include web documents classification, semantic information processing and semantic web.

Pan-KooKim is received the BS degree in computer engineering from Chosun University of Korea and the MS and PhD degrees in computer engineering from Seoul National University of Korea in 1994. He is a full professor in the Department of Computer Engineering at Chosun University. His specific research interests include semantic web techniques, semantic information processing and retrieval, multimedia processing and semantic web.

8

Data gathering using mobile agents for reducing traffic in dense mobile wireless sensor networks

Keisuke Goto, Yuya Sasaki, Takahiro Hara* and Shojiro Nishio
Department of Multimedia Engineering, Graduate School of Information Science and Technology, Osaka University, Osaka, Japan

Abstract. Recently, there has been increasing interest in Mobile Wireless Sensor Networks (MWSNs) that are constructed by mobile sensor nodes held by ordinary people, and it has led to a new concept called *urban sensing*. In such MWSNs, mobile sensor nodes densely exist, and thus, there are basically many sensor nodes that can sense a geographical point in the entire sensing area. To reduce the communication cost for gathering sensor data, it is desirable to gather the sensor data from the minimum number of mobile sensor nodes which are necessary to guarantee the sensing coverage or the quality of services. In this paper, to achieve this, we propose a data gathering method using mobile agents in dense MWSNs. The proposed method guarantees the sensing coverage of the entire area using mobile agents that autonomously perform sensing operations, transmit sensor data, and move between sensor nodes. By gathering only sensor data generated by sensor nodes where mobile agents are running, our proposed method can achieve efficient gathering of sensor data.

Keywords: Mobile wireless sensor networks, data gathering, mobile agent, geo-routing

1. Introduction

Recently, there has been a great deal of interest in Wireless Sensor Networks (WSNs [12,17,22]) because advances in semiconductor and wireless communication technologies have led to the development of small and inexpensive sensor devices. One of typical applications of WSNs is a monitoring application. A lot of sensor nodes sense an environment and send their readings (sensor data) to the sink.

Due to the advances in radio communication and computer technologies, Mobile Ad Hoc Networks (MANETs [1,8,9]), which are constructed by only mobile nodes, have also been actively studied for recent years. In MANETs, every mobile node plays the role of a router, and they communicate with each other. As an integration of WSNs and MANETs, Mobile Wireless Sensor Networks (MWSNs), which are constructed by mobile nodes with sensor devices, have recently attracted much attention.

Urban sensing, in which ordinary people cooperate for data gathering with some incentive, is a typical example of MWSNs. Traditionally, urban sensing generally assumes some infrastructures, i.e., the Internet for communication and data gathering. However, since many applications share and compete for

---

*Corresponding author: Takahiro Hara, Department of Multimedia Engineering, Graduate School of Information Science and Technology, Osaka University, 1-5 Yamadaoka, Suita, Osaka 565-0871, Japan.
E-mail: hara@ist.osaka-u.ac.jp.

limited network resources in the Internet, it is desirable to minimize traffic that MWSNs injects into the Internet. For this aim, MWSNs constructed by mobile sensor nodes (e.g., PDA and smart phones with sensor devices held by ordinary people) have attracted much attention for urban sensing because they do not require any infrastructures [3,13,14,16]. Sensor nodes in such MWSNs automatically sense an environment (e.g., sound and temperature), and a sink gathers the sensor data using multi-hop radio communication between sensor nodes (and the sink).

In MWSNs constructed by sensor nodes held by ordinary people, the number of sensor nodes is generally very large, and thus, there are basically many sensor nodes that can sense (cover) a geographical point in the entire sensing area (i.e., dense MWSNs). From the perspective of applications, a lot of same sensor data are not useful, but just waste limited network bandwidth. Rather, in most cases, applications require a certain geographical granularity of sensing, e.g., sensor data of every 100 [m]$\times 100$ [m] square. In such a situation, if a sink gathers sensor data from all sensor nodes, it unnecessarily wastes the network bandwidth and the battery of sensor nodes. To reduce the data traffic for data gathering, it is desirable to efficiently gather sensor data so that the geographical granularity required from an application can be guaranteed with the minimum number of sensor nodes.

Achieving this is not easy because sensor nodes move freely, and thus, sensor nodes that can sense a specific point dynamically change. A naive method is that a sink determines sensor nodes that perform sensing and send back the data every time of sensing. However, this method requires exchange of requests and replies between the sink and these nodes, which produces large traffic and cause a long delay for message exchanges particularly when the distances between the sink and sensor nodes are long.

In this paper, we propose a data gathering method that efficiently gathers sensor data using mobile agents in dense MWSNs. A mobile agent is an application software that autonomously operates on a sensor node and moves between sensor nodes. In our proposed method, mobile agents are generated by the sink and allocated on sensor nodes located near the sensing points, which are determined from the requirement on geographical granularity of sensing. Each mobile agent moves from the currently located sensor node to the nearest sensor node from the sensing point responding to the movement of sensor nodes. This makes the mobile agent continuously locates near the sensing point. Every time when the sensing time comes, sensor nodes where mobile agents locate perform sensing and send the sensor data to the sink. Our proposed method can reduce the traffic for gathering the sensor data since mobile agents control sensor nodes for sensing operations and transmissions of sensor data. The contributions of this paper are summarized as follows.

- Our proposed method reduces the traffic for sensor data gathering, which is effective in dense MWSNs because the network bandwidth and batteries of sensor nodes are limited.
- It is natural to assume that applications require a certain geographical granularity of sensor data. Thus, in our proposed method, the sink divides the sensing area into lattice-shaped sub-areas according to the required granularity, and determines sensing points as the center points of the sub-areas. Then mobile agents, which are allocated by the sink on sensor nodes located near the sensing points, guarantee the requirement of geographical granularity of sensing by controlling transmissions of sensor data. To the best of our knowledge, this is the first effort that addresses the issue of guaranteeing the geographical granularity of sensing in an efficient manner in dynamic MWSNs.
- Through extensive simulations assuming practical and various situations, we show that our proposed method works very well in terms of traffic reduction and guaranteeing the requirement of geographical granularity of sensor data.

The remainder of this paper is organized as follows. In Section 2, we introduce related works. In Section 3, we describe assumptions in this paper. In Section 4, we explain the details of our proposed method. In Section 5, we show the results of the simulation experiments. Finally, in Section 6, we summarize this paper.

2. Related works

In [10], the authors proposed a hierarchical data gathering method in WSNs. In this method sensor nodes are hierarchically arranged, where a sensor node in a lower level sends the sensor data to a node in the higher level, and the sensor node in the highest level sends the (aggregated) sensor data to the sink. It can reduce the traffic for data gathering since nodes in higher levels aggregate and compress the sensor data which are sent from nodes in lower levels. In [6], the authors proposed a data gathering method using mobile agents in WSNs. Mobile agents are generated by a sink and gather sensor data by visiting sensor nodes. This method can reduce the traffic for data gathering since mobile agents compress the sensor data every time when they visit a sensor node. In [21], the authors proposed a topology control protocol, which splits the sensing area into multiple grid cells. The size of a grid cell is determined based on the communication range of sensor nodes so that any pair of sensor nodes in adjacent two cells can directly communicate with each other. Sensor data are forwarded to the sink through the cells, where one node in each cell plays the role of a router. By doing so, other sensor nodes can reduce the energy consumption.

In [5,7], the authors proposed a duty cycling algorithm, which aims to reduce the cost of topology management by changing the state of redundant sensor nodes to the sleep mode. In the algorithm proposed in [5], a sensor node changes its state to the sleep mode for a certain time in order to save its battery when its neighbor nodes can communicate with each other directly or via another sensor node without itself. In the algorithm proposed in [7], each sensor node periodically sends a HELLO message to its neighbor nodes. Then, sensor nodes that received HELLO messages more than a certain predetermined number change their state to the sleep mode because it means that they have many activate neighbors.

In [4,18,19], the authors proposed a network configuration method that guarantees the coverage in the entire area, i.e., all positions are covered by at least a predetermined number of sensor nodes. In the method proposed in [4], each sensor node periodically sends a HELLO message including the information on its position and state to its neighbor nodes. Sensor nodes with a small amount of remaining battery change their state to the sleep mode until the next period in order to save their battery if their sensing areas are covered by their neighbor nodes. In the method proposed in [18], similar to the method in [4], sensor nodes with less remaining battery move to the sleep mode. In doing so, the network connectivity is also taken into account. Specifically, this method adopts the technique proposed in [5] to guarantee the network connectivity when deciding the transit of mode. In the method proposed in [19], sensor nodes have two different working phases; *initialization phase* and *sensing phase*. Sensor nodes in the initialization phase make a working schedule by sharing the information on their positions and synchronizing times with their neighbors. The working schedule is determined so that it can guarantee that all positions in the entire sensing area are covered by at least one sensor node. In the sensing phase, sensor nodes save their battery by switching their modes (active and sleep) based on the working schedule.

The existing studies presented above do not assume the movements of sensor nodes, and thus, cannot handle the change of network topology. Furthermore, these studies are different from our work that aims data gathering to guarantee the geographical granularity of sensing which is required from an application.

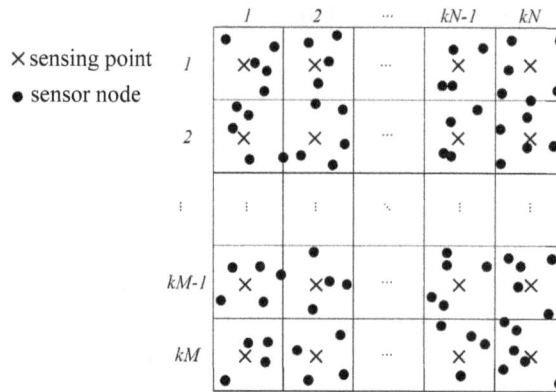

Fig. 1. Sensing area and sensing points.

In [20], the authors assumed location based services and proposed a data gathering and disseminating method in MANETs. The method uses a mobile agent that stays within a certain geographical area by moving between mobile nodes. This work assumes services to disseminate location based information that is generated by disseminating nodes and passed to mobile nodes located near those locations. This work is different from our work which assumes that sensor data generated by sensor nodes are sent to the sink located far from them.

3. Assumptions

In this paper, we assume dense MWSNs constructed by mobile sensor nodes which are held by ordinary people and equipped with a radio communication facility. These sensor nodes periodically observe (sense) a physical phenomenon (e.g., sound, temperature and light), and communicate with each other using multi-hop radio communication. According to the requirement from an application, the sink periodically monitors the sensing area while guaranteeing the geographical granularity of sensing. More specifically, the sink gathers sensor data from sensor nodes located near the sensing points which are determined from the requirement of the geographical granularity at the timing of sensing.

3.1. System environment

The sensing area is assumed to be a two-dimensional plane whose horizontal to vertical ratio is $M{:}N$ (M and N are positive integers). The application specifies its requirement of the geographical granularity of sensing as an integer of $k^2 \cdot M \cdot N$ ($k = 1, 2, \cdots$). Then, the sink divides the sensing area into $k \cdot M \times k \cdot N$ lattice-shaped sub-areas and determines the center point of each sub-area as a sensing point, which is the target of data gathering (see Fig. 1).

As mentioned, we assume MWSNs constructed by mobile sensor nodes held by ordinary people. Since no infrastructure for communication is available in the sensing area, the sink gathers sensor data by using a MANET constructed by the sensor nodes. The communication range of each sensor node is a circle with a radius of r. Each sensor node is equipped with a positioning device such as GPS, and they communicate with each other using multi-hop radio communication based on their positions (i.e., geo-routing described in the next Subsection). The position information is represented as a pair of longitude and latitude.

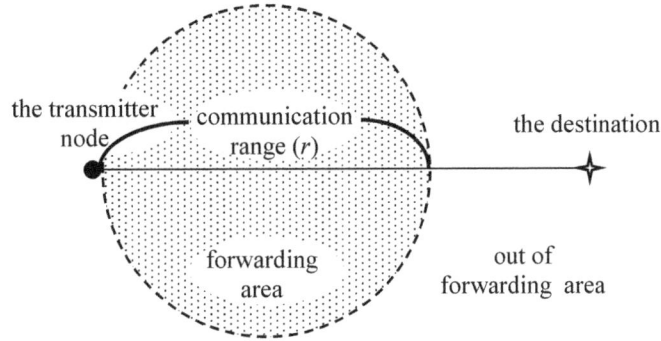

Fig. 2. An example of forwarding area.

Each sensor node freely moves in the sensing area, while the sink is stationary. Since the number of mobile nodes is very large, there are many sensor nodes that can sense (cover) a geographical point in the entire sensing area.

3.2. Geo-routing

Sensor nodes adopt a geo-routing protocol based on that proposed in [11] to transport a message to the destination specified as a position (not a node).

In this protocol, nodes perform a transmission process using the information on positions of the transmitter node and the destination, which is specified in the packet header. Specifically, the transmitter node writes the information on positions of the destination and itself into the packet header of the message, and broadcasts it to its neighboring nodes. Each node that received this message judges whether it locates within the forwarding area. Here, the forwarding area is determined based on the positions of the transmitter node and the destination and the communication range, so that any node in the forwarding area is closer to the destination than the transmitter node and can communicate directly (by one hop) to all nodes in the forwarding area. Figure 2 shows an example of the forwarding area. The forwarding area is represented by the shaded circle whose diameter is the communication range of sensor nodes (r).

The node within the forwarding area sets the waiting time, and then it forwards the message after the waiting time elapses unless it has detected the message sent by another node during the waiting time. Here, the waiting time (WT) is determined by the following equation, where it gets shorter as the distance (P) between the node and the destination gets shorter:

$$WT = Max\_delay \cdot \left(\frac{r - P}{r} \right) \tag{1}$$

Max_delay is the maximum waiting time which is a positive constant. Each node within the forwarding area cancels its transmission process when it detects the message forwarded by another node. By repeating this procedure, the message is forwarded to nodes which are closer to the destination. If the transmitter node is within half of the communication range ($r/2$) from the destination, each node that received the message sends an ACK to the transmitter node after the waiting time elapses instead of forwarding the message. As the result, the nearest node from the destination (that has sent the ACK) can find that it is the nearest one because all nodes within $r/2$ from the destination can detect the ACK sent by the node and cancel to send own ACK. If the transmitter node did not receive an ACK from any node, it also can find that it is the nearest node.

Here, it is very rare but possible that multiple nodes try to reply ACKs at the same timing when their distances to the destination of the message are equivalent. However, even in this case, each node does not send an ACK at the same timing by using a multiple access method (e.g., CSMA/CA) in the MAC layer. Therefore, nodes that receive (overhear) an ACK can stop sending their own ACK, which makes only one node replies an ACK.

4. A data gathering method using mobile agents

In this section, first we describe the design policy of our proposed method and explain about mobile agents. Then, we describe the outline and detailed behaviors of our proposed method.

4.1. Design policy

In MWSNs constructed by mobile sensor nodes held by ordinary people, there are a large number of sensor nodes that move freely, which causes frequent network topology changes. In such an environment, managing topology information and keeping routing tables up-to-date using beacons are impractical because it generates large traffic for exchanging beacons, which consumes a large amount of energy and also might cause frequent packet collisions due to the limited network bandwidth. Therefore, as described in Section 3.2, we adopt a geo-routing protocol, which transmits a packet to the sensor node located near the destination (sensing point) by selecting sensor nodes forwarding the packet based on their positions.

Moreover, it is natural that sensor nodes held by ordinary people do not know the information for data gathering (e.g., the position of a sink and the requirement of geographical granularity of sensing from an application). Therefore a sink has to disseminate the information to sensor nodes located near the sensing points. However, disseminating the information to all of those sensor nodes every time when the sensing time comes is not efficient because it produces large traffic as well as causes some delay for data gathering. To solve this problem, in our proposed method, a sink gathers sensor data using mobile agents each of which operates on a sensor node that is located near a sensing point and covers it. By doing so, sensor data can be gathered from the minimum number of sensor nodes necessary to guarantee the requirement of geographical granularity of sensing. Moreover, it is not necessary to send the information for data gathering from the sink every time when the sensing time comes. As the result, our proposed method can reduce traffic and delay for gathering sensor data.

4.2. Mobile agent

In our proposed method, a mobile agent is an application software that autonomously operates on a sensor node and moves between sensor nodes. A sensor node boots up a mobile agent by referring to the agent data, which consists of the information on geographical granularity of sensing, the sensing cycle and the position of the sink. The role of mobile agents is to transmit a sensor data to the sink at every sensing time. Mobile agents are deployed on sensor nodes so that they can guarantee the requirement from the application regarding the geographical granularity of sensing.

Table 1
Forwarding directions of agent data

Location of the sensor node that previously forwarded the agent data	Forwarding directions of the agent data
– (sink)	up, down, left, right
sub-area next to right	up, down, left
sub-area next to left	up, down, right
sub-area next to down	up
sub-area next to up	down

4.3. Outline of the proposed method

In our proposed method, initially, the sink generates mobile agents and deploys them in the network. Specifically, the sink sends the agent data to $k^2 \cdot M \cdot N$ sensing points that are determined as described in Subsection 3.1. According to Subsection 4.4, the sensor node closest to each sensing point receives the agent data and boots up a mobile agent.

At each sensing time, according to the method described in Subsection 4.5 the mobile agents send sensor data generated by sensor nodes on which they run to the sink. If a mobile agent disappears, the sink and other mobile agents redeploy the disappeared agent and the redeployed mobile agent retransmits the sensor data to the sink, according to the method described in Subsection 4.6. Moreover, if a sensor node on which a mobile agent runs moves away from the sensing point, the mobile agent moves from the sensor node to another node which is the closest to the sensing point, according to the method described in Subsection 4.7.

4.4. Deployment of mobile agents

Our proposed method deploys mobile agents in the network with small traffic. Specifically, the sink sends the agent data along the forwarding tree that is constructed based on the geographical relationships among sensing points. The procedures of the sink and mobile agents to deploy mobile agents are as follows.

1. The sink generates the agent data and sends it to the sensing point in the sub-area in which the sink locates using the geo-routing protocol described in Subsection 3.2.
2. When the sensor node located closest to the sensing point in its existing sub-area receives the agent data, it boots up a mobile agent. As the initial operation, the mobile agent retransmits the agent data to the sensing points in some of sub-areas which are adjacent to its existing sub-area. Table 1 shows the directions to which the mobile agent forwards the agent data in the lattice-shaped sub-areas. For example, if the agent data is transmitted by the sink, the mobile agent forwards the agent data to upper, lower, left and right sub-areas. If the location (sub-area) of the sensor node that forwarded the agent data previously is the sub-area next to left, the mobile agent forwards the agent data to upper, lower and right sub-areas.

    Then, the procedure returns to step 2. Here, mobile agents in sub-areas at the top and bottom edges of the sensing area do not retransmit the agent data.

Through the above procedures, mobile agents dynamically construct a tree structure for forwarding the agent data (e.g., Fig. 3), which we call the *forwarding tree*. The forwarding tree is also used for collecting sensor data, which enables mobile agents to reduce traffic by aggregating sensor data received from child agents and that of itself when they reply the sensor data to their parent agents (describe in the next subsection).

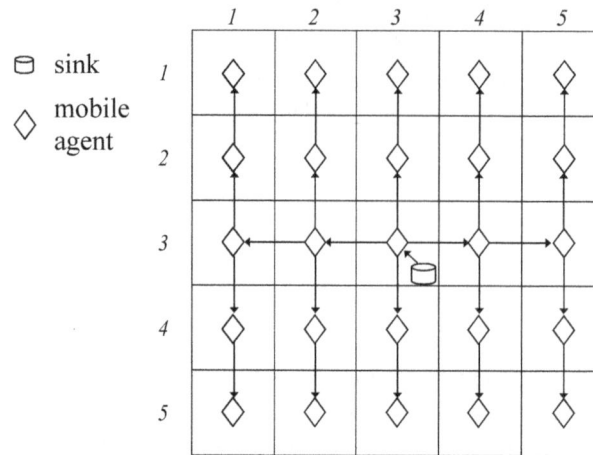

Fig. 3. Example of forwarding agent data between mobile agents.

We show an example that the sink deploys mobile agents to all sensing points using Fig. 3. Let us assume that the sensing area is divided into 5×5 sub-areas and the sink locates in the sub-area (3, 3). Initially, the sink forwards the agent data to the sensing point in the sub-area in which the sink exists. The mobile agent that is booted up by the sensor node closest to the sensing point retransmits the agent data to the four sensing points in the upper, lower, left, and right sub-areas. Then, mobile agents in the four sub-areas are booted up and they retransmit the agent data to directions shown in Table 1. By repeating these procedures, the sink can deploy mobile agents near all the sensing points.

4.5. Transmission of sensor data

Mobile agents deployed at (near) sensing points send the sensor data held by sensor nodes on which these agents operate to the sink at every sensing time. Our proposed method can reduce the traffic of sending sensor data since mobile agents compress the received sensor data that are sent along the forwarding tree in the reverse direction. The detailed procedures of mobile agents are as follows.

1. At every sensing time, mobile agents located in sub-areas at the top and bottom edges of the sensing area start to send their sensor data to the mobile agents that forwarded the agent data to them. In doing so, the geo-routing protocol described in Subsection 3.2 is used. Here, due to the movement of sensor nodes on which mobile agents operate, the mobile agents may become far away from the sensing points. In our geo-routing protocol, sensor data cannot be received by the mobile agent which is responsible for a sensing point if the mobile agent has moved farther than $r/2$ from the sensing point. Therefore, in our method, a mobile agent moves from the sensor node on which it operates to the node closest to the sensing point so that it keeps to locate within $r/2$ from the sensing point. The procedure to achieve this is described in Subsection 4.7. By doing so, mobile agents can always receive the sensor data sent from child nodes in the forwarding tree.

2. When mobile agents located in sub-areas except for that the sink exists receive the sensor data from all mobile agents which are children in the forwarding tree, they aggregate the received sensor data and their own sensor data, and then forward the aggregated data to the mobile agents which are parent in the forwarding tree. The procedure returns to step 2.

3. The mobile agent located in the sub-area where the sink exists aggregates the received sensor data and its own sensor data, and then sends it to the sink.

4.6. Redeployment of mobile agents and retransmission of sensor data

If mobile agents and the sink cannot receive sensor data from their children in a given time it shows that, a packet loss may occur or a mobile agent may disappear. When a packet loss occurs, the mobile agent that cannot receive the sensor data simply requests its child to re-send the data. However, when a mobile agent disappears, the missing mobile agent must be redeployed because otherwise, sensor data cannot be transmitted to the sink. Therefore, a mechanism is necessary to retransmit sensor data and redeploy a mobile agent. The detailed procedures of redeployment of mobile agents and retransmission of sensor data are as follows.

1. At every sensing time, all mobile agents transmit sensor data based on the procedures described in Subsection 4.5.
2. When mobile agents located in sub-areas other than the sub-areas at the top and bottom edges of the sensing area and where the sink exists cannot receive sensor data from any of their children within time period δ, they send a data request message to the child using the geo-routing protocol described in Subsection 3.2.
3. When mobile agents located in sub-areas at the top and bottom edges of the sensing area receive a data request message, they reply by sending the requested sensor data. When mobile agents located in sub-areas other than those at the top and bottom edges of the sensing area receive a data request message, they reply by sending the sensor data unless they have already received the sensor data from all mobile agents which are children in the forwarding tree. If not, then they reply with an existence message instead of the sensor data.
4. Mobile agents that have sent a data request message restart the transmission of sensor data if they receive the sensor data from all mobile agents that are children in the forwarding tree. If they cannot receive the sensor data or an existence message, they determine that the mobile agent that is its child has disappeared, and they send new agent data to the sensing point located in the sub-area where the mobile agent has disappeared.
5. The sensor node located closest to the sensing point in the sub-area in which the mobile agent has disappeared receives the agent data and boots up a new mobile agent to that area. In the initial operation, if a mobile agent in the sub-area at the top and bottom edges of the sensing area is redeployed, it sends its sensor data. If a mobile agent located in a sub-area other than the sub-areas at the top and bottom edges of the sensing area is redeployed, it sends a data request message to all mobile agents that are children in the forwarding tree. The procedure returns to step 3.
   On the other hand, when an existing mobile agent detects the agent data because the transmission of its sensor data has failed (e.g., in packet collisions), the mobile agent replies with an ACK and executes the same process as if it had received a data request message (step 3), and then discards the agent data.

Through the above procedures, mobile agents and the sink send a data request message to mobile agents that are their children in the forwarding tree only if they cannot receive sensor data from the child agents within time period δ. After that, these mobile agents and the sink send agent data only if they cannot receive a response from their children within δ. In this way, our proposed method can redeploy mobile agents and retransmit sensor data without additional traffic in the location (sub-areas) in which mobile agents successfully transmit their sensor data and are still alive.

4.7. Movement of mobile agent

If a sensor node on which a mobile agent operates moves away from the sensing point, it may not be able to sense (cover) the point. Also, as mentioned, it may not be able to receive the sensor data sent from its children in the forwarding tree. Therefore, in our method, to avoid such situations, a mobile agent moves from the current sensor node to the other node which is the closest to the sensing point.

Specifically, a mobile agent starts moving when the distance between the sensing point and itself becomes longer than threshold α. Here, α is a system parameter which is set as a constant smaller than $r/2$ and the range of sensing of each sensor node (sensing coverage), which can guarantee that a sensor node on which a mobile agent operates can communicate with all sensor nodes located near (within $r/2$) from the sensing point and can sense the data at the sensing point.

In order to move to the sensor node closest to the sensing point, the mobile agent broadcasts a message containing the agent data, to neighbor sensor nodes within $r/2$ from the sensing point. Sensor nodes that received this message set the waiting time according to equation (1) and send an ACK to the transmitter node. Since the sensor node closest to the sensing point firstly sends an ACK, it boots up a mobile agent in the same way as in Subsection 3.2. Other sensor nodes detect this ACK and cancel to send own ACK. After receiving the ACK, the original mobile agent stops its operation.

Here, it is possible (but very rare) that there are multiple sensor nodes which locate closest to the sensing point with the same distance when a mobile agent needs to move. Even in this case, since each node sends an ACK at different timings with each other by using a multiple access method in the MAC layer, only the node that first sent an ACK boots up a new agent and the other nodes discard the agent data.

5. Simulation experiments

In this section, we show the results of simulation experiments regarding the performance evaluation of our proposed method. For the simulations, we used the network simulator, Scenargie 1.3 (Subsections 5.2 $\sim$ 5.5) and 1.4 (Subsection 5.6) [15].[1] First, in order to verify the effectiveness of using mobile agents, we evaluate our method without the mechanism of retransmission of sensor data and redeployment of mobile agents described in Subsection 4.6. Then, we evaluate our method with that mechanism in Subsection 5.6.

5.1. Simulation model

There are n mobile sensor nodes ($M_1, \cdots, M_n$) and a sink (S_1) in a two-dimensional field of is 1000 [m] $\times$1000 [m]. The sink is fixed at the point of (580 [m], 580 [m]) from the left and the bottom edges of the sensing field. Each sensor node moves according to the random walk mobility model [2] where it selects a random direction and a random speed from 0 to 1 [m/sec] at intervals of 50 [sec]. The sink and sensor nodes communicate with IEEE 802.11g whose transmission rate is 6 [Mbps] and communication range r is about 100 [m]. Each sensor node continuously senses the field and the valid range of its sensing (i.e., sensing coverage) is 100 [m]. The sink divides the field into G lattice-shaped sub-areas whose the size is $1000/\sqrt{G}$ [m]$\times 1000/\sqrt{G}$ [m] and sets the center point of each sub-area as a sensing point.

The sink deploys a mobile agent at each sensing point after 400 [sec] from the start time of the simulation. The size of the agent data is set as 128 [B], assuming that each sensor node has the source code of mobile agent in advance. The size of sensor data generated at each sensor node is set as D [B].

---

[1]The differences between the two versions (1.3 and 1.4) of Scenargie have no impact on our simulation results.

Table 2
Parameter configuration.

Parameter (Meaning)	Value (Range)	
T (Sensing cycle [sec])	120	$(30 \sim 300)$
G (Number of sub-areas)	25	$(1 \sim 100)$
n (Number of sensor nodes)	2000	$(1000 \sim 3000)$
D (Size of a sensor data item)	32	$(24 \sim 240)$

Table 3
Message size

Method	Roll (Message name)		Size [B]
Our proposed method	Deploying a mobile agent	(Deployment message)	192
	Sending sensor data	(Data message)	$64 + D \cdot i$
	Moving a mobile agent	(Movement message)	160
Comparative method	Requesting sensor data	(Request message)	72
	Sending sensor data	(Data message)	$64 + D$
Common	ACK	(ACK message)	32

The sensing cycle is set as T [sec], i.e., the sensing time is $400 + mT$ [sec] ($m = 0, 1, \cdots$) until the end of the simulation. At every sensing time, the sensor nodes on which mobile agents operate send sensor data to the sink. Moreover, a mobile agent moves to the sensor node which is closest to the sensing point when the distance between the sensing point and itself becomes longer than 47 [m] (i.e., $\alpha = 47$), which is set as an appropriate value according to our preliminary experiments.

For comparison, we also evaluate the performance of the method where the sink sends a request message to all sensing points at every sensing time and collects the sensor data. Specifically, at every sensing time, the sink individually sends messages for requesting a sensor data to each sensing point using the geo-routing protocol described in Subsection 3.2. Here, the sink sets a constant interval of 0.1 [sec] between every pair of request messages, and sends them to the sensing points from the bottom-left to top-right sub-areas in order to avoid packet collisions.

Table 2 shows the four parameters T, G, n and D, and their values are used in the simulation experiments. The parameters are basically fixed to constant values specified by numbers to the left of the parenthetic values, but are varied specified by the parenthetic elements in Table 2 in order to verify the impact on the performance of our proposed method.

In the above simulation model, we performed 50 experiments in which the initial position of each mobile sensor nodes was randomly determined where there was the same number of sensor nodes in all sub-areas. The total simulation time was 4000 [sec], and we evaluated the following four criteria.

1. Delivery ratio: We judge that the sink has succeeded the data gathering if it acquired sensor data from all sensing points. The delivery ratio is defined as the ratio of the number of successes in data gathering to the total number of sensing times occurred during the simulation.

2. Delay: The delay is defined as the average elapsed time from the start of each sensing time to the time that the sink successfully receives all sensor data.

3. Number of packets: The number of packets is defined as the total number of the packets transmitted by the sink and all sensor nodes during the simulation.

4. Traffic: The traffic is defined as the summation of the packets size of all packets sent by the sink and all sensor nodes during the simulation. Table 3 shows messages used in our method and the comparative method and their sizes at the application layer. In this table, i denotes the number of sensor data aggregated in a data message. The message size in the Mac layer is calculated by adding 64 [B] (the header size of a Mac message) to the message size at the application layer.

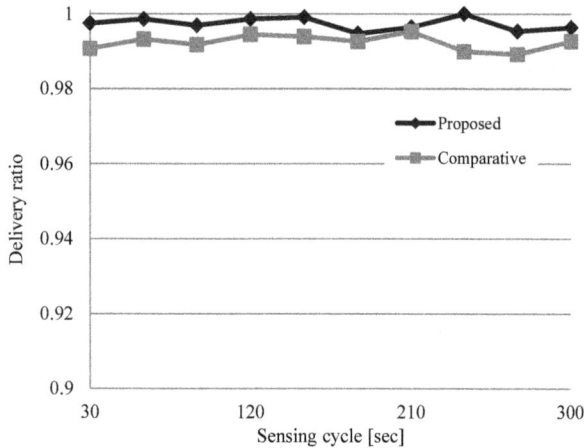

Fig. 4. Effects of sensing cycle (Delivery ratio).

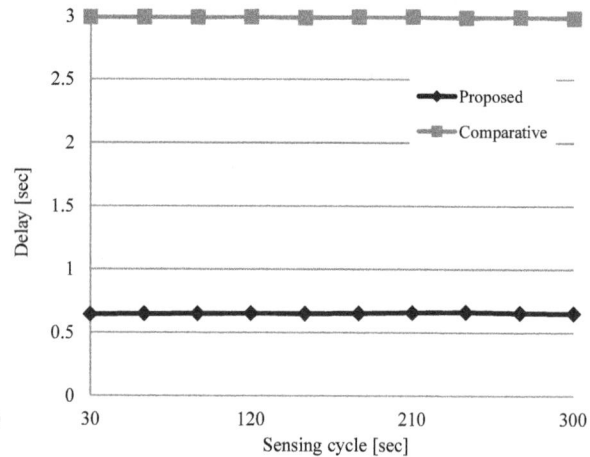

Fig. 5. Effects of sensing cycle (Delay).

5.2. Effects of the sensing cycle

First, we examine the effects of T, the sensing cycle [sec]. Figures 4 through 7 show the simulation results. In these graphs, the horizontal axes indicate T, and the vertical axes indicate the delivery ratio in Fig. 4, the delay in Fig. 5, the number of packets in Fig. 6, and the traffic in Fig. 7.

Figure 4 shows that the delivery ratio in both methods is very high. However, both methods could not achieve 100 [%] of delivery ratio because packet collisions infrequently occurred during geo-routing. In our proposed method, there is another case that data gathering fails. Specifically, a data gathering fail happens when a data message arrives at a mobile agent while the mobile agent moves between sensor nodes. On the other hand, our method produces less traffic than the comparative method (see Fig. 6), and thus, can reduce packet collisions. As the result of these two factors (negative and positive), the delivery ratio in our proposed method is almost same as that in the comparative method.

Figure 5 shows that the delay is not affected by the sensing cycle. It also shows that our proposed method can gather sensor data in much shorter time than the comparative method. In our proposed method, the sensor data are concurrently sent from mobile agents to the sink. On the other hand, in the comparative method, the sensor data are sent from sensor nodes located closest to the sensing points after receiving a request message which is successively sent from the sink with an interval (0.1 [sec]). Here, the longest waiting time until receiving a request is $0.1 \times (G-1)$ [sec]. Although it can be reduced by setting a smaller interval, we have confirmed from our preliminary experiment that packet collisions frequently occur and the delivery ratio heavily decreases if the interval is set as less than 0.1 [sec].

Figure 6 shows that the number of packets in our proposed method is much less than that in the comparative method. In the comparative method, the sink individually sends the request messages to sensing points at every sensing time, and the sensor nodes located closest to the sensing point individually send back their sensor data. On the other hand, in our proposed method, the sink does not issue any request messages because mobile agents autonomously send the sensor data to the sink. In addition, the number of data messages in our proposed method is less than that in the comparative method because mobile agents aggregate the received sensor data and its own sensor data. As the sensing cycle gets longer, the number of packets in both methods decreases because the number of sensing times decreases. When the sensing cycle is long, the difference in the number of packets between our proposed method and the comparative method is small. This is because the impact of deployment and movement messages, which

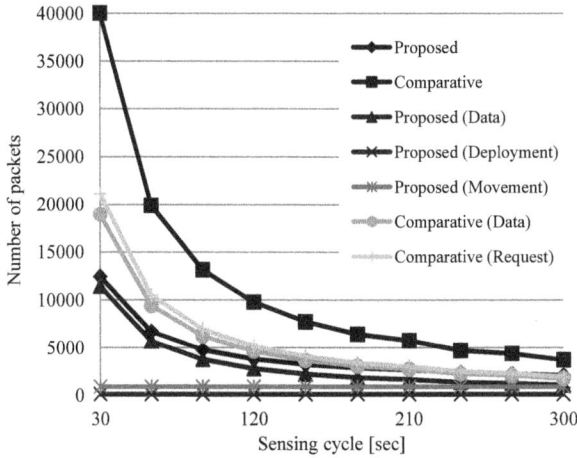

Fig. 6. Effects of sensing cycle (Number of packets).

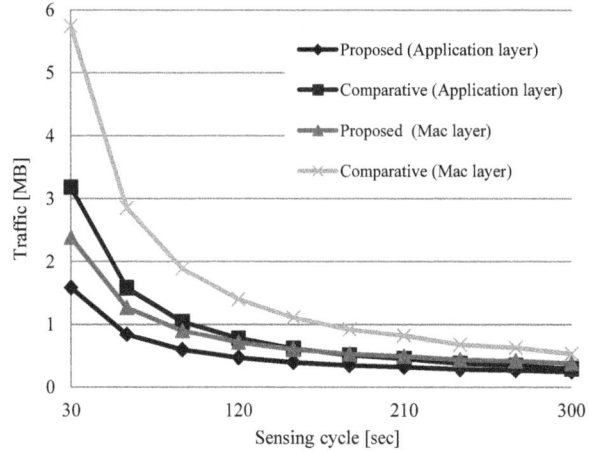

Fig. 7. Effects of sensing cycle (Traffic).

are issued only in our proposed method and not affected by the sensing cycle, becomes more dominant while other packets decrease as the sensing cycle increases.

Figure 7 shows that the traffic in our proposed method is much smaller than that in the comparative method. This is due to the reduction of the number of messages by aggregating sensor data contributes to reduce the traffic. This can be seen from the result that the difference in traffic between our method and the comparative method in the Mac layer is larger than that in the application layer because the proportion of the header size to the total message size is larger in the Mac layer.

As the sensing cycle increases, the traffic in both methods decreases. When the sensing cycle is long, the difference in traffic between our proposed method and the comparative method is small.

5.3. Effects of the number of sub-areas

Next, we examine the effects of G, the number of sub-areas. Figures 8 through 11 show the simulation results. In these graphs, the horizontal axes indicate G, and the vertical axes indicate the delivery ratio in Fig. 8, the delay in Fig. 9, the number of packets in Fig. 10, and the traffic in Fig. 11.

Figure 8 shows that as G increases, the delivery ratio in both methods decreases because the increase of sub-areas means the increase of sensor data collected, and thus, packet collisions occur more often. Moreover, in our proposed method, as G increases, the cases that sensor data are missing while mobile agents move between sensor nodes (as described in the previous subsection) increase. As the result, the delivery ratio in our proposed method becomes smaller than that in the comparative method.

Figure 9 shows that as G increases, the delay in the comparative method increases because the sink successively sends request messages to all of G sensing points with a constant interval. More specifically, it takes at least $0.1 \times (G - 1)$ [sec] for sending request messages to all sub-areas. On the contrary, our proposed method can gather data in a short time regardless of G because our proposed method does not issue any request messages and mobile agents send the sensor data to the sink along the forwarding tree.

Figure 10 shows that as G increases, the number of packets in both methods increases because of the number of sensor data transmitted to the sink increases. However, the number of packets in our proposed method is much fewer than that in the comparative method. This is due to the same reason as that in Fig. 6.

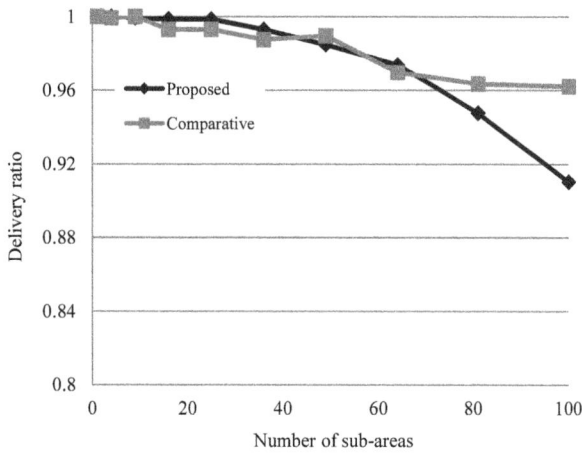

Fig. 8. Effects of number of sub-areas (Delivery ratio).

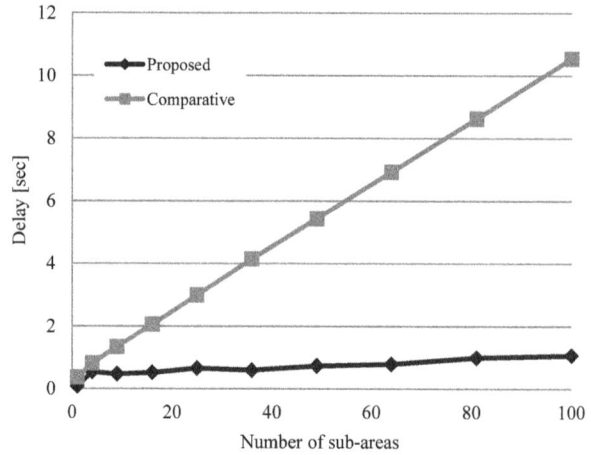

Fig. 9. Effects of number of sub-areas (Delay).

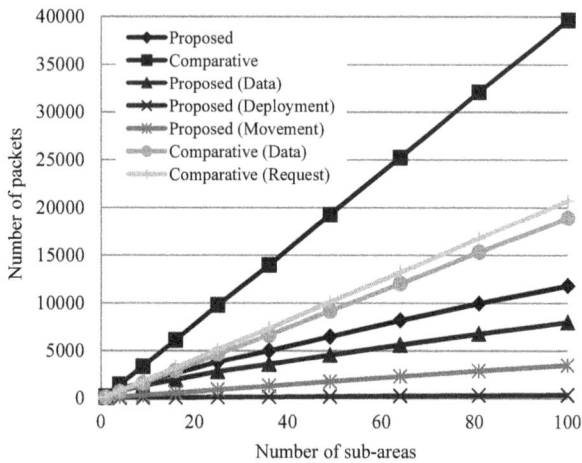

Fig. 10. Effects of number of sub-areas (Number of packets).

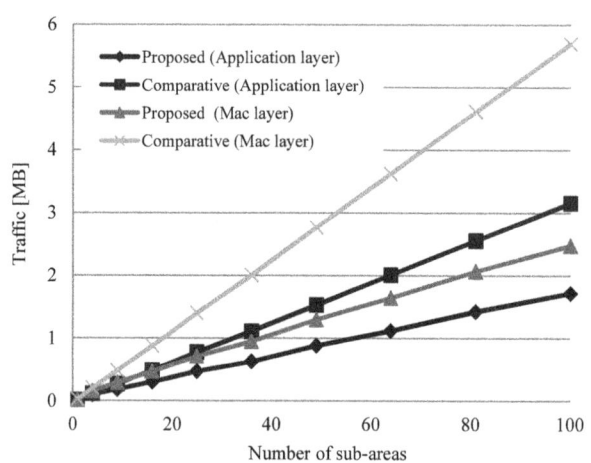

Fig. 11. Effects of number of sub-areas (Traffic).

Figure 11 shows that as G increases, the traffic in both methods increases. However, the traffic in our method is much less than that in the comparative method. This is due to the same reason as that in Figure 7.

5.4. Effects of the number of sensor nodes

We examine the effects of n, the number of sensor nodes. Figures 12 through 15 show the simulation results. In these graphs, the horizontal axes indicate n, and the vertical axes indicate the delivery ratio in Fig. 12, the delay in Fig. 13, the number of packets in Fig. 14, and the traffic in Fig. 15.

Figure 12 shows that when n is very small, the delivery ratio in both methods is small because it happens frequently that no sensor nodes exist in the forwarding area in geo-routing or no sensor nodes exist that can sense a certain sensing point. Moreover, in our proposed method, if there are no sensor nodes within the circle with radius α from the sensing point, the mobile agent which is responsible for that point cannot move to a new sensor node, and thus, the agent for that point disappears from the

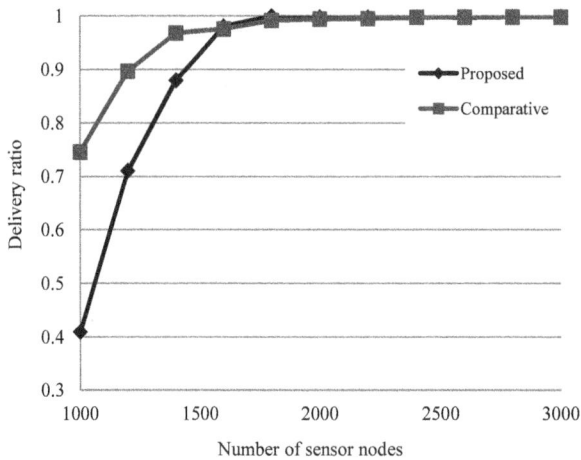

Fig. 12. Effects of the number of sensor nodes (Delivery ratio).

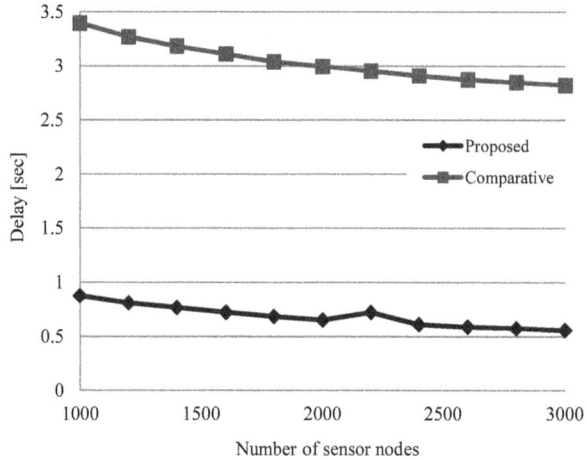

Fig. 13. Effects of the number of sensor nodes (Delay).

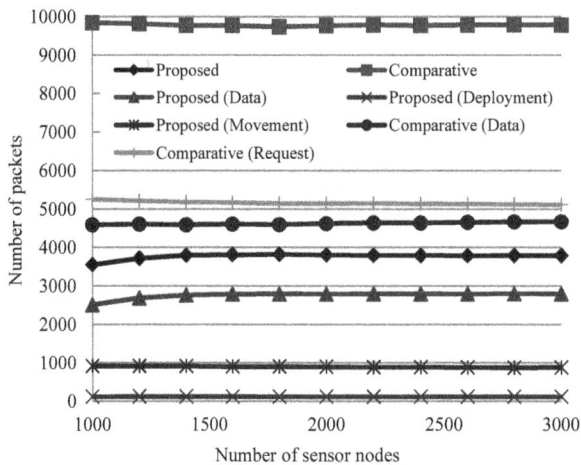

Fig. 14. Effects of the number of sensor nodes (Number of packets).

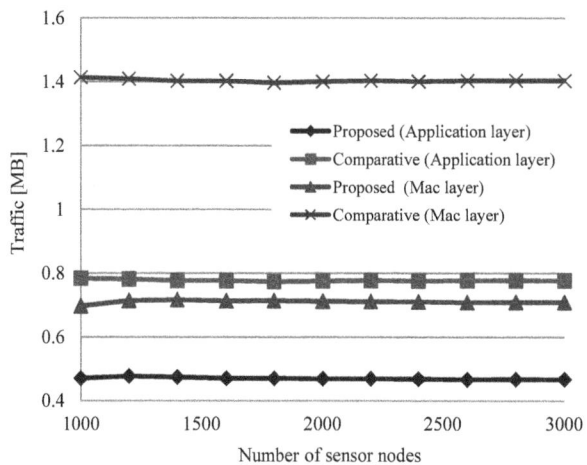

Fig. 15. Effects of the number of sensor nodes (Traffic).

network. As the result, the delivery ratio significantly decreases because the corresponding sensor data cannot be sent to the sink. The effectiveness of handling such situations by the procedures described in Subsection 4.6 is shown in Subsection 5.6.

Figure 13 shows that the delay in our proposed method is much shorter than that in the comparative method due to the same reason as that in Fig. 5. As n increases, the delay in both methods slightly decreases. This is because the distance between the destination and the node near the destination in the forwarding area for geo-routing tends to be shorter due to the increase of node density, which results in reduction of the waiting time for geo-routing (shown in Eq. (1)).

Figure 14 shows that the number of packets in our proposed method is much lower than that in the comparative method due to the same reason as that in Fig. 6. The number of packets in both methods is rarely affected by the number of nodes. In our proposed method, the number of packets slightly decreases as the number of sensor nodes decreases except for the case of Proposed (Movement). Proposed (Movement) gives slightly larger number of packets when the number of sensor nodes is 1000. This is

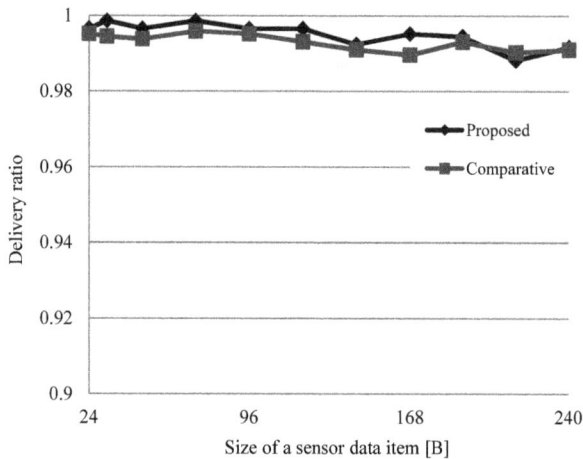

Fig. 16. Effects of size of a sensor data item (Delivery ratio).

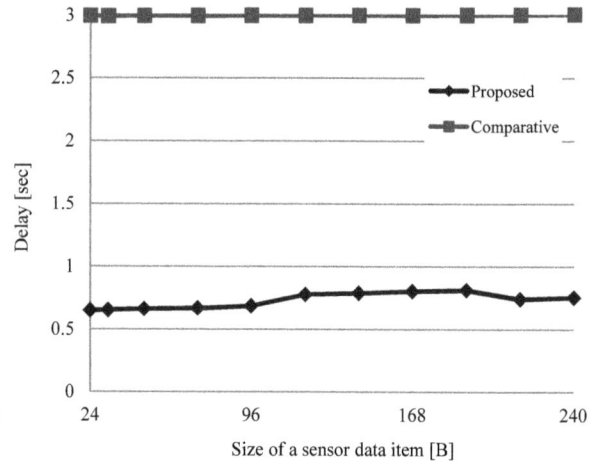

Fig. 17. Effects of size of a sensor data item (Delay).

because when the number of sensor nodes is small, the average distance between the sensing point and the nearest node from the sensing point becomes long, and thus, the mobile agent moves more frequently between sensor nodes.

Figure 15 shows that the traffic in our proposed method is much less than that in the comparative method due to the same reason as that in Fig. 7. The traffic in both methods is rarely affected by the number of nodes. This is because only one node sends the sensor data for each sensing point regardless of the number of sensor nodes.

5.5. Effects of the size of a sensor data

We examine the effects of D, the size of a sensor data item. Figures 16 through 19 show the simulation results. In these graphs, the horizontal axes indicate D, and the vertical axes indicate the delivery ratio in Fig. 16, the delay in Fig. 17, the number of packets in Fig. 18, and the traffic in Fig. 19.

Figure 16 shows that the delivery ratio in both methods is very high regardless of D. As the size of sensor data increases, the delivery ratio slightly decreases in both methods because packet losses occur a bit more frequently. However, the impact of this is small because the size of sensor data is still small. (In this paper, we assume that the size of a sensor data item is small, such as with temperature data.)

Figure 17 shows that the delay in our proposed method is much lower than that in the comparative method. The delay in our proposed method slightly increases as the size of sensor data increases, whereas it is almost constant in the comparative method. This is because the time for sending sensor data along the forwarding tree increases as the size of sensor data becomes larger. In our proposed method, because the sensor data are aggregated, the effect is greater than in the comparative method.

Figure 18 shows that the number of packets in our proposed method is much lower than that in the comparative method. The number of packets in both methods is rarely affected by the size of sensor data, which means that packet losses were rare in this simulation setting.

Figure 19 indicates that as the size of sensor data increases, the traffic in both methods also increases. Here, the increment in traffic in our proposed method is larger than that in the comparative method. This is because in the comparative method sensor data are sent directly to the sink (along the shortest paths), while in our proposed method the sensor data are sent along the forwarding tree (not the shortest paths), i.e., the path length is longer. Since our proposed method does not issue request messages and it aggregates sensor data, the size of sensor data is a dominant factor in traffic.

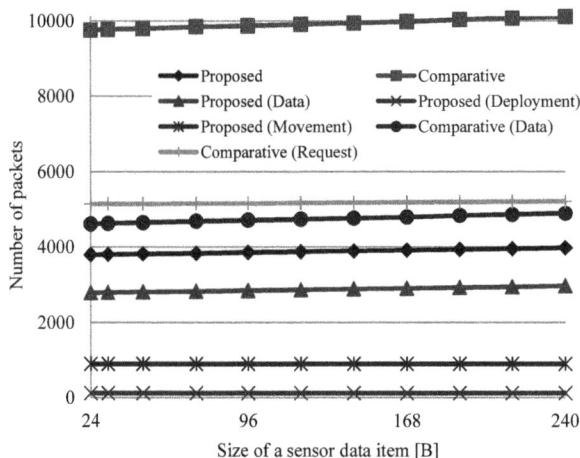

Fig. 18. Effects of size of a sensor data item (Number of packets).

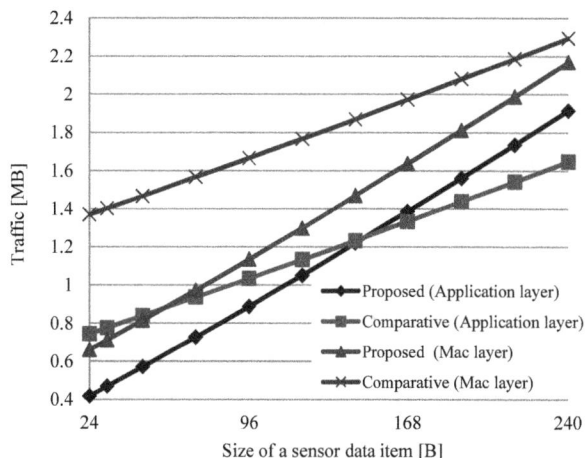

Fig. 19. Effects of size of a sensor data item (Traffic).

5.6. Effects of node failure

Finally, we evaluate the effectiveness of applying mechanism of redeployment of mobile agents and retransmission of sensor data described in Subsection 4.6. Each sensor node moves according to the random waypoint mobility model with a home area [2]. Specifically, it selects a random destination in the sub-area that it is assigned to (home area) with the probability of 90% or a destination in the entire sensing area with 10%, and moves at a constant speed uniformly determined from 1 [m/sec] to 2 [m/sec]. It stops at the destination for 60 [sec]. Additionally, each sensor node disappears from the network with $F\%$ every 50 [sec], and the disappeared nodes return to the network after 100 [sec]. The sink and sensor nodes communicate using the IEEE 802.11a protocol. In our proposed method with the mechanism in Subsection 4.6 (extended method), when mobile agents and the sink cannot receive sensor data from any of their children within 2 [sec] ($\delta = 2$), they redeploy the missing mobile agents and request them to retransmit the sensor data. For retransmission of sensor data in the comparative method, the sink sends request messages at intervals of 0.1 [sec] to the sensing points whose sensor data the sink cannot receive within 2 [sec]. We evaluated the extended method, the comparative method, and our proposed method without the mechanism of retransmission of sensor data and redeployment of mobile agents (basic method).

Figures 20 through 23 show the simulation results. In these graphs, the horizontal axes indicate F (ratio of node failure), and the vertical axes indicate the delivery ratio in Fig. 20, the delay in Fig. 21, the number of packets in Fig. 22, and the traffic in Fig. 23.

Figure 20 shows that the delivery ratio in the extended and comparative methods is very high regardless of F. With the basic method, however, as the ratio of node failure increases, the delivery ratio drastically decreases because in this method the sink only initially deploys mobile agents, and data are not gathered until the end of the simulation after a mobile agent disappears (the sensor node on which a mobile agent operates fails). In the extended method, the delivery ratio does not decrease because when a mobile agent disappears, its parent in the forwarding tree redeploys it and the redeployed agent retransmits the sensor data. In the comparative method, the ratio of node failure does not affect the delivery ratio because the comparative method does not use mobile agents.

Figure 21 shows that as the ratio of node failure increases, the delay increases in the extended method because the ratio of redeployment of mobile agents and the retransmission of sensor data increases, and

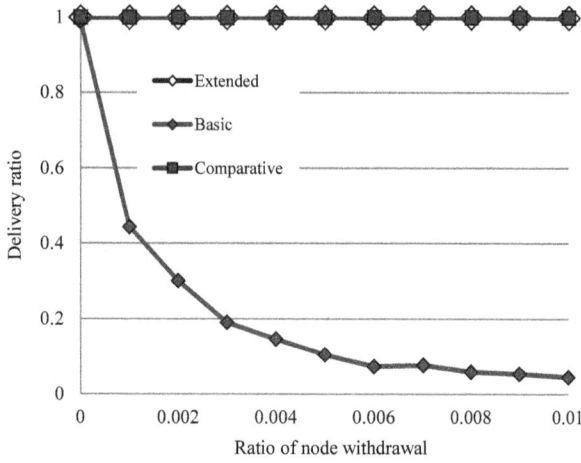

Fig. 20. Effects of ratio of node withdrawal (Delivery ratio).

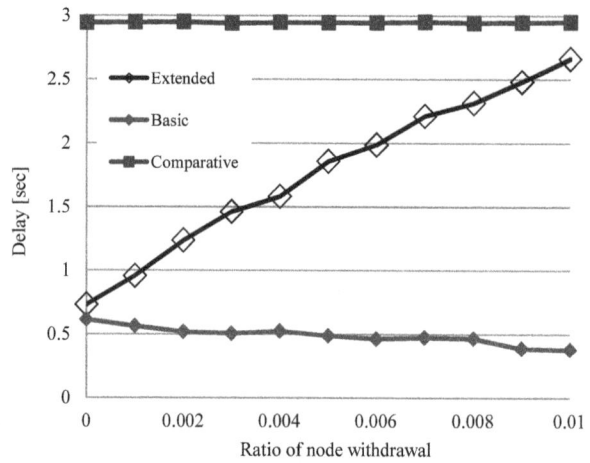

Fig. 21. Effects of ratio of node withdrawal (Delay).

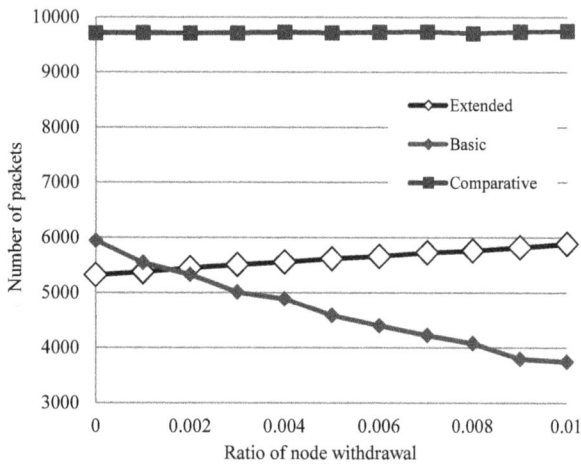

Fig. 22. Effects of ratio of node withdrawal (Number of packets).

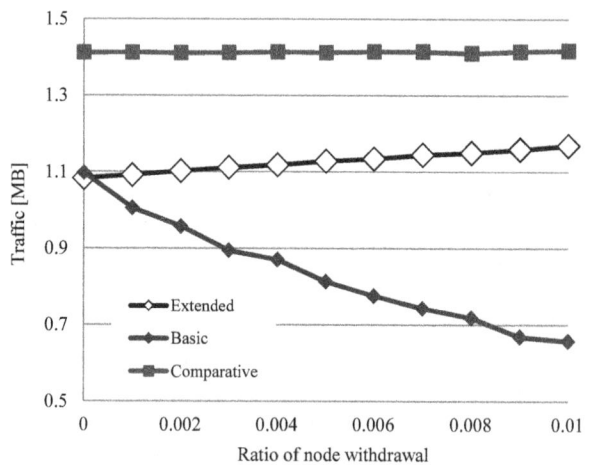

Fig. 23. Effects of ratio of node withdrawal (Traffic).

the time for these procedures adds to the delay. In the basic method, the ratio of node failure does not affect the delay because the basic method does not retransmit sensor data. The ratio of node failure also does not affect the delay in the comparative method because this method does not use mobile agents, thus, retransmission of sensor data does not happen very often.

Figures 22 and 23 show that as the ratio of node failure increases, the number of packets and the traffic increase in the extended method because the ratio of redeployment of mobile agents and retransmission of sensor data increases, and the messages for these procedures increase. In the basic method, as the ratio of node failure increases, the number of packets and the traffic decrease because the disappeared agents and their ancestors in the forwarding tree do not transmit their sensor data after the mobile agents disappear. The result for the comparative method shows the same delay tendency.

6. Conclusions

In this paper we propose a data gathering method using mobile agents in dense MWSNs. In this method, the sink deploys mobile agents on sensor nodes located near from the sensing points, and the mobile agents autonomously send sensor data to the sink at every sensing time. Our proposed method can reduce the traffic for sensor data gathering since it does not issue any request messages for data gathering and mobile agents can aggregate the sensor data while relaying the data. The results of the simulation experiments show that our proposed method can gather sensor data in short time and with small traffic.

In this paper, for the purpose of simplicity, we do not assume sensing errors (i.e., sensors generate incorrect data). However, in real environments, sensing errors do occur. Therefore, we plan to extend our proposed method to support sensing errors. A possible approach is that neighboring mobile agents exchange their sensor data to check the accuracy before sending the readings to the sink.

Acknowledgment

This research is partially supported by the Grant-in-Aid for Scientific Research (S)(21220002), (B)(24300037) of MEXT, Japan.

References

[1] M. Abolhasan, T. Wysocki and E. Dutkiewicz, A review of routing protocols for mobile ad hoc networks, *Ad Hoc Networks* **2**(1) (2004), 1–22.

[2] T. Camp, J. Boleng and V. Davies, A survey of mobility models for ad hoc network research, *Wireless Communications and Mobile Computing* **2**(5) (2002), 483–502.

[3] A.T. Campbell, S.B. Eisenman, N.D. Lane, E. Miluzzo, R.A.Peterson, H. Lu, X. Zheng, M. Musolesi, K. Fodor and G. Ahn, The rise of people-centric sensing, *IEEE Internet Computing* **12**(4) (2008), 12–21.

[4] J. Carle and D. Simplot, Energy-efficient area monitoring for sensor networks, *IEEE Computer* **37**(2) (2004), 40–46.

[5] B. Chen, K. Jamieson, H. Balakrishnan and R. Morris, Span: An energy efficient coordination algorithm for topology maintenance in ad hoc wireless networks, *ACM Wireless Networks* **8**(5) (2002), 481–494.

[6] M. Chen, T. Kwon, Y. Yuan and V.C.M. Leung, Mobile agent based wireless sensor networks, *JCP* **1**(1) (2006), 14–21.

[7] B. Godfrey and D. Ratajczak, Naps: Scalable, robust topology management in wireless ad hoc networks, In Proceedings of ACM/IEEE International Conference on Information Processing in Sensor Networks (IPSN 2004), 2004, pp. 443–451.

[8] J. Haillot and F. Guidec, A protocol for content-based communication in disconnected mobile ad hoc networks, *Mobile Information Systems* **6**(2) (2010), 123–154.

[9] A.M. Hanashi, I. Awan and M.E. Woodward, Performance evaluation with different mobility models for dynamic probabilistic flooding in MANETs, *Mobile Information Systems* **5**(1) (2009), 65–80.

[10] W.R. Heinzelman, A. Chandrakasan and H. Balakrishnan, Energy-efficient communication protocol for wireless microsensor networks, In Proceedings of Hawaii International Conference on System Sciences (HICSS-33), 2000, pp. 1–10.

[11] M. Heissenbüttel, T. Braun, T. Bernoulli and M. Wälchli, BLR: Beacon-less routing algorithm for mobile ad hoc networks, *Computer Communications* **27**(11) (2004), 1076–1086.

[12] Y.S. Lee, J.W. Park and L. Barolli, A localization algorithm based on AOA for ad-hoc sensor networks, *Mobile Information Systems* **8**(1) (2012), 61–72.

[13] S. Reddy, D. Estrin, M.H. Hansen and M.B. Srivastava, Examining micro-payments for participatory sensing data collections, In Proceedings of ACM International Conference on Ubiquitous Computing (UbiComp 2010), 2010, pp. 33–36.

[14] S. Reddy, V. Samanta, J. Burke, D. Estrin, M.H. Hansen and M.B. Srivastava, MobiSense – Mobile network services for coordinated participatory sensing, In Proceedings of International Symposium on Autonomous Decentralized Systems (ISADS 2009), 2009, pp. 231–236.

[15] Scenargie Base Simulator, Space-Time Engineering, http://www.spacetime-eng.com.

[16] J. Shi, R. Zhang, Y. Liu and Y. Zhang, PriSense: Privacy-preserving data aggregation in people-centric urban sensing systems, In Proceedings of IEEE Infocom, 2010, pp. 758–766.

[17] B. Wang, Coverage problems in sensor networks: A survey, *ACM Computing Surveys* **43**(4) (2011), 32:1–32:53.

[18] X. Wang, G. Xing, Y. Zhang, C. Lu, R. Pless and C. Gill, Integrated coverage and connectivity configuration in wireless sensor networks, In Proceedings of ACM Conference on Embedded Networked Sensor Systems (Sensys 2003), 2003, pp. 28–39.

[19] T. Yan, T. He and J.A. Stankovi, Differentiated surveillance for sensor networks, In Proceedings of ACM Conference on Embedded Networked Sensor Systems (Sensys 2003), 2003, pp. 51–62.

[20] T. Yashir, A new paradigm of V2V communication services using nomadic agent, In Proceedings of International Workshop on Vehicle-to-Vehicle Communications (V2VCOM 2006), 2006, pp. 1–6.

[21] Y. Xu, J.S. Heidemann and D. Estri, Geography-informed energy conservation for ad hoc routing, In Proceeding of Annual International Conference on Mobile Computing and Networking (MobiCom 2001), 2001, pp. 70–84.

[22] J. Yick, B. Mukherjee and D. Ghosa, Wireless sensor network survey, *Computer Networks* **52**(12) (2008), 2292–2330.

Keisuke Goto Keisuke Goto received the B.E. degree from Kyoto Institute of Technology and the M.I.S. degree from Osaka University, Japan, in 2010 and 2012, respectively. He is currently a research student at the Graduate school of Information Science and Technology of Osaka University. His research interests include Mobile Wireless Sensor Networks.

Yuya Sasaki Yuya Sasaki received the B.E. and M.I.S. degrees from Osaka University, Japan, in 2009 and 2011, respectively. He is currently pursuing his Ph.D. at the Graduate school of Information Science and Technology of Osaka University. His research interests include Mobile Ad Hoc Networks and Vehicular Ad Hoc Networks.

Takahiro Hara Takahiro Hara received the B.E., M.E., and Dr. E. degrees in Information Systems Engineering from Osaka University, Osaka, Japan, in 1995, 1997, and 2000, respectively. Currently, he is an Associate Professor of the Department of Multimedia Engineering, Osaka University. He has published more than 300 international Journal and conference papers in the areas of databases, mobile computing, peer-to-peer systems, WWW, and wireless networking. He served and is serving as a Program Chair of IEEE International Conferences on Mobile Data Management (MDM'06 and 10) and Advanced Information Networking and Applications (AINA'09), and IEEE International Symposium on Reliable Distributed Systems (SRDS'12). He guest edited IEEE Journal on Selected Areas in Communications, Sp. Issues on Peer-to-Peer Communications and Applications. His research interests include distributed databases, peer-to-peer systems, mobile networks, and mobile computing systems. He is a senior member of IEEE and ACM and a member of three other learned societies.

Shojiro Nishio Shojiro Nishio received his B.E., M.E., and Ph.D. degrees from Kyoto University in Japan, in 1975, 1977, and 1980, respectively. He has been a full professor at Osaka University since August 1992. He served as a Vice President and Trustee of Osaka University from August 2007 to August 2011. He also acted as the Program Director in the Area of Information and Networking, Ministry of Education, Culture, Sports, Science and Technology (MEXT), Japan from April 2001 to March 2008. His research interests include database systems and multimedia systems for advanced networks such as broadband networks and mobile computing environment. Dr. Nishio has co-authored or co-edited more than 55 books, and authored or co-authored more than 560 refereed journal or conference papers. He served as the Program Committee Co-Chairs for several international conferences including DOOD 1989, VLDB 1995, and IEEE ICDE 2005. He has served and is currently serving as an editor of several international journals including IEEE Trans. on Knowledge and Data Engineering, VLDB Journal, ACM Trans. on Internet Technology, and Data & Knowledge Engineering. He is also a fellow of IEICE and IPSJ, and is a member of six learned societies, including ACM and IEEE.

9

Performance evaluation of WMN-GA for different mutation and crossover rates considering number of covered users parameter

Tetsuya Oda[a,*], Admir Barolli[b], Fatos Xhafa[c], Leonard Barolli[d], Makoto Ikeda[d] and Makoto Takizawa[b]

[a]*Graduate School of Engineering, Fukuoka Institute of Technology (FIT), Higashi-Ku, Fukuoka, Japan*
[b]*Department of Computers and Information Science, Seikei University, Musashino-Shi, Tokyo, Japan*
[c]*Department of Languages and Informatics Systems, Technical University of Catalonia, Barcelona, Spain*
[d]*Department of Information and Communication Engineering, Fukuoka Institute of Technology (FIT), Higashi-Ku, Fukuoka, Japan*

Abstract. Node placement problems have been long investigated in the optimization field due to numerous applications in location science and classification. Facility location problems are showing their usefulness to communication networks, and more especially from Wireless Mesh Networks (WMNs) field. Recently, such problems are showing their usefulness to communication networks, where facilities could be servers or routers offering connectivity services to clients. In this paper, we deal with the effect of mutation and crossover operators in GA for node placement problem. We evaluate the performance of the proposed system using different selection operators and different distributions of router nodes considering number of covered users parameter. The simulation results show that for Linear and Exponential ranking methods, the system has a good performance for all rates of crossover and mutation.

1. Introduction

With the emergence of several new networking paradigms such as Ad-hoc networks, sensor networks, vehicular networks, cellular networks, optimization modelling and resolution turns out to be crucial to achieve optimized performance networks [1–3]. One such networking paradigm that is requiring the resolution of optimization problems is that of WMNs. At the heart of WMNs are the issues of achieving network connectivity and stability as well as QoS in terms of user coverage. These issues are very closely related to the family of node placement problems in WMNs, such as mesh router nodes placement. Node placement problems have been long investigated in the optimization field due to numerous applications in location science and classification (clustering). In such problems, we are given a number of potential facilities to serve to costumers connected to facilities aiming to find locations such that the cost of serving

---

[*]Corresponding author. E-mail: oda.tetuya.fit@gmail.com.

to all customers is minimized. In traditional versions of the problem, facilities could be hospitals, polling centers, fire stations serving to a number of clients and aiming to minimize some distance function in a metric space between clients and such facilities.

WMNs [4,5] infrastructures are currently used in developing and deploying medical, transport and surveillance applications in urban areas, metropolitan, neighboring communities and municipal area networks [6]. WMNs are based on mesh topology, in which every node (representing a server) is connected to one or more nodes, enabling thus the information transmission in more than one path. The path redundancy is a robust feature of this kind of topology. Compared to other topologies, mesh topology needs not a central node, allowing networks based on such topology to be self-healing. These characteristics of networks with mesh topology make them very reliable and robust networks to potential server node failures. In WMNs mesh routers provide network connectivity services to mesh client nodes. The good performance and operability of WMNs largely depends on placement of mesh routers nodes in the geographical deployment area to achieve network connectivity, stability and user coverage. The objective is to find an optimal and robust topology of the mesh router nodes to support connectivity services to clients.

For the mesh router node placement problem, are given a grid area arranged in cells where to distribute a number of mesh router nodes and a number of mesh client nodes of fixed positions (of an arbitrary distribution) in the grid area. The objective is to find a location assignment for the mesh routers to the cells of the grid area that maximizes the network connectivity and client coverage. Network connectivity is measured by the size of the giant component of the resulting WMN graph, while the user coverage is simply the number of mesh client nodes that fall within the radio coverage of at least one mesh router node.

For most formulations, node placement problems are shown to be computationally hard to solve to optimality [7–10], and therefore heuristic and meta-heuristic approaches are useful approaches to solve the problem for practical purposes. Several heuristic approaches are found in the literature for node placement problems in WMNs [11–15].

Genetic Algorithms (GAs) have been recently investigated as effective resolution methods. In this work, we present the results of the effect of mutation and crossover operators in GA for mesh router nodes placement problem. The study aims to identify the mutation and crossover types that work best for instances of different characteristics. A bi-objective optimization, namely, the maximization of the size of the giant component in the network (for measuring network connectivity) and that of user coverage are considered. However in this paper, we use only the number of covered users as a parameter for the system evaluation. In the simulations, we have used different distributions of mesh node clients (Uniform, Normal, Exponential and Weibull).

The rest of the paper is organized as follows. In Section 2 is presented Genetic Algorithm template and its application to mesh router nodes placement. We focus further on mutation types and mutation rate values in Section 3. The crossover operator is described in Section 4. The simulation results are given in Section 5. We end the paper in Section 6 with some conclusions.

2. Genetic algorithms

Genetic Algorithms (GAs) [16,17] have shown their usefulness for the resolution of many computationally combinatorial optimization problems. For the purpose of this work, we have used the *template* given in Algorithm 1.

We present next the particularization of GAs for the mesh router nodes placement in WMNs (see [20] for more details).

Algorithm 1 Genetic Algorithm template

Generate the initial population P^0 of size μ;
Evaluate P^0;
while not termination-condition **do**
   Select the parental pool T^t of size λ; $T^t := Select(P^t)$;
   Perform crossover procedure on pairs of individuals in T^t with probability p_c; $P_c^t := Cross(T^t)$;
   Perform mutation procedure on individuals in P_c^t with probability p_m; $P_m^t := Mutate(P_c^t)$;
   Evaluate P_m^t;
   Create a new population P^{t+1} of size μ from individuals in P^t and/or P_m^t;
   $P^{t+1} := Replace(P^t; P_m^t)$
   $t := t + 1$;
end while
return Best found individual as solution;

2.1. Encoding

The encoding of individuals (also known as chromosome encoding) is fundamental to the implementation of GAs in order to efficiently transmit the genetic information from parents to offsprings.

In the case of the mesh router nodes placement problem, a solution (individual of the population) contains the information on the current location of routers in the grid area as well as information on links to other mesh router nodes and mesh client nodes. This information is kept in data structures, namely, `pos_routers` for positions of mesh router nodes, `routers_links` for link information among routers and `client_router_link` for link information among routers and clients (matrices of the same size as the grid area are used.) Based on these data structures, the size of the giant component and the number of users covered are computed for the solution.

It should be also noted that routers are assumed to have different radio coverage, therefore any router could be linked to a number of clients and other routers. Obviously, whenever a router is moved to another cell of the grid area, the information on links to both other routers and clients must be computed again and links are re-established.

2.2. Selection operators

In the evolutionary computing literature we can find a variety of selection operators, which are in charge of selecting individuals for the pool mate. The operators considered in this work are those based on *Implicit Fitness Re-mapping* technique. It should be noted that selection operators are generic ones and do not depend on the encoding of individuals.

- *Random Selection*: This operator chooses the individuals uniformly at random. The problem is that a simple strategy does not consider even the fitness value of individuals and this may lead to a slow convergence of the algorithm.
- *Best Selection*: This operator selects the individuals in the population having higher fitness value. The main drawback of this operator is that by always choosing the best fitted individuals of the population, the GA converges prematurely.
- *Linear Ranking Selection*: This operator follows the strategy of selecting the individuals in the population with a probability directly proportional to its fitness value. This operator clearly benefits the selection of best endowed individuals, which have larger chances of being selected.
- *Exponential Ranking Selection*: This operator is similar to Linear Ranking but now probabilities of ranked individuals are weighted according to an exponential distribution.

– *Tournament Selection*: This operator selects the individuals based on the result of a tournament among individuals. Usually winning solutions are the ones of better fitness value but individuals of worse fitness value could be chosen as well, contributing thus to avoiding premature convergence. Particular cases of this operator are the *Binary Tournament* and $N-Tournament\ Selection$, for different values of N.

2.3. Crossover operators

The crossover operators are the most important ingredient of GAs. Indeed, by selecting individuals from the parental generation and interchanging their *genes*, new individuals (descendants) are obtained. It is very important to stress that the crossover operators depend on the chromosome representation. The crossover operator should thus take into account the specifics of mesh router nodes encoding.

2.4. Mutation operators

Mutation operator is one of the GA ingredients. Unlike crossover operators, which achieve to transmit genetic information from parents to offsprings, mutation operators usually make some small local perturbation of the individuals, having thus less impact on newly generated individuals.

Crossover is "a must" operator in GA and is usually applied with high probability, while mutation operators when implemented are applied with small probability. The rationale is that a large mutation rate would make the GA search to resemble a random search. Due to this, mutation operator is usually considered as a secondary operator.

3. Mutation in genetic algorithms

3.1. Mutation operators issues

Many studies in the literature have shown that mutation when effectively combined with selection operators can improve the performance of GAs. Some of the important issues behind the use of mutation operators are as follows:

– *The design of different versions of mutation operators and their evaluation in GA in conjunction with selection operators*: In general, it is difficult to know *a priori* if a simple mutation, such as bit-flip mutation, would produce desired effect. Therefore different types of mutation operators should be designed and evaluated. In fact, their evaluation should be done in conjunction with selection operators.
– *Mutation Rate*: This is among most critical aspects in applying mutation operators. In most applications, mutation rate is kept low and constant through the search. However, the concrete mutation rate has to be tuned. In some works the use of dynamic mutation rates have also been proposed.
– *Effect of Mutation Rate on GA Evolution*: In GAs one critical aspect is keeping diversity of the population and avoid premature convergence to local optima. Certainly, mutation, through small perturbation of the individuals can contribute to maintaining the diversity of the population, if appropriate mutation rates are achieved.

– *Generic vs. Ad Hoc Problem-oriented Mutation Operators*: In general, one could consider the mutation operator as generic, problem-independent operator. This is so because in most cases, small local perturbation to individuals such as bit flip, swap, and so on can be implemented similarly for most combinatorial structures. However, it might be interesting to explore the characteristics of the problem at hand and design some *ad hoc* problem-dependent operators.

3.2. Setting mutation rates

Setting the values of parameters in GAs has been and is still a major issue in GA research field. One such parameter is the mutation rate. The complexity of tuning GA parameters, and the mutation rate in particular, comes from the side-effect of one operator on the rest of operators. Ideally, tuning should be able to explore the synergies among different operators.

3.2.1. Population size, crossover and mutation rate

Among the different GA parameters, population size, crossover rate and mutation rate have considerable effect on GA convergence and, consequently, the quality of solution found. Some research studies reported in the literature suggest that there are direct relations among selection strategies, crossover rate and mutation rate. Regarding population size vs. mutation rate, the issue here is whether it would be more beneficial for the genetic search to use: (a) a small population size and rather large mutation rate, or (b) large population sizes and low mutation rate. The choice could have a significant impact on the GA convergence.

3.2.2. Mutation rate

As can be seen from above, setting the mutation value is difficult, in part due to the many other GA parameters to be set and the existing synergies among them. In the literature we would find essentially two ways to set the mutation rate: (a) using constant values and (b) using variable mutation rate. In the former, considered as the easiest way, the mutation rate is kept unchanged during the whole genetic search process. In this case, the value of the mutation rate can be fixed either *a priori* (based on an experimental study) or computed based on some problem size parameter. For instance, some authors have explored the advantages and limitations of using the $1/L$ heuristic, in which the mutation rate is set to $1/L$, where L stands for the chromosome string length [18].

In the second group, mutation rate is changed during the genetic search process. The difficulty here is how to find an appropriate decreasing function of mutation rate. One such possible function is linear function, that is, mutation rate linearly decreases along the search process. One implementation would then be to keep mutation rate constant during a certain number of generations and then decrease it linearly. Some more advanced schemes proposed in the literature suggest using self-adaptive mutation rates either by splitting the population into sub-populations and using specific mutation rates to sub-populations or by using a ranking of the individuals in the population and then assign to individuals mutation according to the rank [19].

3.3. Mutation operators for mesh routers nodes placement in WMNs

In the case of mesh routers node placement, the matrix representation is chosen for the individuals of the population, in order to keep the information on mesh router nodes positions, mesh client positions, links among routers and links among routers and clients. The definition of the mutation operators is therefore specific to matrix-based encoding of the individuals of the population. Several specific mutation operators were considered in this study, which are move-based and swap-based operators.

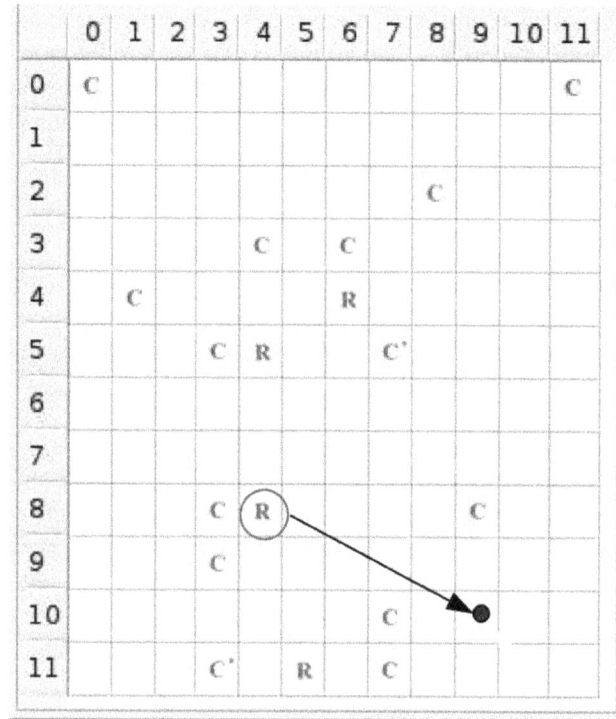

Fig. 1. Single mutate operator.

3.3.1. SingleMutate

This is a move-based operator. It selects a mesh router node in the grid area and moves it to another cell of the grid area (see Fig. 1).

3.3.2. RectangleMutate

This is a swap-based operator. In this version, the operator selects two "small" rectangles at random in the grid area, and swaps the mesh routers nodes in them (see Fig. 2).

3.3.3. SmallMutate

This is a move-based operator. In this case, the operator chooses randomly a router and moves it a small (*a priori* fixed) number of cells in one of the four directions: up, down, left or right in the grid (see Fig. 3). This operator could be used a number of times to achieve the effect of SingleMutate operator.

3.3.4. SmallRectangleMutate

This is a move-based operator. The operator selects first at random a rectangle and then all routers inside the rectangle are moved with a small (*a priori* fixed) numbers of cells in one of the four directions: up, down, left or right in the grid (see Fig. 4).

3.3.5. Time efficiency of mutation and selection operators

Simple mutation operators, such as bit-flip or simple move are very efficient as only small changes are done in the combinatorial structure. However, mutation based on making larger perturbations of the individual could be time costly given they are applied a total expected number of $p_m \cdot population\_size,$

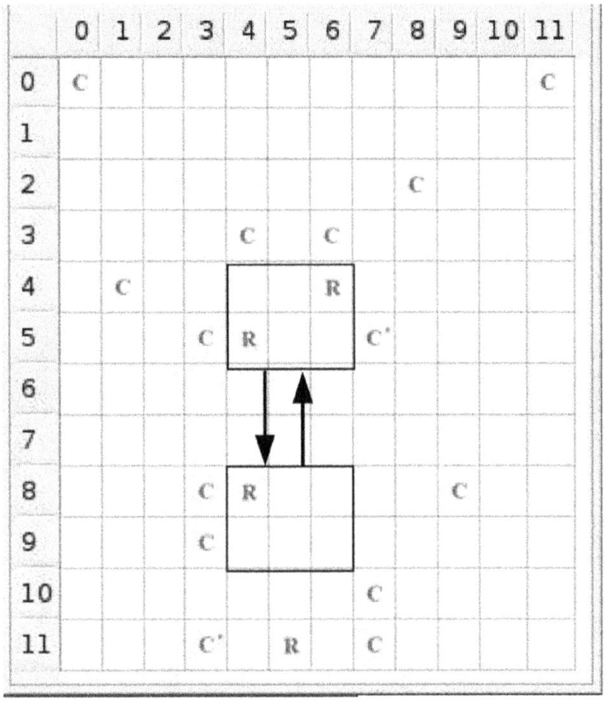

Fig. 2. Rectangle mutate operator.

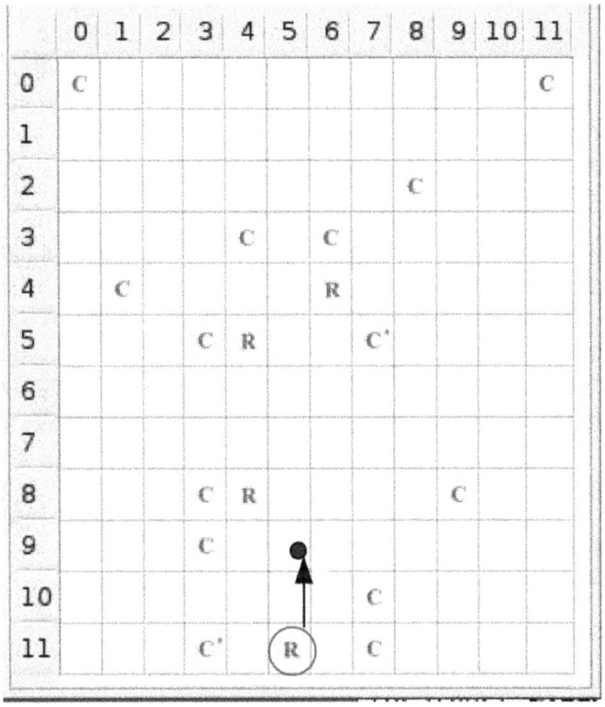

Fig. 3. Small mutate operator.

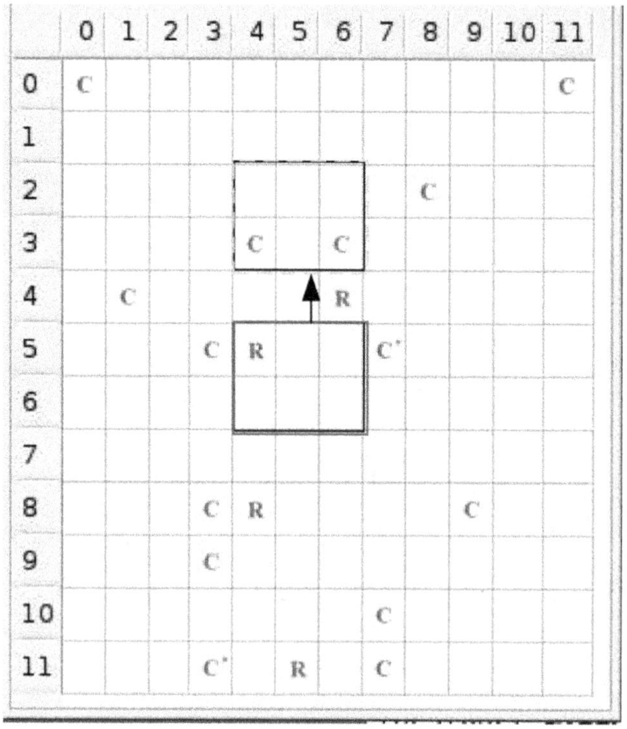

Fig. 4. Small rectangle mutate operator.

where p_m denotes the mutation probability. This amount is to be multiplied by the time cost of performing one mutation, which in case of rectangle mutate and small rectangle mutate are $O(grid\_side \times grid\_side)$. The computational effort is even larger for the crossover operators as in this case the grid area has to be explored to find the sparsest and densest areas.

3.4. Crossover operators

The crossover operator selects individuals from the parental generation and interchanging their *genes*, thus new individuals (descendants) are obtained. The aim is to obtain descendants of better quality that will feed the next generation and enable the search to explore new regions of solution space not explored yet.

There exist many types of crossover operators explored in the evolutionary computing literature. It is very important to stress that crossover operators depend on the chromosome representation. This observation is especially important for the mesh router nodes problem, since in our case, instead of having strings we have a grid of nodes located in a certain positions. The crossover operator should thus take into account the specifics of mesh router nodes encoding. We have considered the following crossover operator, called *intersection operator* (denoted CrossRegion, hereafter), which take in input two individuals and produce in output two new individuals (see Algorithm 2).

4. Proposed and implemented WMN-GA system

In this section, we present WMN-GA system. Our system can generate instances of the problem using different distributions of client and mesh routers.

Algorithm 2 Crossover Operator

1:**Input**: Two parent individuals P_1 and P_2; values H_g and W_g for height and width of a small grid area;

2:**Output**: Two offsprings O_1 and O_2;

3:Select at random a $H_g \times W_g$ rectangle RP_1 in parent P_1. Let RP_2 be the same rectangle in parent P_2;

4:Select at random a $H_g \times W_g$ rectangle RO_1 in offspring O_1. Let RO_2 be the same rectangle in offspring O_2;

5:Interchange the mesh router nodes: Move the mesh router nodes of RP_1 to RO_2 and those of RP_2 to RO_1;

6:Re-establish mesh nodes network connections in O_1 and O_2 (links between mesh router nodes and links between client mesh nodes and mesh router nodes are computed again);

7:**return** O_1 and O_2;

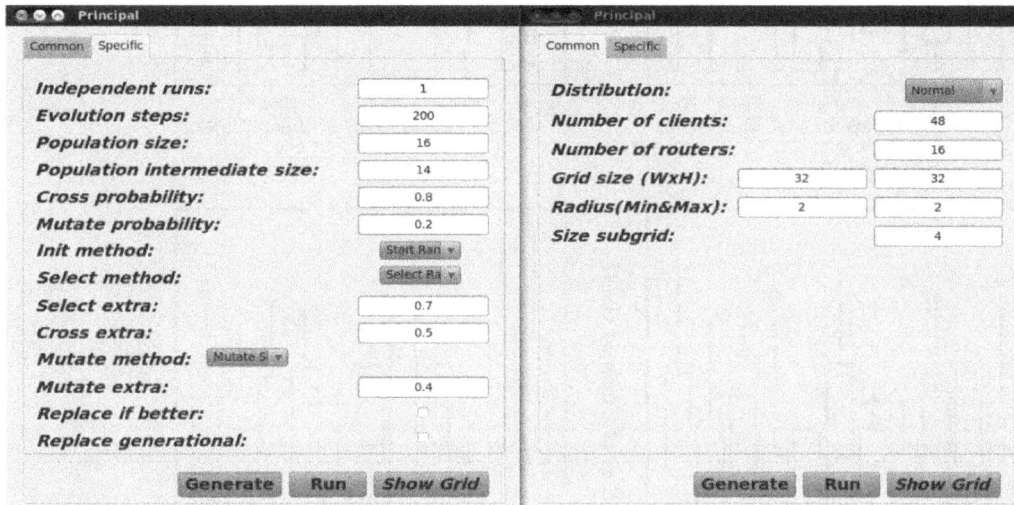

Fig. 5. GUI tool for WMN-GA system.

The GUI interface of WMN-GA is shown in Fig. 5. The left side of the interface shows the GA parameters configuration and on the right side are shown the network configuration parameters.

For the network configuration, we use: distribution, number of clients, number of mesh routers, grid size, radious of transmission distance and the size of subgrid.

For the GA parameter configuration, we use: number of independent runs, GA evalution steps, population size, population intermediate size, crossover probability, mutation probability, initial methods, select method.

5. Simulation results

We carried out many simulations to evaluate the performance of WMNs using WMN-GA system.

In this work, we considered the number of covered users parameter. The number of Mesh Routers is considered 16. We take in consideration four distribution methods: Exponetial, Normal, Uniform and Weibull. Six selection operators are used: Best, Exponential Ranking, Linear Ranking, Binary Tournament, N Tournament and Random.

In Figs 6, 7, 8 and 9 are used box plots to analyse the range of data values. Each box plot shows the median of each simulation data (the median is noted by the notch), the upper quartile (25 th and 75 th percentile, respectively) and the outliers. The whiskers rapresent the lowest data which is still within

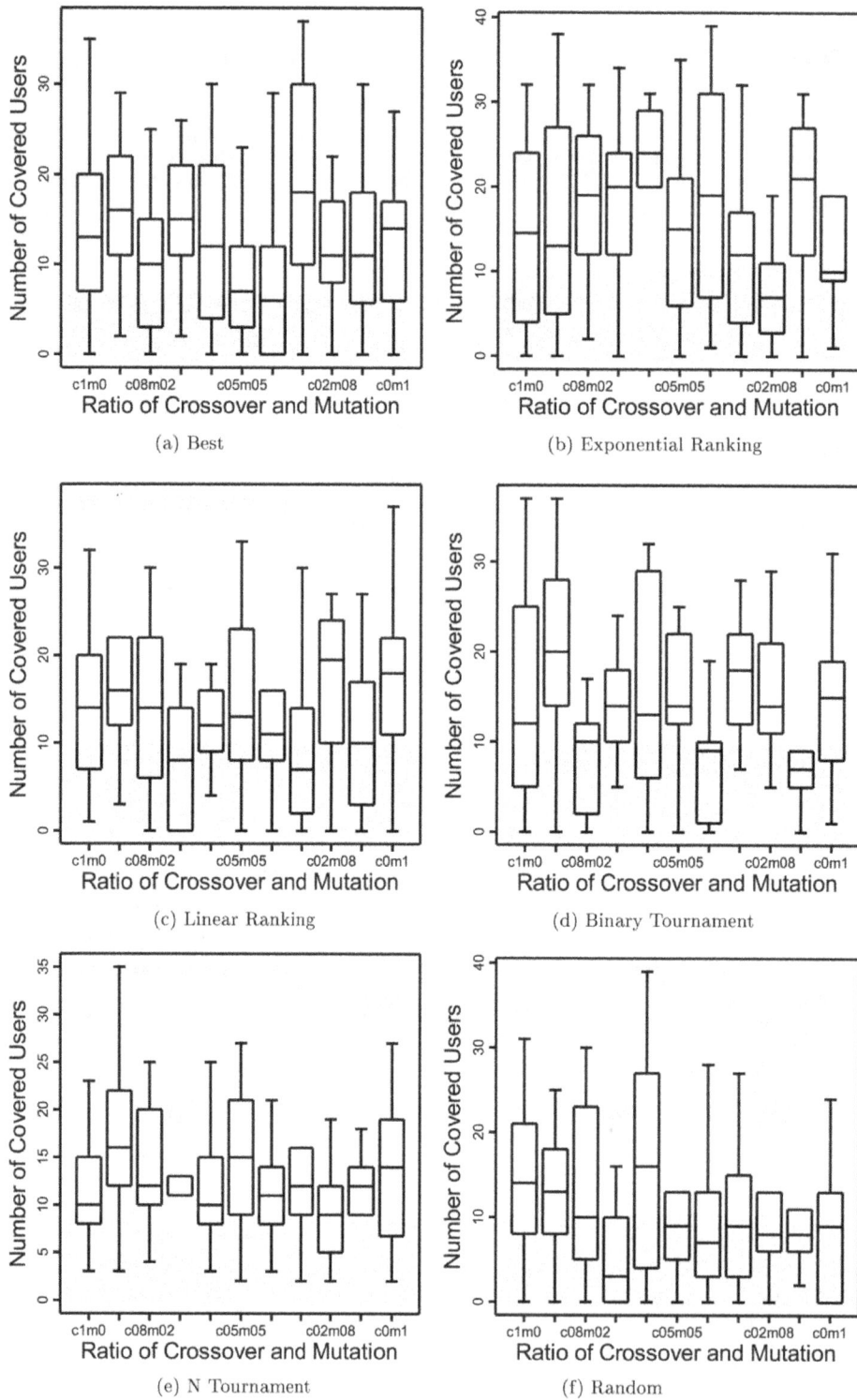

(a) Best

(b) Exponential Ranking

(c) Linear Ranking

(d) Binary Tournament

(e) N Tournament

(f) Random

Fig. 6. Number of covered users vs. ratio of crossover and mutation for exponential distribution.

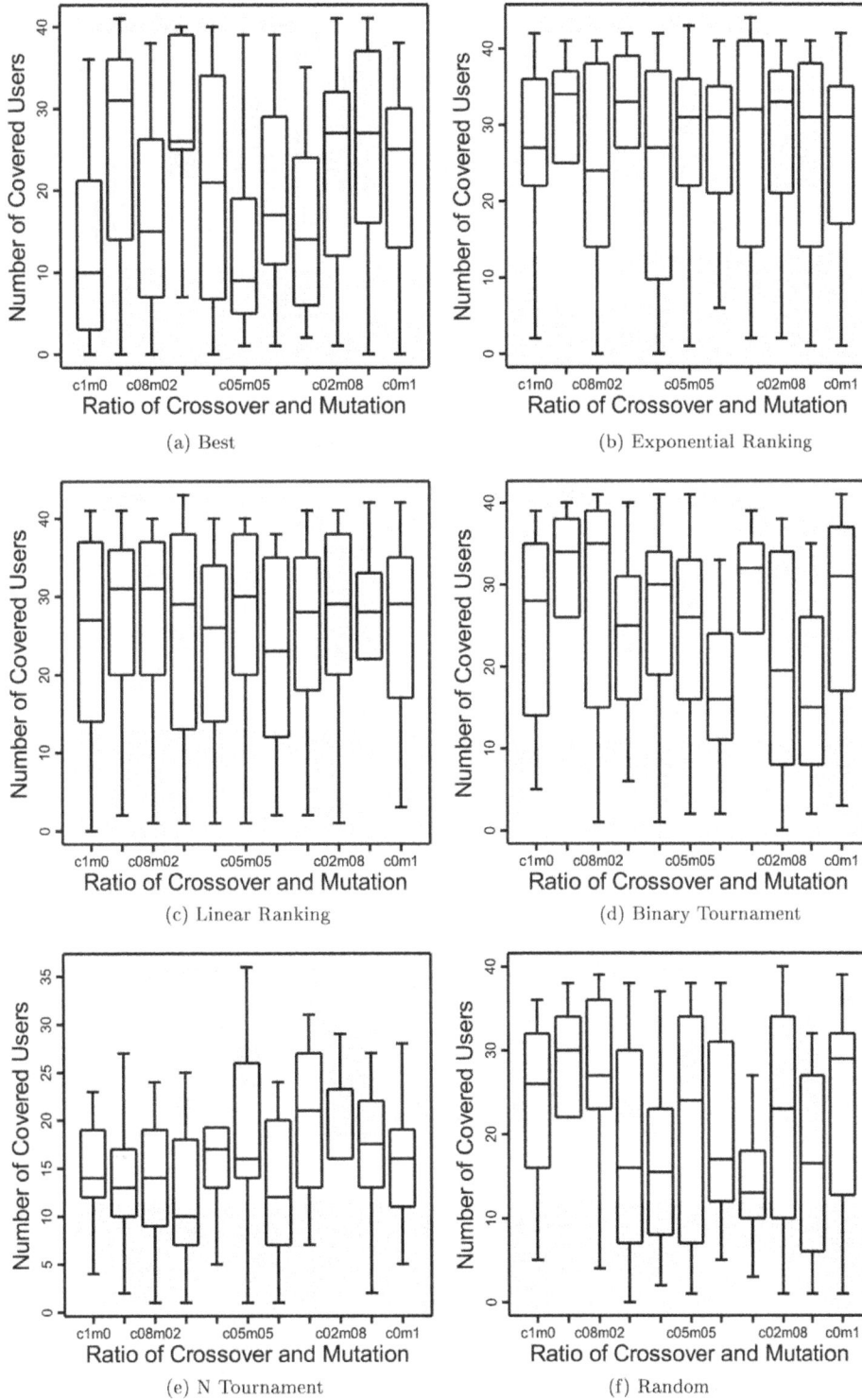

Fig. 7. Number of covered users vs. ratio of crossover and mutation for normal distribution.

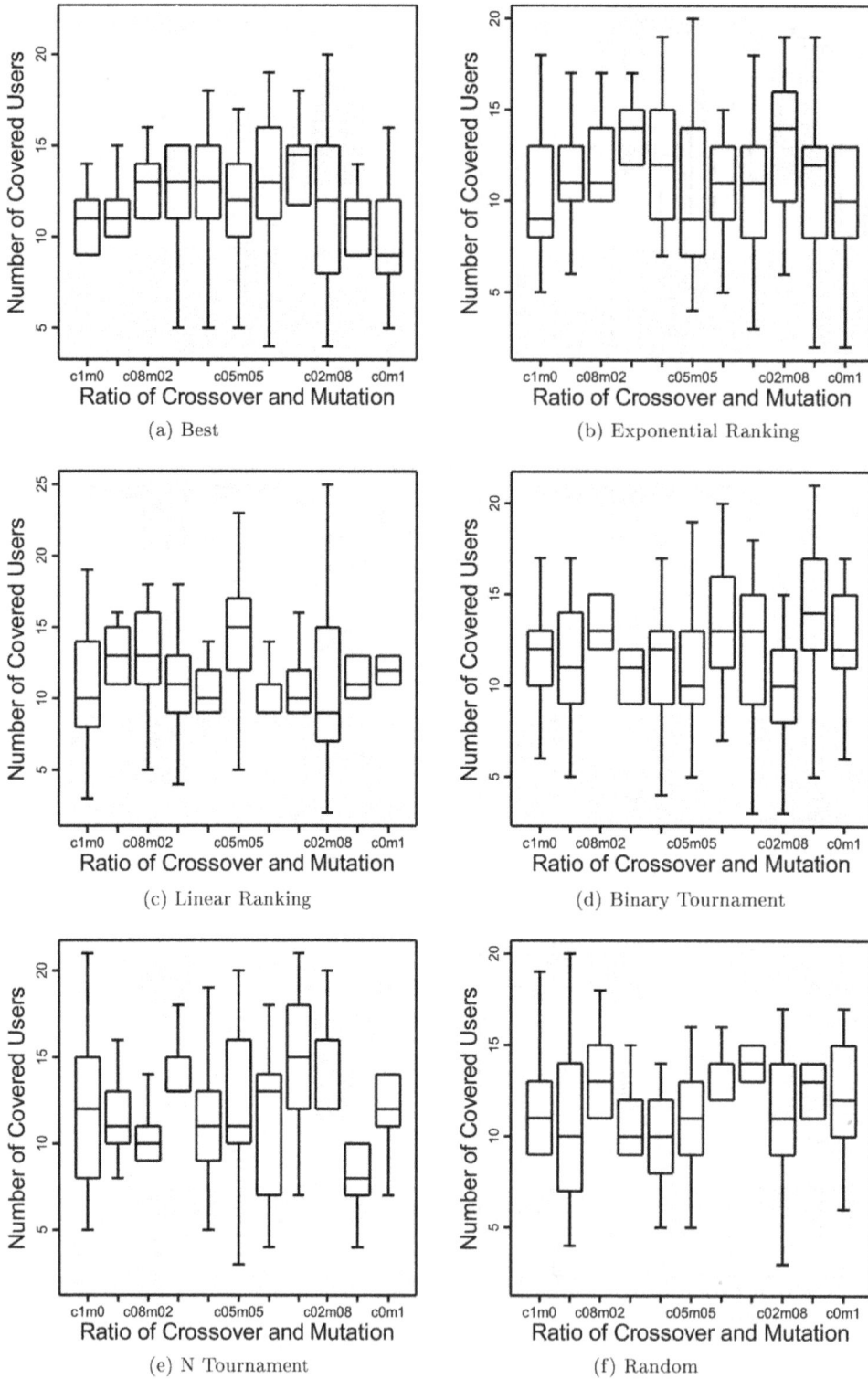

Fig. 8. Number of covered users vs. ratio of crossover and mutation for uniform distribution.

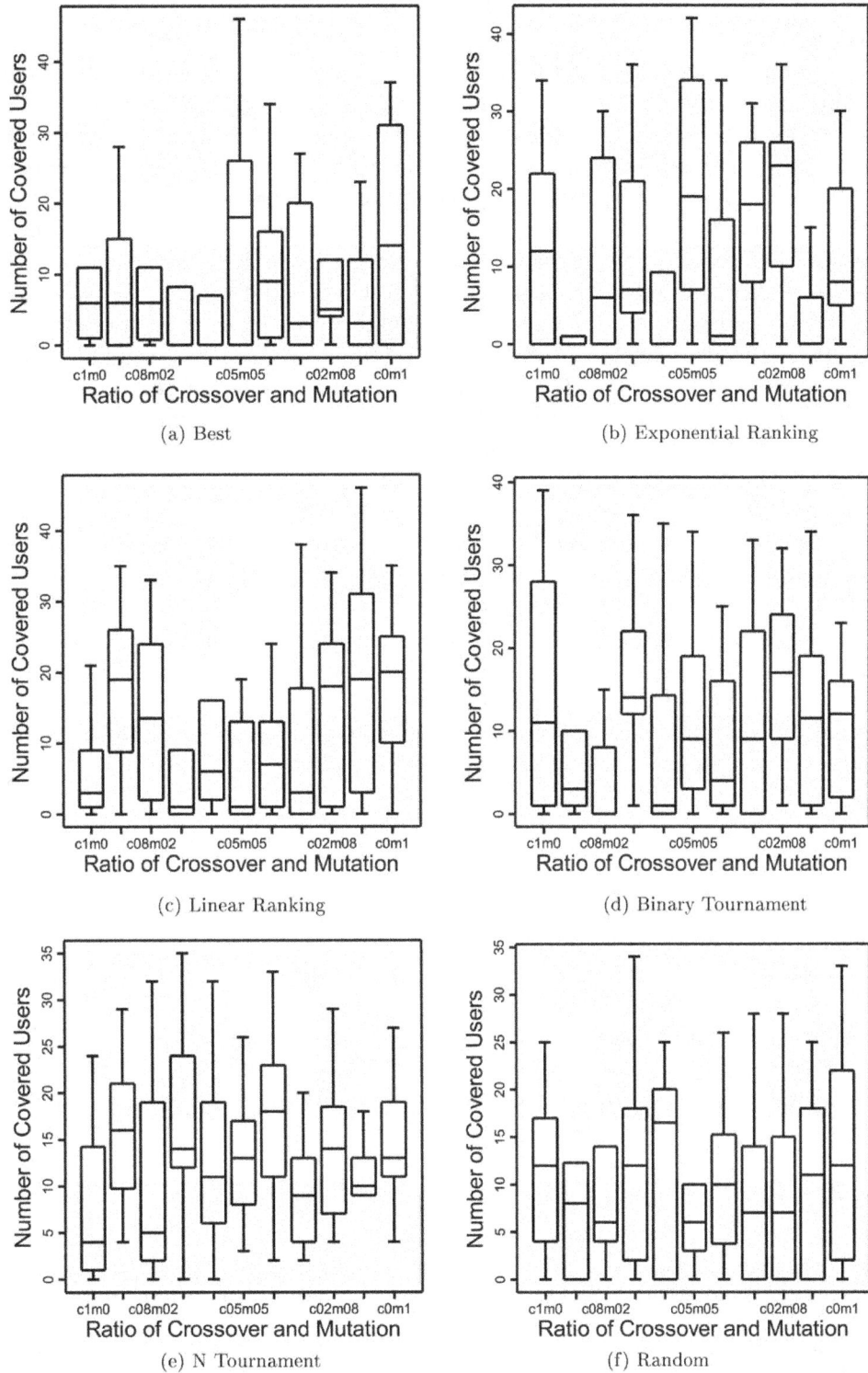

(a) Best

(b) Exponential Ranking

(c) Linear Ranking

(d) Binary Tournament

(e) N Tournament

(f) Random

Fig. 9. Number of covered users vs. ratio of crossover and mutation for weibull distribution.

1.5 inter-quartile range of the lowest quartile and the highest data which is still within 1.5 inter-quartile range of the upper quartile.

In Fig. 6 is shown the number of covered users vs. ratio of crossover and mutation for exponential distribution. The maximal number of covered users among these selection operators is achieved by Exponential Ranking operator. The maximum number of covered users is 25. From the results, we can see that Exponential Ranking has better performance. Also, Linear Ranking and Binary Tournament have better behaviour than Best, N Tournament and Random selection operators.

In Fig. 7 is shown the number of covered users vs. the ratio of crossover and mutation for normal distribution. From the results, we can see that Exponential Ranking and Linear Ranking operators have almost the same performance, the number of covered users is almost stable for different rates of crossover and mutation, but for $c = 90\%$ and $m = 10\%$ the Exponential Ranking has a bigger number of covered users (35 users). Best operator reaches the best coverage for $c = 90\%$ and $m = 10\%$ but for the other rates it covers only few users. For the Binary Tournament the maximal number of covered users is the same with Exponential Ranking but in Exponential Ranking the number of covered users is more stable for all rates of crossover and mutation. N Tournament operator has a bad coverage, the number of covered users is very small. For Random operator the number of covered users is bigger for $c = 90\%$ and $m = 10\%$ but for other rates it covers only a small number of users.

In Fig. 8 is shown the number of covered users vs. the ratio of crossover and mutation for uniform distribution. As can be seen, the performance for all these selection operators is almost the same and it is not good because the number of the covered users is less than 15 users.

In Fig. 9 is shown the number of coverd users vs. the ratio of crossover and mutation for weibull distribution. The maximum number of covered users is 23 users and is achieved when $c = 20\%$ and $m = 80\%$. Other operators offer a small number of covered users.

From all simulation results, we can see that for normal distribution, different rates of crossover and mutation, and different selection operators, the best performance is for Exponential Ranking.

6. Conclusions

In this work, we have proposed and implemented a system based on GAs to solve the connectivity problem in WMNs. We call this system WMN-GA.

We evaluated the performance of the proposed system using different genetic operators and different distributions of router nodes considering number of covered users parameter.

The simulation results show that for Exponential and Linear Ranking methods, the system has a good performance.

In the future work, we would like to compare the performance of WMN-GA system with other optimization algorithms.

References

[1] E. Kulla, M. Hiyama, M. Ikeda, L. Barolli, V. Kolici and R. Miho, MANET Performance for Source and Destination Moving Scenarios Considering OLSR and AODV Protocols, *Mobile Information Systems (MIS), IOS Press* **6**(4) (2010), 325–339.

[2] A. Durresi, M. Durresi and L. Barolli, Secure Authentication in Heterogeneous Wireless Networks, *Mobile Information Systems (MIS)* **4**(2) (2008), 119–130.

[3] L. Barolli, A Speed-Aware Handover System for Wireless Cellular Networks Based on Fuzzy Logic, *Mobile Information Systems (MIS), IOS Press* **4**(1) (2008), 1–12.

[4] I.F. Akyildiz, X. Wang and W. Wang, Wireless Mesh Networks: A Survey, *Computer Networks* **47**(4) (2005), 445–487.

[5] N. Nandiraju, D. Nandiraju, L. Santhanama, B. He, J. Wang and D. Agrawal, Wireless mesh networks: Current challenges and future direction of web-in-the-sky, *IEEE Wireless Communications*, 2007, pp. 79–89.

[6] C. Chen and C. Chekuri, Urban Wireless Mesh Network Planning: The Case of Directional Antennas, *Tech Report* No. UIUCDCS-R-2007-2874, 2007.

[7] M.R. Garey and D.S. Johnson, Computers and Intractability -A Guide to the Theory of NP-Completeness, *Freeman*, San Francisco, 1979.

[8] A. Lim, B. Rodrigues, F. Wang and Zh. Xua, $k-$Center Problems with Minimum Coverage, *Theoretical Computer Science* **332**(1–3) (2005), 1–17.

[9] E. Amaldi, A. Capone, M. Cesana, I. Filippini and F. Malucelli, Optimization Models and Methods for Planning Wireless Mesh Networks, *Computer Networks* **52** (2008), 2159–2171.

[10] J. Wang, B. Xie, K. Cai and D.P. Agrawal, Efficient Mesh Router Placement in Wireless Mesh Networks, *MASS*, Pisa, Italy, 2007, pp. 9–11.

[11] S.N. Muthaiah and C. Rosenberg, Single Gateway Placement in Wireless Mesh Networks, *In Proc. of 8th International IEEE Symposium on Computer Networks*, Turkey, 2008, pp. 4754–4759.

[12] P. Zhou, B.S. Manoj and R. Rao, A Gateway Placement Algorithm in Wireless Mesh Networks, *In Proc. of the 3rd International Conference on Wireless Internet*, Austin, Texas, 2007, pp. 1–9.

[13] M. Tang, Gateways Placement in Backbone Wireless Mesh Networks, *International Journal of Communications, Network and System Sciences* **2**(1) (2009), 45–50.

[14] A. Antony Franklin and C. Siva Ram Murthy, Node Placement Algorithm for Deployment of Two-Tier Wireless Mesh Networks, *In Proc. of IEEE GLOBECOM-2007*, Washington, USA, 2007, pp. 4823–4827,

[15] T. Vanhatupa, M. Hännikäinen and T.D. Hämäläinen, Genetic Algorithm to Optimize Node Placement and Configuration for WLAN Planning, *In Proc. of 4th International Symposium on Wireless Communication Systems*, 2007, pp. 612–616.

[16] J. Holland, Adaptation in Natural and Artificial Systems, *University of Michigan Press*, Ann Arbor, 1975.

[17] A. Barolli, E. Spaho, L. Barolli, F. Xhafa and M. Takizawa, QoS Routing in Ad-hoc Networks Using GA and Multi-objective Optimization, *Mobile Information Systems (MIS), IOS Press* **7**(3) (2011), 169–188.

[18] G. Ochoa, Setting The Mutation Rate: Scope And Limitations Of The 1/L Heuristic, *In Proc. of the Genetic and Evolutionary Computation Conference (GECCO-2002)*, USA, 2002, pp. 495–502.

[19] J. Cervantes and C.R. Stephens, Limitations of Existing Mutation Rate Heuristics and How a Rank GA Overcomes Them, *IEEE Transactions on Evolutionary Computation* **13**(2) (2009), 369–397.

[20] F. Xhafa, Ch. Sanchez and L. Barolli, Genetic Algorithms for Efficient Placement of Router Nodes in Wireless Mesh Networks, *Proc. of AINA 2010*, 2010, pp. 465–472.

Tetsuya Oda received BE in Information and Communication Engineering from Fukuoka Institute of Technology (FIT), Japan in 2010. Presently, he is a Master Student at Graduate School of Engineering, FIT, Japan. His current research interests include wireless networks, mesh networks, ad-hoc networks and sensor networks.

Admir Barolli was graduated from Agricultural University of Tirana, Albania. He received his Diploma Degree in April 2008. From October 2009 to June 2010, he was a Visiting Researcher at Curtin University of Technology, Australia. Presently, he is a Visiting Researcher at the Department of Computer and Information Science, Seikei University, Japan and working toward his PhD Degree. He has been working for many years in Agriculture Sectors in Albania. His research interest are in genetics, genetic algorithms, agricultural engineering, intelligent algorithms, climate change, global warming, computer networks, ad-hoc networks, mesh networks and P2P systems.

Fatos Xhafa joined the Department of Languages and Informatics Systems of the Technical University of Catalonia as an Assistant Professor in 1996 and is currently an Associate Professor and member of the ALB- COM Research Group of this department. His current research interests include parallel algorithms, combinatorial optimization, approximation and meta-heuristics, distributed programming, grid and P2P computing. He has published in leading international journals and conferences and has served in the organizing committees of many conferences and workshops. He is also a member of the editorial board of several international journals.

Leonard Barolli is a Professor at the Department of Information and Communication Engineering, Fukuoka Institute of Technology (FIT), Japan. He received BE and PhD Degrees from Tirana University and Yamagata University in 1989 and 1997, respectively. He has published about 350 papers in Journals, Books and International Conference. He has served as a Guest Editor for many Journals. He was PC Chair of IEEE AINA-2004 and ICPADS-2005. He was General Co-Chair of IEEE AINA-2006, AINA-2008 and AINA-2010. He is Steering Committee Chair of CISIS and BWCCA International Conferences. His research interests include, P2P, intelligent algorithms, ad-hoc and sensor networks. He is a member of IEEE, IEEE Computer Society, IPSJ and SOFT.

Makoto Ikeda is an Assistant Professor, at Fukuoka Institute of Technology (FIT), Japan. He was an Assistant Research Fellow at the Center for Asian and Pacific Studies, Seikei University, Japan from April 2010 to March 2011. He received BE, MS and PhD degrees in Information and Communication Engineering, Fukuoka Institute of Technology (FIT), Japan, in 2005, 2007, and 2010, respectively. He was a Japan Society for the Promotion of Science (JSPS) Research Fellow from April 2008 to March 2010. Dr. Ikeda has widely published in peer reviewed international journals and international conferences proceedings. He has served as PC Members for many international conferences. He is a member of IEEE, ACM, IPSJ and IEICE. His research interests include wireless networks, mobile computing, high-speed networks, P2P systems, mobile ad-hoc networks, wireless sensor networks and vehicular networks.

Makoto Takizawa is a Professor at the Department of Computer and Information Science, Seikei University. He was a Professor and the Dean of the Graduate School of Science and Engineering, Tokyo Denki University. He was a Visiting Professor at GMD-IPSI, Keele University, and Xidian University. He was on the Board of Governors and a Golden Core member of IEEE CS and is a fellow of IPSJ. He received his DE in Computer Science from Tohoku University. He chaired many international conferences like IEEE ICDCS, ICPADS, and DEXA. He is the founder of IEEE AINA. His research interests include distributed systems and computer networks.

10

BGN: A novel scatternet formation algorithm for bluetooth-based sensor networks

Yongjun Li[a], Hu Chen[a], Rong Xie[a] and James Z. Wang[b,*]

[a]*School of Computer Science & Engineering, South China University of Technology, Guangzhou, 510640, China*

[b]*School of Computing, Clemson University, Clemson, SC 29634, USA*

Abstract. To address the unreliability and inefficiency of the existing algorithms for Bluetooth scatternet formation, this paper proposes a new efficient algorithm, Bluetooth Growing Network (BGN) algorithm, which constructs Bluetooth scatternet by adding reserve links among its branch nodes so that the resulting scatternet can maintain a high degree of connectivity while avoiding exccessive redundant links. Extensive simulation results demonstrate that the scatternet constructed by BGN algorithm is efficient in terms of the number of bridge nodes, average data transmission distance, network reliablity and overall network traffic.

Keywords: Wireless sensor network, BGT, BGN, scatternet

1. Introduction

In recent years, advances in wireless technology and mobile applications have made mobile devices an essential part of people's daily life. Many mobile applications, such as mobile databases and navigation [1–5], have been widely adopted. Recent advances in wireless ad-hoc and sensor networks further expand the mobile application arena because mobile devices can form personal area networks (PANs) without using the base stations and wired networks. More and more mobile applications [6–9], such as healthcare, fitness, gaming, etc. are now based on wireless ad-hoc and sensor networks. Bluetooth is a wireless protocol utilizing short-range communication technologies to facilitate data transmission over short distances between fixed and/or mobile devices. Because of its excellent wireless properties, such as low cost, low power consumption and good anti-jamming performance, and its advantages in network facilitation, Bluetooth has become an attractive communication protocol for wireless ad-hoc and sensor networks.

To construct a Bluetooth network using Bluetooth protocol, devices are first grouped into piconets, each of which consists of no more than eight devices due to a 3-bit address space. These piconets then interconnect with each other by one member which is also another piconet's member. The device participating in two piconets forms a "bridge" between these two piconets and relays data between their members. All interconnected piconets form a scatternet to support communication among devices in the network.

---

*Corresponding author. E-mail: jzwang@cs.clemson.edu.

A tough challenge for a scatternet formation algorithm is to find an optimal network configuration in terms of the number of bridges and the connectivity degree of each bridge to achieve the best network performance without affecting the reliability and connectivity of the network. In general, the number of bridges and the connectivity degree of each bridge affect the performance and survival time of the Bluetooth-based sensor networks. With more bridges, the network is usually more reliable and the average data transmission distance is shorter. However, the bridge node is responsible of forwarding packets between two or more piconets in a time-shared manner. Thus, the bridge nodes usually have shorter survival time because they consume more energy due to the extra communication responsibilities. Therefore, too many bridge nodes often result in the decline of the network overall survival time.

To address the weaknesses of the existing algorithms, this paper proposes a novel algorithm, Bluetooth Growing Network (BGN) algorithm. This algorithm constructs the scatternet from a tree structure by adding reserve links among its branch nodes. It has the merits of both the mesh-based and the tree-based Scatternet formation algorithms. The resulting scatternet can maintain a high degree of connectivity while avoiding excessive redundant links. Our simulation studies demonstrate that this algorithm outperforms existing algorithms in terms of the number of bridges, reliability, and average transmission distance.

The rest of this paper is organized as follows. We first discuss related work in Section 2. Then, in Section 3, we define the Bluetooth Growing Tree (BGT) and how to construct the BGT, and propose the BGN scatternet formation algorithm. In Section 4, we compare the performance of our BGN algorithm with the tree-based and mesh-based algorithms through extensive simulation studies. Finally, we give concluding remarks and discuss the future work in Section 5.

2. Related work

Recently, researchers proposed many scatternet formation algorithms which can be roughly divided into three categories: tree-based algorithms [10–13], ring-based algorithms [14–16] and mesh-based algorithms [17–22].

The scatternet constructed by a tree-based algorithm has a single root node and descendant tree nodes. A bottleneck can easily form at the trunk node of the tree, although such scatternet has a logarithmic average path length and its data routing algorithm is very simple. Zaruba et al. [10] proposed a tree-based algorithm (BlueTree Algorithm), which utilizes tree structure characteristics (N devices only need N-1 links to interconnect with each other) to minimize the number of communication links. However, Not only such a tree structure has poor reliability, but also bridges can easily become the communication bottleneck. Rashid Bin Muhammad [11] proposed a method for range assignment problem on the Steiner tree based topology, but the method should throw away nodes that are too far from neighbor nodes which lead to too many dense nodes for the connectivity. Tang et al. [12] proposed a tree-based heuristic algorithm for mobile and pervasive computing based on special network processors. Saginbekov et al. [13] also proposed a tree-based shortest path algorithm, trying to realize the energy saving through minimizing the energy consumption during communication.

Ring-based algorithms try to alleviate the bottleneck problems in tree-based algorithms while maintaining simple scatternet routing properties. However, they suffer from network partition problems and significantly longer average path length. Huang et al. [16] presented a ring structure called Solidring and a routing algorithm to lower transmission time delay. But the assumption of devices can communicate directly with one another having strict restriction.

Thomas et al. [17] proposed an algorithm (BlueNet Algorithm) based on a mesh topology to construct a smooth network architecture to enhance network reliability. But due to the large number of piconets and redundant communication links, excessive interference among piconets often leads to low communication performance of the self-organized network. Song et al. [18] adopt the well-known de Bruijn graph [19] to form the backbone of Bluetooth scatternet, in which for every pair of nodes it can find a path with at most O(log n) hops without any routing table and the congestion of every node is at most $O(\frac{\log n}{n})$. Chang et al. [20] proposed a traffic-aware restructure protocol for partially restructuring a piconet or a pair of two neighboring piconets by applying role switch mechanism to reduce the route path length, however, the frequent role switch will result in more transmission time delay. Zaguia et al. [21] introduced a new protocol based on maximal independent sets (BlueMis) attempt to simplify the BlueMesh procedure [22], but the number of slaves for each node is not limited, and in some cases, e.g. complete graph, one node may be selected as slave to all other nodes.

3. The BGN algorithm

In this section, after defining the Bluetooth Growing Tree and discussing the basic ideas on how to construct the scatternet using BGT, we propose a distributed implement of our BGN algorithm.

3.1. Definitions

Definition 1 (Piconet): A piconet is a network consisting of a master node and at most seven slave nodes dominated by this master node. A piconet P_i can be represented as:

$$P_i = \{(M_i, S_{i,j}) | 1 \leqslant j \leqslant 7\},$$

where M_i is the master node and $S_{i,j}$ is the slave nodes dominated by M_i.

Definition 2 (Scatternet): A scatternet is formed by two or more independent, asynchronous piconets overlapped in time and space. Formally, a scatternet can be presented as:

$$S = \{\cup P_i\}, i \in [1, n], \quad n \geqslant 2,$$

where $\forall k, j, k \neq j \wedge \exists P_k \xrightarrow{path} P_j$, i.e., given any two piconets P_k and P_j in S, there exists a path from P_k to P_j.

Definition 3 (Bridge): A node in a scatternet S is a bridge node if and only if it is shared by two or more piconets. It receives data from one piconet and forwards the data to another piconet. A bridge node can be slave nodes in two separate piconets (slave/slave bridge), or a master node in one piconet and a slave node in another piconet (master/slave bridge). A set of bridge nodes in a scatternet S can be presented as: $B = \{t_k | t_k \in P_i \cap P_j\}$, where $i \neq j$ and $P_i, P_j \in S$.

As shown in Fig. 1, piconet P1, P2 and P3 are partially overlapped in time and space, where M_2 is the master node in piconet P2 and it controls the slave node S_1 and bridge nodes B_{21} and B_{23}.

3.2. Growing tree based bluetooth scatternet

In this subsection, we introduce the concept of Bluetooth Growing Tree and discuss its construction process.

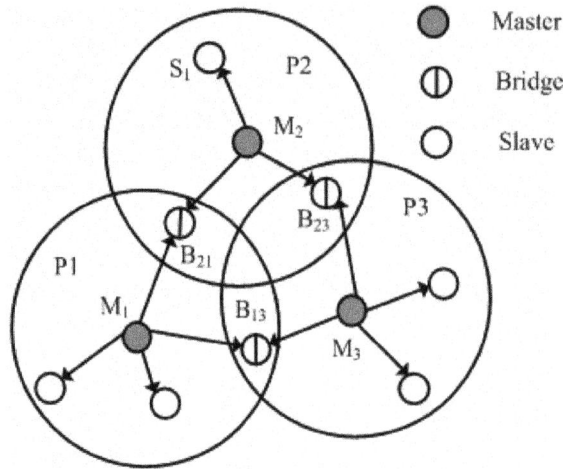

Fig. 1. The topology of a Bluetooth scatternet.

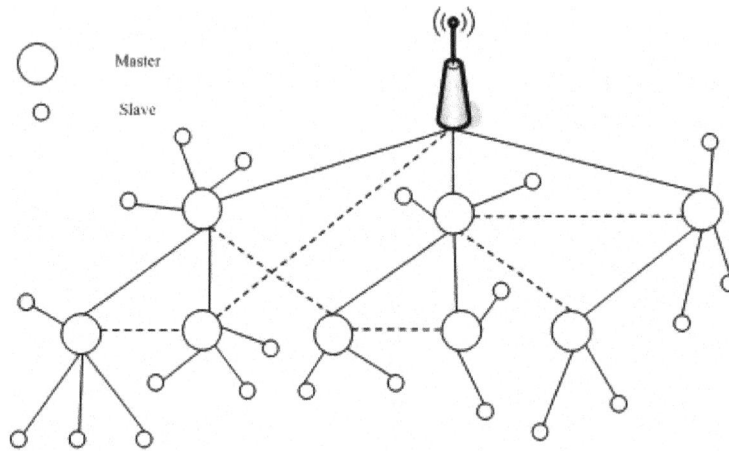

Fig. 2. The structure of a Bluetooth growing tree.

3.2.1. Bluetooth growing tree construction

The BGT construction process starts with a random selection of a seed node. This seed node first selects up to 7 of its neighboring nodes to become its children. These child nodes then serve as seed nodes to find their neighboring nodes which are not yet selected by other nodes to be their children respectively. Once a node is selected to be a child by another one, it becomes a seed node and can compete with other nodes to select available nodes as its children. This process continues until all nodes are connected. Because this construction process essentially grows a tree through constantly bifurcating, we call the resulting tree as a Bluetooth Growing Tree (BGT), as shown in Fig. 2.

In Fig. 2, the topology formed by solid lines and their associated nodes is a BGT, in which all leaf nodes are slave nodes in their associated Piconet. A branch node plays two roles. It is a slave node in its parent Piconet, which is formed by its parent node during the growing tree process. On the other hand, it is a master node of the Piconet formed by this branch node when it grows the tree to its direct neighbors.

3.2.2. Scatternet formation through BGT

Using the Bluetooth protocol, one device may directly connect with up to seven neighboring devices. Therefore, it is possible that some branch nodes may only used a portion of their available wireless capacities during the BGT construction. Therefore, it is possible to add more links between the branch nodes as shown by the dotted lines in Fig. 2.

Adding links between the branch nodes makes the BGT a mesh-like scatternet, which is much more reliable than the BGT. In addition, edges between the branch nodes can help the traffic routing in the scatternet. Since this scatternet is constructed through a Bluetooth Growing Tree, we call it a Bluetooth Growing Network (BGN).

After adding the links between branch nodes, the scatternet grows into a mesh structure from the original BGT structure. Although, the number of backbone nodes remains the same, the reliability and throughput of the scatternet has significantly improved over the original BGT structure.

Definition 4: Given graph $G(V, E)$ and its Growing Tree $T(V', E')(V' \subseteq V, E' \subseteq E)$, the BGN is defined as:

$$BGN = \{T \cup e(u, v) | e(u, v) \in E - E', u, v \in M,$$
$$\deg ree(u) \leqslant 7, \deg ree(v) \leqslant 7\}$$

where M is the set of branch nodes of T, and $degree(u)$, $degree(v)$ are the numbers of connections associated with the branch nodes u and v respectively.

Definition 5: Candidate Edge: Given a Growing Tree T, the set of candidate edges for a branch node u is:

$$ce_u = \{e_{u,v} | length(e_{u,v}) \leqslant R, u \in M, v \in M\},$$

where R is the communication radius of the node, and M is the set of branch nodes.

3.2.3. Link capacity reservation

There are some restrictions for BGT growing into BGN. Based on Definition 5, as long as the wireless distance between two branch nodes is less than or equal to R, there exists a candidate edge. However, the maximum link capacity of each node is 7. If any branch node has reached its link capacity, no candidate edge can be added between this node and any other branch nodes. Since the BGT construction process is a competing process, it is highly possible a branch node has connected with 7 child nodes during the BGT growing process. This situation happens more often when the node density is high, limiting the number of candidate edges that can be added to the BGN.

To allow a BGT to grow into a BGN, we can limit the number of links of a branch node during the BGT growing process. That is, we reserve some link capacity for a branch node to add links to other branch ones. A smaller link capacity reservation for a branch node means it has less capacity to add links to other branch nodes and the resulting BGN is less reliable. On the contrary, a larger link capacity reservation for a branch node means it has more capacity to add links to other branch nodes. Having more links among the branch nodes will lead to more bridge nodes and affect the survival time of the BGN.

3.3. BGN algorithm

The BGN algorithm consists of two phases: (1) constructing the BGT, and (2) adding candidate edges between branch nodes.

The Bluetooth growing tree construction starts from a root node, called blueroot. This root node pages its neighbors one by one and becomes the master of the nodes that respond the paging messages. If a node is paged and is not yet a part of any piconet, it responds the paging message and becomes the slave of the paging node; otherwise, it may not respond to the page or may respond the paging with an "already taken" message to stop further paging messages sent to it. Once a node becomes a slave in a piconet, it starts paging all its neighbors one by one. This procedure is recursively repeated until all nodes join a piconet [23].

After completing the Bluetooth growing tree construction, each branch node runs the following BGN procedures to add candidate edges [24,25]. These procedures are essentially a competition process. The delay time for the branch node i during the competition is defined as:

$$ t = \frac{1 + |B_i|}{2T} \tag{1} $$

where T is competition cyclic time and B_i is the number of piconets connected of nodes i.

Procedure for all nodes during time period T_q
s.1 if the number of slave $\geqslant$ 7 then exit;
s.2 $t \leftarrow (1+Bi)/2T$
s.3 schedule a delay in t second
s.4 if receive any QUERY(ALL,sender) msg in t second then
        cancel the scheduler
        parent←sender
        role←loser
    else
        broadcast Query(ALL,myaddr) msg
        role←winner;
s.5 wait until T_j period arrives

Procedure for nodes where role = winner during time period T_j
s.1 if receive JOIN(parent, sender) msg && sender not in B_i
        add a link to the sender
        $B_i \leftarrow B_i \cup$ sender
        Send Accept(sender,myaddr) msg
s.2 if receive none of JOIN msg then
        exit
    else
        wait until next T_q period arrive

Procedure for nodes where role = loser during time period T_j
s.1 $t \leftarrow$ rand(0,T_j/2)
s.2 delay t second
s.3 send JOIN(parent,myaddr) msg
s.4 if receive ACCEPT msg from parent then
        add a link to parent
        $B_i \leftarrow B_i \cup$ sender
s.5 wait until next T_q period arrives

As shown in the above procedures, the BGN algorithm is a competition-based algorithm. It contains a query period T_q and a join period T_j in each competition cycle. Usually we have $T_q = T_j$. A branch node competes for sending the QUERY message in T_q and the JOIN message in T_j.

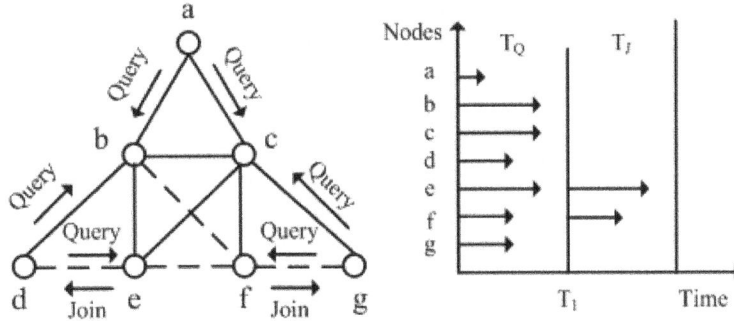

Fig. 3. An example of competition-based candidate edge adding.

During the query period T_q, each branch node computes its delay time t using Formula Eq. (1). If a branch node does not receive any QUERY message from its neighboring branch nodes in its delay time t, it is the winner in competition and will broadcast QUERY messages to its neighboring nodes.

Nodes receiving QUERY messages from other nodes are losers and will not broadcast QUERY messages to other nodes. Instead, these losers will send JOIN messages to the winner during period T_j. The winner will send an ACCEPT message to the node that sends a JOIN message to it the first time and is not subordinate to it. It will ignore other JOIN requests. A node receiving an ACCEPT message will add itself to the Piconet mastered by the node that sends the ACCEPT message, forming a link between two branch nodes. Each branch node repeats this procedure until the note reaches its link capacity or no JOIN message is received from its neighboring nodes.

Figure 3 shows an example of adding candidate edges among branch nodes. In Fig. 3, a solid line means that the connection is already established, while the dotted line represents a candidate edge. As shown in the right part of Fig. 3, each branch node computes its delay time according to Formula 1 during period T_q. Node a has the shortest delay time, and nodes b, c, e all have the longest delay time. During period T_q, only nodes a, d, f and g succeed in competition. Other nodes receive QUERY messages broadcasted by nodes a, d, f and g and will not send QUERY messages. During period T_j, nodes that received QUERY messages will check whether they have connected to source nodes, if not they will send JOIN message to source nodes. In Fig. 3, after competition, candidate edge (d, e) and (f, g) will be added to the BGN.

4. Performance evaluation

In this section, we use simulation to evaluate the performance of our BGN algorithm. We compare the scatternets constructed by our algorithm with those constructed by Bluetree [10] and BlueNet [13] algorithms respectively in terms of the number of candidate edges, the number of backbone nodes, the average transmission distance, network reliability and throughput. We conduct the simulation using the NS-2 network simulator. We assume there are N nodes randomly distributed in a 100 m × 100 m square area and each node has communication radius R = 40 m. We vary N from 70 to 250 and observe the characteristics of the scatternets constructed by three different algorithms. For each fixed N, we repeat the simulation for 100 times and report the average values of these 100 simulation runs.

4.1. The optimal link capacity reservation

We first find the optimal link capacity reservation value θ and investigate its impact to the performance of our proposed BGN algorithm.

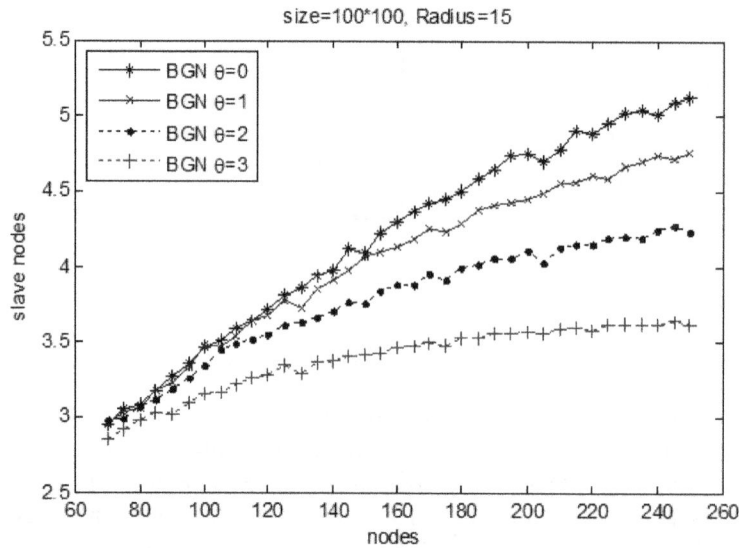

Fig. 4. The number of average slave nodes of piconet of Bluetooth growing tree.

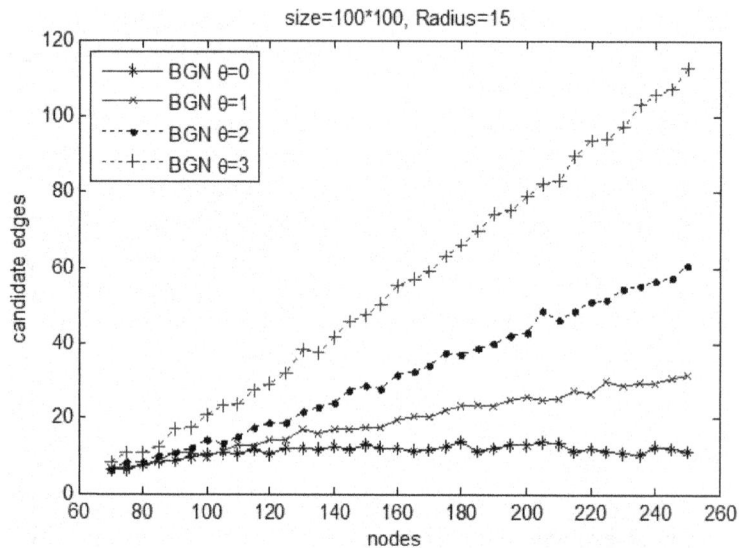

Fig. 5. The number of candidate edges of Bluetooth growing tree.

Figure 4 depicts the average number of slave nodes per piconet in the BGT under different θ values. Although the average number of slave nodes increases with the node density increases, this number does not exceed 5.5 when total number of network nodes in a 100 m × 100 m area reaches 250. This means there is still available wireless capacity in branch nodes to add candidate edges. In general, the greater the θ value, the more available wireless links are reserved for candidate edges.

Figure 5 shows the number of candidate edges in the BGN under different θ values. When the node density is low (N ⩽ 120), the available candidate edges are limited because the physical distances between branch nodes are high. Therefore, the numbers of candidate edges added to the BGN are very close when $\theta = 0$, 1 and 2. But as the node density increases, the physical distances between the branch

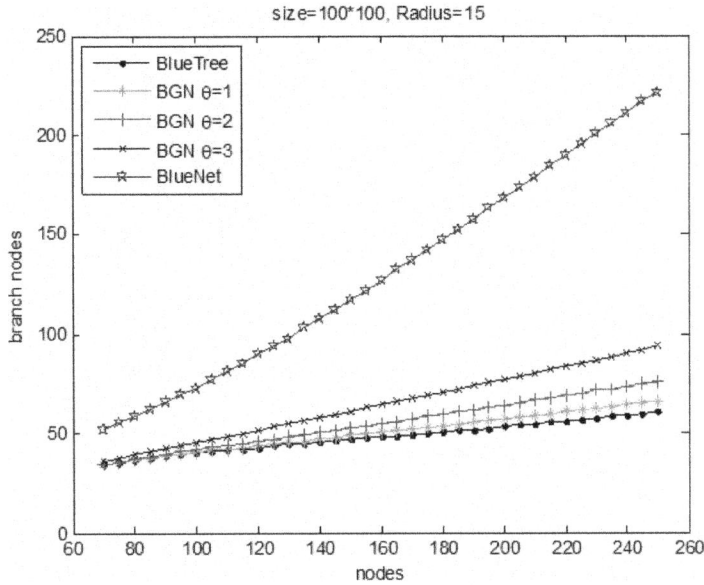

Fig. 6. The number of branch nodes of BlueTree, BlueNet and BGN.

nodes decrease, resulting more available candidate edges among the branch nodes. Therefore, with a higher link capacity reservation, more candidate edges can be added into the BGN. For example, when $N = 250$, given a link capacity reservation value $\theta = 3$, 112 candidate edges can be added. Meanwhile, when $\theta = 2$ and $\theta = 1$, only 60 and 38 candidate edges can be added respectively. Thus, when the node density is high, the link capacity reservation value θ has a significant impact on the number of candidate edges that can be added to the BGN. If we don't reserve any link capacity, i.e., $\theta = 0$, almost all branch nodes become "full" during the BGT growing process, only about 10 candidate edges can be added to the BGN in our simulation. Therefore, it is very important to reserve the link capacity for branch nodes to add candidate edges among them.

Now we compare the numbers of branch nodes in the Bluetooth Scatternets constructed by BGN, BlueNet and BlueTree algorithms respectively.

As shown in Fig. 6, the number of branch nodes in the Scatternet constructed by BlueNet algorithm is the largest, while the BlueTree algorithm requires the least number of branch nodes when constructing the Scatternet. The number of branch nodes in the Scatternet constructed by our BGN algorithm is very close to that constructed by BlueTree algorithm. For instance, when $N = 200$, BlueTree algorithm requires 39 branch nodes as shown in Fig. 6. Since it is a tree structure, there are only 38 edges among those branch nodes. When $\theta = 1$, the number of branch nodes in the Scatternets constructed by our BGN algorithm is only 42. Therefore, the number of edges in the BGT is 41 since BGT is also a tree structure. However, the number of candidate edges in the constructed scatternet is 19. It means the total number of edges in the scatternet constructed by BGN algorithm can be 62% more than that constructed by BlueTree algorithm with only a slight increase on the number of branch nodes. Therefore, BGN algorithm constructs a much better scatternet than BlueTree algorithm does in terms of the number of links in the backbone network. Thus, the resulting scatternet has better network throughput and reliability.

The results in Figs 5 and 6 indicate that setting the link capacity reservation value θ at 1 or 2 for BGN algorithm is better choices.

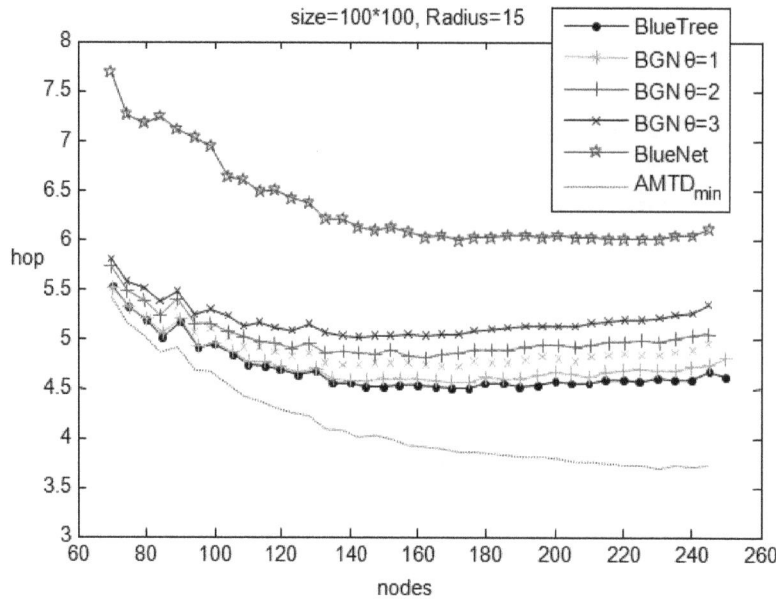

Fig. 7. Average minimum transmission distance under various node densities.

4.2. Average transmission distance

Let min_hop(i, 0) represent the minimal number of hops from node i to the base station. Then the average minimum transmission distance (AMTD) is computed as: $AMTD = \left(\sum_{i=1}^{N} \text{min\_hop}(i,0) \right) / N$. Obviously, *AMTD* determines the energy cost and data transmission delay of a BGN. Given an arbitrary connected network, it is not hard to calculate the optimal *AMTD*.

Figure 7 shows the *AMTD* values of the scatternets constructed by BGN, BlueNet and BlueTree algorithms under different node densities. The simulation results show that the BlueTree algorithm has the best performance in terms of AMTD while BGN algorithm achieves very similar performance when $\theta = 1$. The scatternet constructed by BlueNet algorithm has the largest *AMTD*, worst among all algorithms evaluated. We also calculated the lower bound of *AMTD*, demonstrated as $AMTD_{\text{min}}$ in Fig. 7.

4.3. Network reliability

Traditionally, the reliability of a network is measured by the number of dividing nodes in this network. Given a connected network, if cutting edges connecting to a node in the network will divide the network into two non-intersected sub-networks, this node is a dividing node of the network. In general, the more dividing nodes in a network, the less reliable it is. The scatternet constructed by BlueTree algorithm is a tree structure. All of its branch nodes are dividing nodes. The ratio of dividing nodes to branch nodes is 1 in such a network. This network is least reliable.

Figure 8 depicts the number of dividing nodes among the branch nodes of the scatternets constructed by different algorithms under various node densities. Figure 9 shows the ratio of dividing nodes to branch nodes in the scatternets constructed by different algorithms under various node densities.

As shown in Figs 8 and 9, the number of dividing nodes and the ratio of dividing nodes to branch nodes decrease as the node density increases. Although the number of dividing nodes and the ratio of

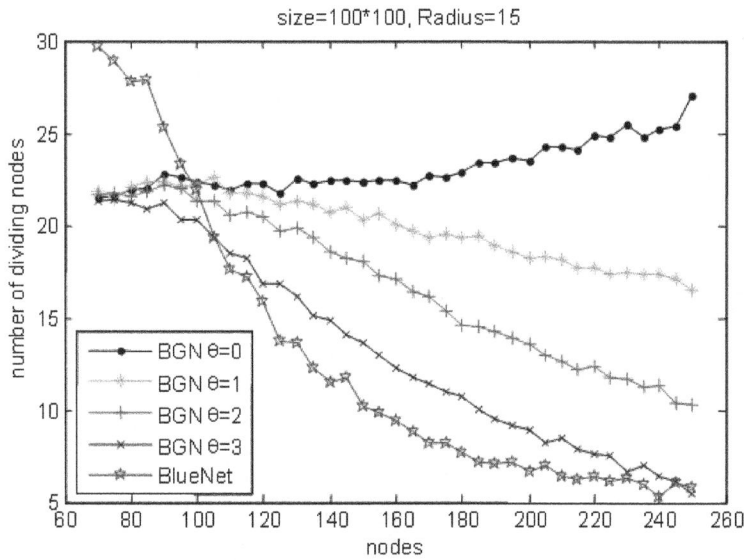

Fig. 8. The number of dividing nodes in branch nodes.

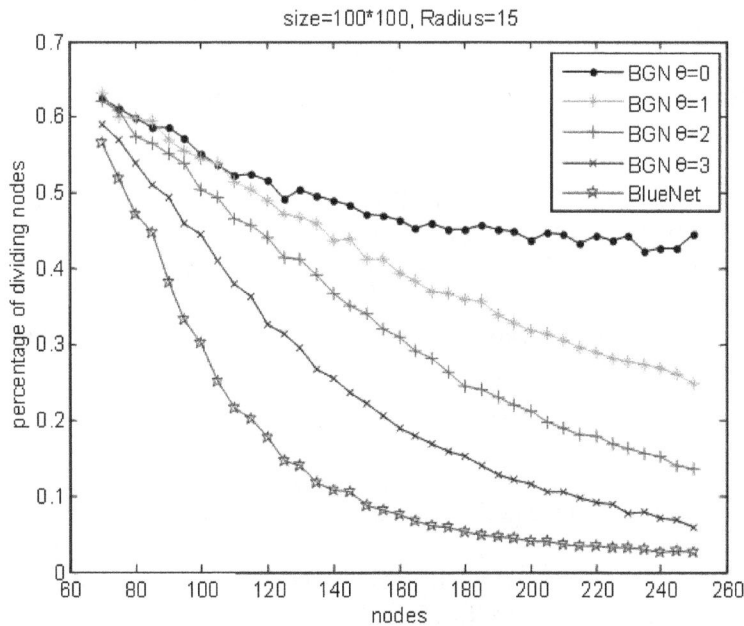

Fig. 9. The percentage of dividing nodes among the branch nodes.

dividing nodes to branch nodes in the scatternet constructed by BlueNet algorithm are slightly less than those by BGN algorithms, the number of branch nodes in the scatternet constructed by BGN algorithm is far less than that by BlueNet algorithm. Therefore, BGN algorithm can maintain a higher degree of network connectivity while avoiding excessive redundant links.

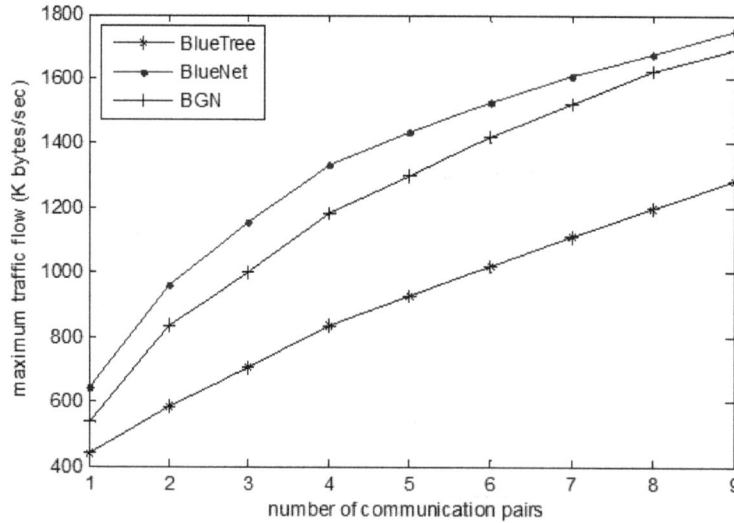

Fig. 10. Maximum traffic flows under various numbers of concurrent flows.

4.4. The maximum traffic flows

In most wireless sensor networks, nodes need to send the collected data to the clustering node in a short period of time. As wireless sensor networks often have only one clustering node, node resource constraints and link capacity limitations limit the amount information a scatternet can transfer within a unit time. Therefore, it is important to study the maximum information can be transmitted by the scatternet within a unit time. We call this the Maximum Traffic Flow.

Figure 10 presents the Maximum Traffic Flows of Scatternets constructed by BlueTree, BlueNet and BGN algorithms under different numbers of concurrent flows. Experimental environment is set as follows: There are 40 nodes randomly distributed in a square area of 20 m × 20 m. We assume the communication link radius $R = 10$ m and each node has a transmission bandwidth of $B_0 = 1000$ Kbps. For a branch node, the bandwidth cost of maintaining a link with its subordinate node is $B_1 = 10$ Kbps and the bandwidth cost for maintaining a bridge between two Piconets is 100 Kbps. Thus, the communication capacity of a node i can be computed as:

$$C_i = \begin{cases} B_0 - n_s^i B_1 - (n_b^i - 1)B_2 & \text{i is a branch node} \\ B_0 & \text{otherwise} \end{cases} \tag{2}$$

where n_s^i denotes the number of the subordinate nodes for branch node i, and n_b^i denotes the number of piconets connected to this branch node.

As shown in Fig. 10, the scatternet constructed by BlueTree algorithm has the smallest maximum traffic flow while the scatternet constructed by BlueNet algorithm has the largest maximum traffic flow. The maximum traffic flow in the scatternet constructed by our BGN algorithm is only slightly smaller than that by BlueNet algorithm.

5. Conclusion and future studies

This paper proposed a novel algorithm, BGN algorithm, to construct Scatternet for Bluetooth-based sensor networks. By adding links among the branch nodes of a Bluetooth Growing Tree, the scatternet constructed by our BGN algorithm can maintain a high degree of connectivity while avoiding excessive redundant links among the branch nodes. Because our BGN algorithm has the merits of both BlueTree algorithm and BlueNet algorithm, the constructed scatternet demonstrates well balanced performance in terms of average minimum transmission distance, network reliability, and maximum traffic flow.

We are currently designing an efficient routing protocol for wireless sensor networks based on our BGN algorithm. We will also study the self-stabilization protocol of constructing the scatternet in a dynamic wireless sensor network environment where some sensor nodes may fail due to power failure or environment impact, or some sensor nodes may move around the targeted sensing area.

Acknowledgement

James Z. Wang's work is partially supported by NSF grant DBI-0960586. Yongjun Li's work is partially supported by Guangdong Science and Technology Department grant 2008B080703004, 2009A030302009 and Guangdong Administration of Work Safety grant x2jsB2080910.

References

[1] K. Xuan, G. Zhao and D. Taniar, *Bala Srinivasan: Continuous Range Search Query Processing in Mobile Navigation*, ICPADS, 2008, 361-368.

[2] Agustinus Borgy Waluyo, Bala Srinivasan, David Taniar: A Taxonomy of Broadcast Indexing Schemes for Multi Channel Data Dissemination in Mobile Database. AINA (1) 2004: 213-218

[3] Agustinus Borgy Waluyo, Bala Srinivasan, David Taniar: Research in mobile database query optimization and processing. Mobile Information Systems 1(4): 225-252 (2005)

[4] James Jayaputera, David Taniar: Data retrieval for location-dependent queries in a multi-cell wireless environment. Mobile Information Systems 1(2): 91-108 (2005)

[5] James Z. Wang, Zhidian Du and Pradip K Srimani. Cooperative Proxy Caching for Wireless Base Stations, International Journal of Mobile Information Systems, Volume 3, Number 1, pp. 1-18, 2007.

[6] X. H. Wang, M Iqbal, Bluetooth: Opening a blue sky for healthcare, Mobile Information Systems, vol 2 (2006), pp. 151-167.

[7] Barolli, L., Hsu, H., and Shibata, Y. Mobile Systems and Applications. Mobile Information Systems. vol 4, (2008), pp. 77-79.

[8] Vojislav B. Misic, Jelena V. Misic: Improving sensing accuracy in cognitive PANs through modulation of sensing probability. Mobile Information Systems 5(2): 177-193 (2009)

[9] A.B. Waluyo, R. Hsieh, D. Taniar and J.W. Rahayu, Bala Srinivasan: Utilising Push and Pull Mechanism in Wireless E-Health Environment, *EEE* (2004), 271–274.

[10] G.V. Zaruba, S. Basagni and I. Chlamtac, Bluetrees- Scatternet Formation to Enable Bluetooth-Based Ad Hoc Networks, *Proceedings of IEEE International Conference on Cominunications*, Helsinki, Finland, June 2001, pp. 273–277.

[11] R.B. Muhammad, Range assignment problem on the Steiner tree based topology in ad hoc wireless networks, *Mobile Information Systems* 5(1) (2009), 53–64.

[12] F. Tang, I. You, M. Guo, S. Guo and L. Zheng, Balanced bipartite graph based register allocation for network processors in mobile and wireless networks, *Mobile Information Systems* 6(1) (2010), 65–83.

[13] S. Saginbekov and I. Korpeoglu, *An Energy-Efficient Scatternet Formation Algorithm for Bluetooth-Based Sensor Networks*, Proceedings of the 2nd European Workshop on Wireless Sensor Networks, Istanbul, 2005, pp. 207–216.

[14] T. lin, Y. Tseng, K. Chang and C. Tu, Formation, Routing, and Maintenance Protocols for the BlueRing Scatternet of Bluetooths, *Proceedings of the 36th Annual Hawaii International Conference on System Sciences (HICSS03)*, January 2003, pp. 313–322.

[15] T. Hassan, *Ring of Masters (ROM) A New Ring Structure for Bluetooth Scattemets*, Master' s thesis, American University of Beirut, Department of Electrical and Computer Engineering, February 2004.

[16] B. Huang, S. Ngah, H. Zhu and T. Baba, Solidring: A Novel Bluetooth Scatternet Structure, *International Journal of Computer Science and Network Security* 8(9) (Sep 2008), 93–103.

[17] Z. Wang, R.J. Thomas, and Z.J. Haas, *Bluenet- a new scatternet formation scheme, Proceedings of 35th Hawaii International Conference on System Science (HICSS-35)*, Big Island, Hawaii, Jan 2002, pp 61–69.

[18]   W.Z. Song, X.Y. Li, Y. Wang and W. Wang, dBBlue: low diameter and self-routing Bluetooth scatternet, *Journal of Parallel and Distributed Computing* **65**(2) (2004), 178–190.

[19]   Ralston, A. de Bruijn Sequences-A Model Example of the Interaction of Discrete Mathematics and Computer Science, *Math Mag* **55** (1982), 131–143.

[20]   C.Y. Chang, C.T. Chang, S.C. Lee and C.H. Tseng, TARP: A Traffic-Aware Restructuring Protocol for Bluetooth Radio Networks, *International Journal of Computer and Telecommunications Networking* **51**(14) (2007), 4070–4091.

[21]   N. Zaguia, Y. Daadaa and I. Stojmenovic, Simplified bluetooth scatternet formation using maximal independent sets, *Integr Comput -Aided Eng* **15**(3) (2008), 229–239.

[22]   C. Petrioli, S. Basagni and I. Chlamtac, BlueMesh: Degreeconstrained multi-hop scatternet formation for Bluetooth networks, *Mobile Networks and Applications* **9** (2004), 33–47.

[23]   I. Al-Kassem, S. Sharafeddine and Z. Dawy, *BlueHRT: Hybrid Ring Tree Scatternet Formation in Bluetooth Networks*, Computers and Communications, 2009. ISCC 2009, IEEE Symposium on 5–8 July 2009, pp. 165–169.

[24]   Y. Li and J.Z. Wang, Cost analysis and optimization for IP multicast group management, *Computer Communications* (8) (2007), 1721–1730.

[25]   R. Xie, D.Y. Qi, Y.J. Li and J.Z. Wang, A novel distributed MCDS approximation algorithm for wireless sensor networks, *Wireless Communications and Mobile Computing* **Issue 3**(9) (2009), 427–437.

Yongjun Li received his B.E. of science, M.E. of engineering and Ph. D. of science degrees from East China Normal University, National University of Defense Technology, and Sun Yat-sen University, respectively. He has been conducting postdoctoral research since 2004 at South China University of Technology (SCUT). He worked first at Hunan Normal University from Jul., 1991 to Oct.,1999, and then at Guangzhou Maritime College from 1999 to 2002, later at Guangzhou Radio & TV University from 2002 to 2005. Now he is an associate professor and continues his teaching and research work at the College of Computer Science & Engineering of SCUT. His research interests include new generation of network technology, intelligent initiative network and computer application technology. He served as the principle investigator in several projects supported by National and Province Science Fund and participated in many software development projects. He has published three text books and more than twenty papers, and received one patent.

Rong Xie received his B.S. M.S. and Ph.D. degrees in Computer Science from South China Normal University, Guangdong University of Technology, and South China University of Technology. He is with ZhaoQing University and currently an Assistant Professor in the School of Computer. His current research interests include computer networks, wireless networks, and next generation Internet.

Hu Chen is now studying for his M.S. degree in South China University of Technology.

James Z. Wang received the B.S. and M.S. degrees in Computer Science from University of Science and Technology of China, and the Ph.D. degree in Computer Science from University of Central Florida. He is currently an associate professor in the School of Computing at Clemson University. He previously worked as senior software engineer in Veritas Corp. and Computer Associate. His research interests include multimedia systems, database, distributed computing, Internet technologies, data mining, information retrieval and bioinformatics. He has published nearly 60 papers in refereed journals and conference proceedings. He is a senior member of IEEE and a senior member of ACM. He is an associate editor for International Journal of Data Mining and Bioinformatics.

11

A web-based application of TELOSB sensor network

Tarek R. Sheltami[a,*], Elhadi M. Shakshuki[b] and Hussein T. Mouftah[c]

[a]*Computer Engineering Department, King Fahd University of Petroleum and Minerals, Dhahran, Saudi Arabia*

[b]*Jodrey School of Computer Science, Acadia University, Wolfville, Nova Scotia, B4P 2R6, Canada*

[c]*School of Information Technology and Engineering, University of Ottawa, ON, Canada*

Abstract. Sensor network can be used in a numerous number of applications. However, implementing wireless sensor networks present new challenges compared with theoretical networks. In addition, implementing a sensor network might provide results different from that derived theoretically. Some routing protocols when implemented might fail to perform. In this paper, we implement three routing protocols, namely: Dynamic MANET on-demand, Collection Tree and Dissemination protocols. To compare the performance of these protocols, they are implemented using a Telosb sensor network. Several performance metrics are carried out to demonstrate the pros and cons of these protocols. A telemedicine application is tested in top of the implemented Telosb sensor network at King Fahd University of Petroleum and Minerals Clinic in Saudi Arabia, utilizing Alive ECG sensors.

Keywords: TELOSB, MANET, wireless sensor networks, alive ECG sensor

1. Introduction

Wireless Sensor Network (WSN) is a network of devices that have one or more types of sensors connected together via wireless communication to monitor cooperatively different environmental conditions such as temperature, humidity, sound, etc. in different locations [4,10]. Lately, WSNs are deployed in a variety of applications such as hazard detection applications, healthcare applications, smart homes applications, and industrial applications traffic control [11,14,15] WSN has unique characteristics such as ability to withstand harsh environmental conditions, ability to cope with nodes failures, communication failures and heterogeneity of nodes. There are other challenges like mobility of nodes, dynamic network topology, deployment and synchronization. A node or a mote, in a WSN, consists of sensors, a microcontroller, a transceiver and batteries. Motes use special operating systems that are designed for wireless embedded sensor networks, such as TinyOS [24,25] and contiki [7]. Vendors use different communication protocols for networking; however, the most commonly used protocol in WSNs is Zigbee [5,30]. Motes may have different interfaces, such as USB and RS-232, and they are mainly used for programming and data transfer. The work presented in this paper tests Telosb motes [22] from Crossbow shown in Fig. 1. These motes are equipped with USB interface, TinyOS operating system and Zigbee communication protocol.

---

*Corresponding author: Tarek R. Sheltami, Computer Engineering Department, King Fahd University of Petroleum and Minerals, Dhahran, Saudi Arabia. E-mail: tarek@kfupm.edu.sa.

Fig. 1. Telosb mote.

This paper discusses different versions of TinyOS and investigates different multi-hop network monitoring protocols. In addition, a testbed of 20 Telosb motes is used for performance evaluations. Finally, a real world application using Alive ECG sensors in top of the developed Telsob network is tested on Electrocardiography medical environment. Some of the Quality of Service (QoS) issues [1–3] are being adopted in our ECG application.

2. TinyOS 1.x vs. TinyOS 2.x

There are several differences exist between TinyOS 1.x [26] and TinyOS 2.x [27] operating systems. For example, TinyOS 1.x has very limited interfaces and structures. These limitations have motivated and encouraged several embedded systems operating systems designers to produce TinyOS 2.x to overcome the problems of the previous version (i.e., TinyOS 1.x) that led to easy programming practices of sensor networks, especially for new programmers. TinyOS 2.x differs from Tinyos 1.x in several aspects. Some of these aspects are: booting and initialization, task scheduling, timers, arbitration, debugging and network protocols. In booting and initialization, TinyOS 2.x has a different boot sequence than TinyOS 1.x. TinyOS 2.x boot sequence do only initialization of the components that are connected to it in the beginning. After it finishes the initialization, the boot sequence signals boot event. Conversely, in TinyOS 1.x any component connected to the boot sequence will be powered up once the boot sequence starts even if the initialization is not complete.

For task scheduling, both versions of TinyOS follow a non-preemptive FIFO algorithm, but TinyOS 2.x offers more flexibility, simplicity and organized task scheduling than Tinyos 1.x. In TinyOS 2.x, every task has its own place in the task queue and can only be posted once by the component. If the task is needed to be posted more than once, the task can repost itself once it finishes. The post operation fails only if the task is already posted. In TinyOS 1.x, the task queue is shared among all tasks, and the tasks can be posted multiple times. This sometimes may cause the task queue to be full, which makes any further post operation a failure. In addition to this disadvantage of TinyOS 1.x, programmers do not have the freedom to use their own scheduling algorithms as in TinyOS 2.x.

For timers, TinyOS 2.x offers more timer interfaces than TinyOS 1.x. It also makes it possible for component to check how much time left before a timer is fired and to make timers to be fired in the future. Moreover, it can make different timers to start asynchronously.

For arbitration, or access to shared resources, TinyOS 1.x lacks in providing a mechanism for managing access to these resources. This problem is eliminated in TinyOS 2.x by the Resource interface. Resource interface provides a policy for managing shared resource accessing.

For debugging, TinyOS 1.x gives only two types of error codes: SUCCESS and FAIL. These two types restrict programmers from being able to understand the cause of the errors. This can be problematic for programmers who want to know the cause of the error. On the other hand, TinyOS 2.x provides four types of error codes. These error codes are SUCCESS, FAIL, EBUSY and ECANCEL.

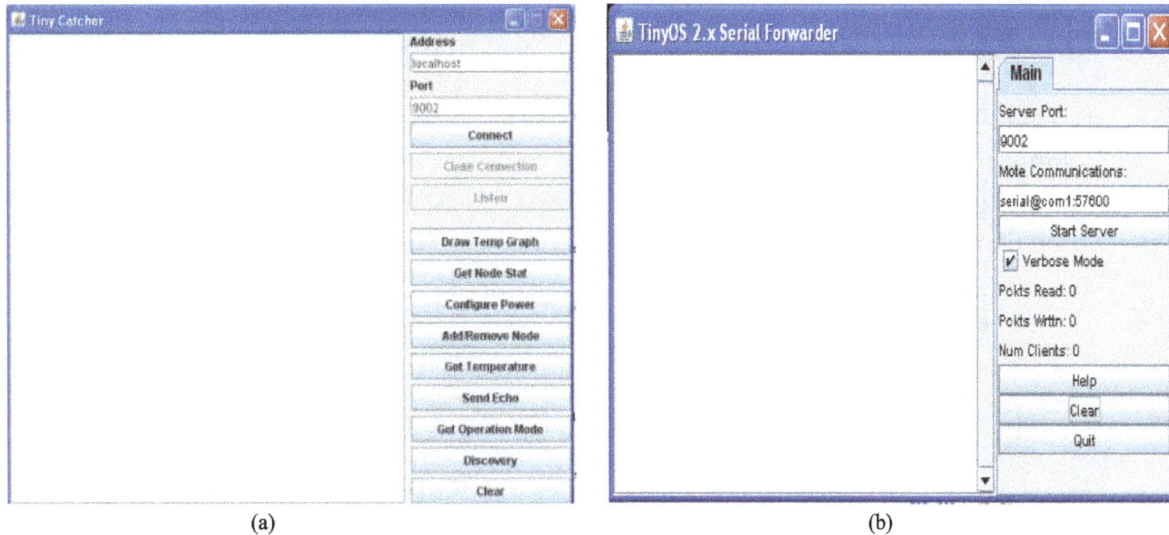

Fig. 2. (a) TinyCatcher interface, (b) SerialForwarder interface.

For network protocols, Tinyos 2.x has updated versions of the two basic network protocols: Dissemination and Collection. These two protocols provide support for more applications with improved results than TinyOS 1.x can provide.

The aforementioned aspects describe a subset of advantages that makes TinyOS 2.x surpasses TinyOS 1.x. TinyOS 2.x have provided great improvements in power management, sensors and communication in a wide range of platform support. TinyOS 2.x advantages made it natural for many researchers in wireless sensor networks to be adapted for their application environments.

3. Graphical user interfaces

To facilitate the interaction between the user and the application environment, a web-based application is built utilizing two Java application interfaces, namely: TinyCatcherand and SerialForwarder [25]. These two interfaces are shown in Figs 2-a and 2-b respectively.

TinyCatcher is a tool that allows the user to control the nodes in the sensor network by reading the messages received from different nodes. It records these messages in a database for analysis purposes. It consists of the following:

Address: address of the base-station PC that has SerialForwarder running and the localhost if Tiny-Catcher is on the same PC as SerialForwarder.

- Port: port number used by SerialForwarder.
- Connect/Close Connection: for connecting and disconnecting to/from the SerialForwarder. A certain connection protocol to SerialForwarder must be followed.
- Listen: start listening on TCP/IP port for messages after establishing the connection.
- Draw Temp Graph: creates a graph for temperature reading.

In our application, we are only interested in temperature mission. TinyCatcher is designed to interpret the temperature readings; however, it could be designed to interpret other phenomenon.

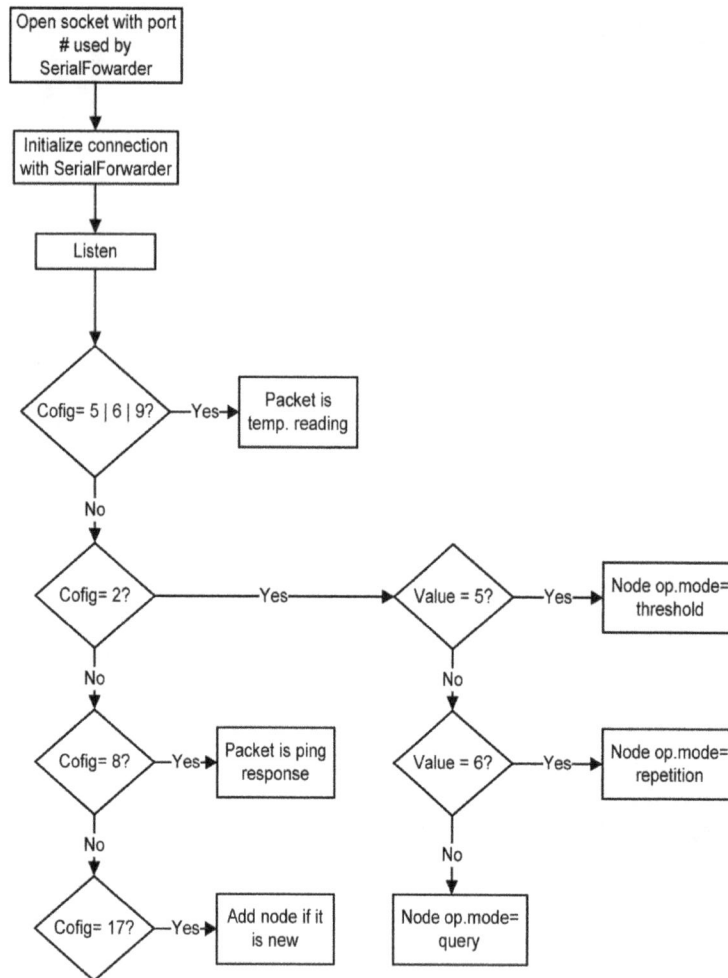

Fig. 3. TinyCatcher operation.

TinyCatcher depends on the SerialForwarder, which is a Java tool built in TinyOS. This tool is used for reading messages that arrive through the serial port from the gateway node (base-station). This allows other applications to share the connection over TCP/IP, which provides an advantage of remote monitoring. In other words, SerialForwarder acts as a proxy for reading and writing packets from and to the sensor network. Instead of reading data directly from the serial port, TinyCatcher can connect to SerialForwarder. The operation flow of TinyCatcher is demonstrated in Fig. 3.

Socket programming is used in the operation of the TinyCatcher. At the beginning, a socket is opened and a connection is initialized with the Serial forwarder. We set the ports 5, 6 and 9 for temperature readings, port 2 for control packets such as threshold, repetition and query, port 8 for ping response and port 17 to add new nodes.

The SerialForwarder has a very simple interface which is shown in Fig. 2-b. It consists of the following:

– Server Port: port of the socket the SerialForwarder opens.
– Mote Communication: specify serial port number and baud rate of the mote. It should follow this syntax [serial@(serial port):(baud rate)]. Serial port depends on the platform type, e.g. Win-

dows/Cygwin or Linux. Baud rate can be a numerical value or mote type, e.g. Telosb.
- Pckts Read: number of packets read from the serial port.
- Pckts Wrttn: number of packets sent through the serial port.
- Num Clients: number of clients connected to the SerialForwarder.

4. Sensor node configuration

In Telosb motes, the transmission power is divided into 31 power levels, where level 1 is the minimum and level 31 is the maximum. Many algorithms are introduced to solve both the transmission and the reception problem [6], some of which are described below.

In the Local Mean Algorithm (LMA) [8], every node starts with the same transmission power and then periodically broadcasts a life message including the transmitter's identity. After successfully receiving the life messages, the receivers respond by sending a life acknowledgement message that contains the address of life message transmitter. Therefore, the transmitter is able to count the number of reachable neighbors at a particular power. If the number of received acknowledgements is less than a predefined threshold value, the transmitter increases its power by a certain predefined factor to allow more nodes to respond. Moreover, the degree of power increases for a single step does not exceed a specified value. The minimum value allocated to both thresholds is used as the transmission power for the next operation cycle. The power is decreased if the number of acknowledged messages is greater than the threshold value.

The Transmission Power Control and Blacklisting (PCBL) [8] algorithm has two main objectives. The first objective is to investigate the behavior of low-power wireless communication links with respect to varied transmission power under different settings. The second objective is to propose a new scheme on power control that is capable of removing low quality links. At the beginning, each node measures the quality of each link by determining Packet Reception Rate (PRR) at various transmission power levels. In order to set a threshold value for the required link reliability, an optimal unicast transmission power for each link is set. The threshold is set to the minimum value of the observed PRR if the PRR is greater than the threshold. Otherwise, it is set to the maximum transmission power. Another threshold, which is known as the blacklist threshold, is set to consider whether a link can be converted to a good link. If not, then that link is blacklisted and no longer used.

The Transmission Power Control mechanism proposed in [16] uses dynamic transmission power adaptation. This mechanism starts with a node transmitting a beacon message. A receiving node with a good link quality, records the ID of the sender. The list of the IDs is piggybacked on the beacon message. Finally, a node knows the number of its neighbors by determining its ID in the incoming messages. The number of neighbors is then compared to the predefined threshold value and the transmission power control mechanism is applied to achieve the goal. The transmission power is increased if the observed number of neighbors is less than a predefined threshold value. The mechanism is capable of reducing the degree of adjustment as the number of neighbors converges to the targeted value.

The Adaptive Transmission Power Control (ATPC) approach described in [20] is designed based on the concept of changing a pair-wise transmission power level over time. As a result, each node assigns a different minimum transmission power for each link. Two main ideas behind its design is a neighbors table which is maintained by each node and a closed loop for transmission power control which runs between each pair of sensors. The entries of the table include Node ID, proper transmission power levels defined as the minimum power which provides a good link quality, and several parameters used

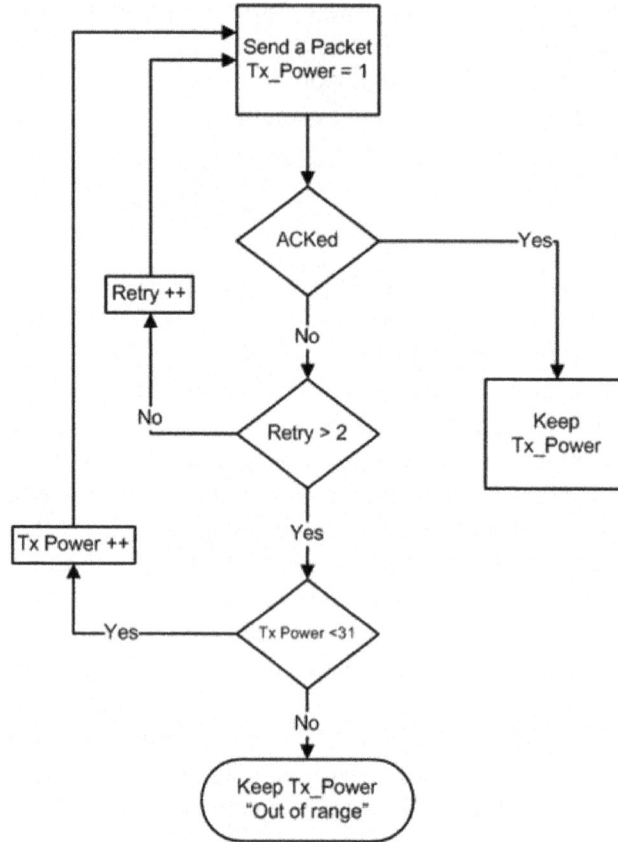

Fig. 4. Proposed power control flow.

for linear predictive models of transmission power control. The closed loop feedback is used to obtain the minimum transmission power by gradually adjusting the power.

According to the reviewed transmission power control schemes, most of them demonstrate two similar procedures. The first procedure can be described as: a transmitting node discovers which or how many active neighbors it has by broadcasting messages. The second procedure can be described as: a feedback or acknowledgement system is used after the neighbors have successfully received the messages.

In our approach, the nodes attempt to discover the minimum required transmission power. A packet is sent with minimum transmission power and then waits for an acknowledgment message. The acknowledgments are generated by the CC2420 radio transceiver as specified in the IEEE 802.15.4 protocol. If a packet is acknowledged then the node keeps its current transmission power. If the transmission fails after three attempts in a row, the node increases its power by one until it gets an acknowledgment. If no acknowledgement is received and reaches level 31, then this means that the other node is out of range. This is described in Fig. 4.

The Telosb sensor node can be configured for different power levels from 1 to 31 (1 is the lowest). Figure 5 depicts Java graphical user interface that allows users to set the power level for any given node using NodeID.

Figure 6 shows an interface that allows a user to add or remove nodes at the base-station.

Through the interfaces shown in Fig. 7, the user is allowed to set the desired period, threshold value and query. These operation modes can be described as follows:

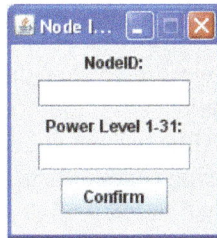

Fig. 5. Power configuration interface.

Fig. 6. Node add/remove interface.

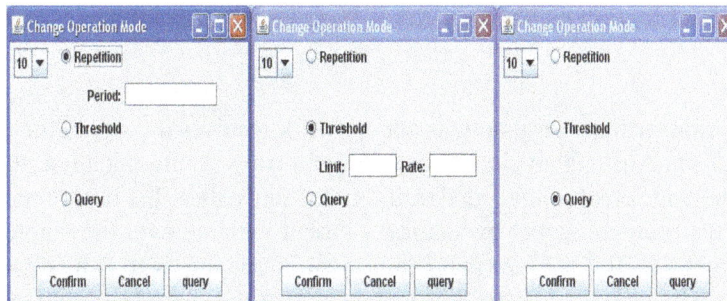

Fig. 7. Operation modes interface.

- Periodic: messages are sent periodically by the selected node.
- Threshold: specify the desired minimum and maximum temperature thresholds for the node to start sending messages when temperature reaches the threshold. Rate of the incoming messages is specified.
- Query: for sending a query to the specified node.

5. Implementation

Three routing protocols: Dynamic MANET On-demand (DYMO) Routing Protocol [12,13], Collection Tree Protocol (CTP) [9,10,19,23], and Dissemination Protocol [9,21,28,29] are implemented using Telsob motes with TinyOS2 operating system. We test and evaluate each of these protocols and compare their advantages and disadvantages. Typically, the base-station (gateway) acts as a bridge between the computer application side and motes. Thus, the data about temperature is collected at the base-station. At the base-station, the payload packet is extracted and forwarded it via the serial port to the computer.

5.1. DYMO routing protocol

After testing the implemented DYMO protocol, we notice that the protocol is unstable. DYMO protocol shows some unpredictable behaviors. In many cases, the nodes stop responding or refuse to discover new routes when old route(s) is removed. Although we could have improved the results with troubleshooting and using different methods, our decision is geared towards investigating other protocols due to time limitation. It should be noted that the tested operation is bi-directional, i.e. the base-station and sensor nodes are using the same routing protocol.

5.2. CTP protocol

A tree around a root is formed for CTP protocol to function. Nodes chose their parents based on the Link Quality Indicator (LQI), which could be calculated using Received Signal Strength Indicator (RSSI) and packet drop rate. Some nodes are equipped with Chipcon CC2420 radio chip that provides such statistics. This resulted in a new CTP called CTP-LQI. The only difference between CTP and CTP-LQI is in how the link quality calculated. Since Telosb motes use the CC2420 chip, it is more appropriate to use the CTP-LQI implementation. We observe that CTP routing protocol works much better than DYMO protocol. Unlike DYMO protocol, CTP data is limited to the size of the tree. This means that the data can only be sent from the edges towards the base-station.

5.3. Dissemination protocol

The Dissemination protocol ensures that every node in the network receives a copy of the data. This protocol is used by the base-station. Although it can be used for both ways communication, the fact that every node has to receive the message creates unneeded traffic in the network. This in turn has an effect on the battery performance. This protocol works by sharing a global variable over the whole network. However, this variable needs to be of a limited size that can be sent in one message. The dissemination protocol ensures, over time, the whole network shares the same value of the shared variable. Such value code is any data type of the default TinyOS data types or a user defined C structure that the user can select [24].

6. Performance evaluation

This section discusses the measurement of the RSSI of Telosb motes, the power field distribution, the throughput, and delay of transmitting packets.

6.1. RSSI measurement

The RSSI of Telosb motes is measured by fixing the power level for different distances. As mentioned earlier, the power levels of a Telosb mote range from 1 to 31 levels. Where, 1 refers to the weakest signal power level and 31 refers to the strongest signal level. RSSI is measured in dBm unit ranging from 0 to -155 dBm. We measure the RSSI for a certain power level using the following steps:

1. Fix the power level to have the variables as Distance and RSSI.
2. Place two motes away from each other by a certain distance X.
3. Send 200 packets by the first mote to the second.

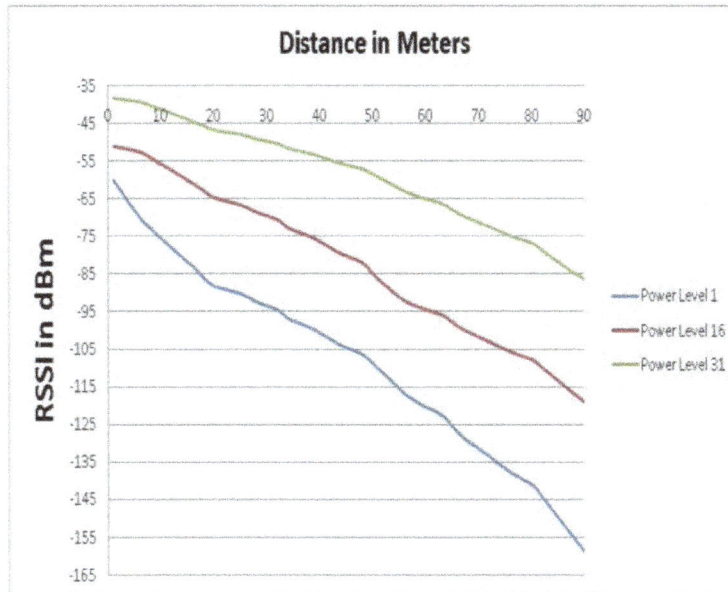

Fig. 8. RSSI readings.

4. Measure the signal strength for each packet received by the second mote.
5. Calculate the average RSSI.

Figure 8 depicts different readings for RSSI for levels 1, 15 and 31. Although the results are intuitive and expected, the main reason of this experiment is to find specific readings of the sensor nodes.

6.2. Power field distribution

In many research papers, it is assumed that the transmission power is in circular shape. We assume the same and performed the below experiment to find the power field distribution, using the following algorithm:

1. Place two motes away from each other by 3 meters.
2. Place the receiver to the right of the sender.
3. Face the sender mote towards north (the USB port of the mote is directed towards north).
4. Send 100 packets to be received by the second mote.
5. Measure the RSSI of all received packets
6. Calculate the average RSSI of the received packets.
7. Repeat the previous steps three more times, by having the sender facing south, east and then west.

At the completion of the experiment, we obtained the measurements of RSSI for Telosb mote when it is facing all four directions: north, south, east and west. All derived values are listed in Table 1 and depicted in Fig. 9.

From this experiment, we noticed that the transmission power close to elliptical shape. Also, we found that the RSSI is best when the sender is facing east (the receiver) with the USB port facing the receiver. Hence, it is best to have the USB interface of the sender mote facing towards the receiver.

Table 1
Power reading in different directions

Direction	RSSI (dBm)
North	−80.86
South	−81.12
East	−70.04
West	−83.45

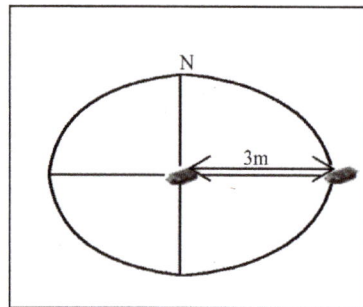

Fig. 9. Power field distribution.

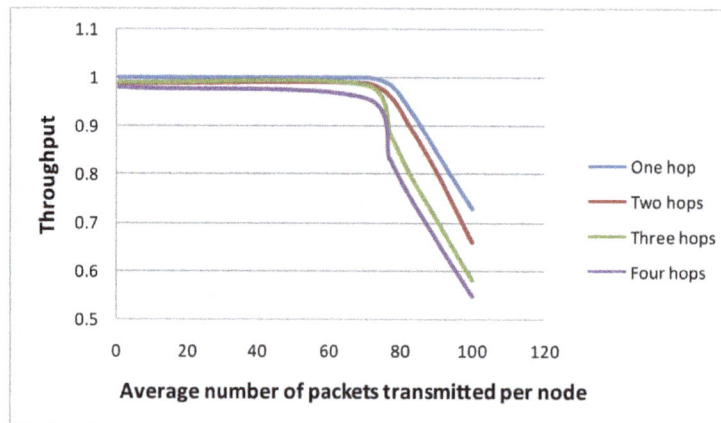

Fig. 10. Throughput vs. average number of transmitted packets.

6.3. Throughput

We utilize one, two, three and four hops topologies to measure the throughput of the transmitted data from and to the base-station. The following algorithm describes how we performed this experiment:

1. Fix the time period of transmitting packets to 1 sec.
2. Set the number of packets to be transmitted.
3. Calculate the throughput (number of received packets over transmitted packets in one second period).
4. Repeat the previous steps for different number of packets.

Figure 10 depicts the throughput of the network vs. the average number of packets transmitted. We notice that the performance of the networks starts to degrade when the average number of packets reaches 75 with sharp degradation when it reaches 80 packets. This is due packets collision.

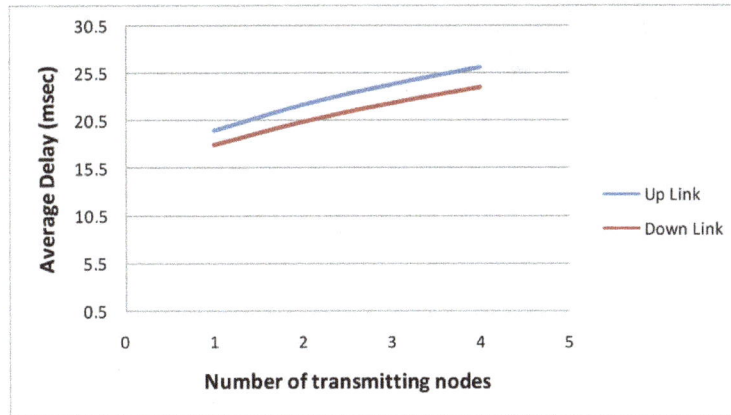

Fig. 11. Average delay (per packet) vs. number of transmitting nodes.

6.4. Delay

In order to find the delay of the transmitted packets between any two motes, many messages are transmitted between two specific nodes and then the average delay is calculated. The sender mote incorporates a timestamp with the packet for the receiver to calculate the time of transmission. The delay is compared with the number of nodes transmitted to the base-station. Figure 11 depicts average delay of the transmitted packets versus the number transmitting nodes. As expected, the more the number transmitting nodes the more the delay.

6.5. A Medical application: Electrocardiography (ECG)

Telemedicine system can be defined as the delivery of medical information over a distance by using a telecommunication means. This system helps and serves the patients and allows their physicians to check them without the actual physical contact, independent of their geographical locations. In this regard, we design a web-based interface between ECG device and any personal computer for the elderly and disabled people at home. Then, the application analyzes the data coming from the ECG device and according to some parameters, it decides whether the patient has an abnormal case or not. As soon as the patient has some difficulties and abnormalities, an email and/or SMS message is sent to the physician defining the patient ID and his situation.

Accurate identification of a heart disorder is vital to saving a life. A physician or an expert like the cardiologist always looks for tools that help in detecting certain heart disorder within the patient. An ECG has provided them the insight to detect these disorders by identifying variations in the functions of the heart. The main parts of the ECG are described in the following paragraphs.

6.5.1. ECG signal

An electrocardiogram (ECG) is a recording of the electrical activity of the heart over time produced by an electrocardiograph, usually in a noninvasive recording via skin electrodes. Its name is composed of three different parts: electro, because it is related to electrical activity, cardio, Greek for heart, and gram, a Greek root word meaning "to write". Electrical impulses in the heart originate in the senatorial node and travel through the heart muscle where they impart electrical initiation of systole or contraction of the heart. The electrical waves can be measured at selectively placed electrodes (electrical contacts) on the skin.

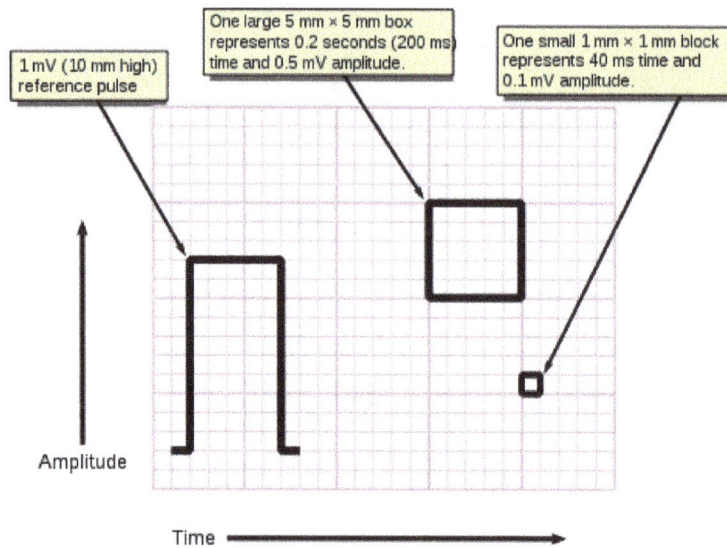

Fig. 12. One second of ECG graph paper.

Electrodes on different sides of the heart measure the activity of different parts of the heart muscle. An ECG displays the voltage between pairs of these electrodes, and the muscle activity that they measure, from different directions, also understood as vectors. This display indicates the overall rhythm of the heart and weaknesses in different parts of the heart muscle. It is the best way to measure and diagnose abnormal rhythms of the heart.

6.5.2. ECG graph paper

Timed interpretation of an ECG was once incumbent to a stylus and paper speed. Computational analysis now allows considerable study of heart rate variability. A typical electrocardiograph runs at a paper speed of 25 mm/s, although faster paper speeds are occasionally used. Each small block of ECG paper is 1 mm^2, as depicted in Fig. 12. At a paper speed of 25 mm/s, one small block of ECG paper translates into 0.04 s (or 40 ms). Five small blocks make up one large block, which translates into 0.20 s (or 200 ms). Hence, there are five large blocks per second. A diagnostic quality 12 lead ECG is calibrated at 10 mm/mV, so one mm translates into 0.1 mV. A calibration signal should be included with every record. A standard signal of 1 mV must move the stylus vertically 1 cm, which are two large squares on ECG paper.

6.5.3. Normal ECG waves and intervals

A typical ECG tracing of a normal heartbeat (or cardiac cycle) consists of a P wave, a QRS complex and a T wave. A small U wave is normally visible in 50 to 75% of ECGs. The baseline voltage of the electrocardiogram is known as the isoelectric line. Typically, the isoelectric line is measured as the portion of the tracing following the T wave and preceding the next P wave, as depicted in Fig. 13.

6.5.4. P wave

During normal atrial depolarization, the main electrical vector is directed from the sinoatrial (SA) node towards the atrioventricular (AV) node, and spreads from the right atrium to the left atrium. This turns into the P wave on the ECG, which is upright in II, III, and augmented vector foot (aVF) (since the general electrical activity is going toward the positive electrode in those leads), and inverted in augmented vector

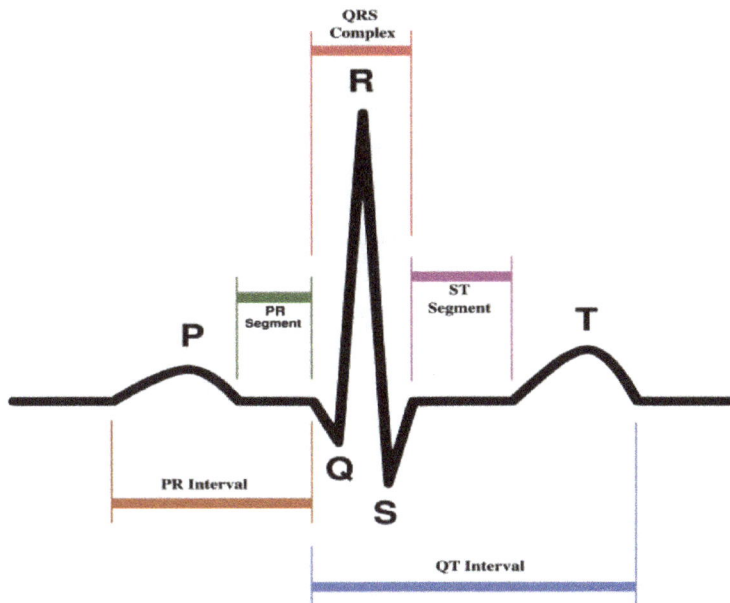

Fig. 13. Schematic representation of normal ECG.

right (aVR) (since it is going away from the positive electrode for that lead). A P wave must be upright in leads II and aVF and inverted in lead aVR to designate a cardiac rhythm as Sinus Rhythm.

- The relationship between P waves and QRS complexes helps distinguish various cardiac arrhythmias.
- The shape and duration of the P waves may indicate atrial enlargement.

6.5.5. QRS complex

The QRS complex is a structure on the ECG that corresponds to the depolarization of the ventricles. Because the ventricles contain more muscle mass than the atria, the QRS complex is larger than the P wave. In addition, because the His/Purkinje system coordinates the depolarization of the ventricles, the QRS complex tends to look "spiked" rather than rounded due to the increase in conduction velocity. A normal QRS complex is 0.06 to 0.10 sec (60 to 100 ms) in duration represented by three small squares or less, but any abnormality of conduction takes longer, and causes widened QRS complexes.

6.5.6. T wave

The T wave represents the repolarization (or recovery) of the ventricles. The interval from the beginning of the QRS complex to the apex of the T wave is referred to as the absolute refractory period. The last half of the T wave is referred to as the relative refractory period (or vulnerable period). In most leads, the T wave is positive. However, a negative T wave is normal in lead aVR. Lead V1 may have a positive, negative, or biphasic T wave. In addition, it is not uncommon to have an isolated negative T wave in lead III, aVL, or aVF.

To demonstrate our approach on a real world application, a Telsob network is tested on Electrocardiography medical environment. This environment consists of Alive ECG sensor [18], Telosb sensor network, application server, patient PC and database, as shown in Fig. 14. The Alive ECG sensor is connected to a patient and sends signals to the Telosb sensor network. The Alive ECG software is tested and ECG data is measured after a Bluetooth connection is established with a Bluetooth enabled device.

Fig. 14. Network topology of ECG environment.

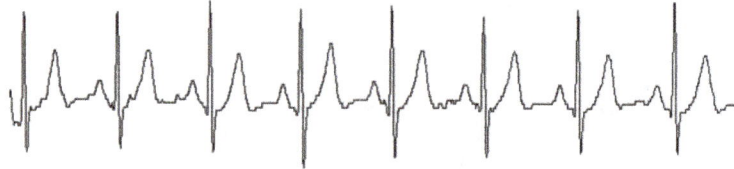

Fig. 15. A normal ECG wave.

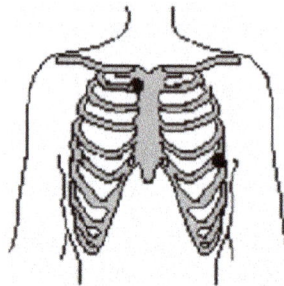

Fig. 16. Normal configure of ECG probes.

We interface one of the Telosb sensors with a Bluetooth connection. This allows us to receive the ECG signals from the Alive sensor, and then the signal is sent via the Telosb network to the gateway. Doctors and nurse are able to access gateway through the Internet.

We perform our experiments at the clinic of King Fahd University of Petroleum and Minerals (KFUPM), in Dhahran city, Saudi Arabia. Our patients are connected to the Alive ECG sensor and each patient sends his/her data via the Telosb network to the gateway.

During this experiment, 150 samples of ECG are collected over a period of one year using ECG Alive sensor. Figure 15 shows a normal (i.e. healthy) ECG signals that are captured by ECG Alive sensor.

ECG along with Heart Rate (HR) and three axes accelerometers readings are sent wirelessly using Telosb network. The normal configuration of the probes is shown in Fig. 16. The ECG Alive sensor is good enough for a doctor to detect any abnormal behavior in the ECG and specially for detecting Arrhythmia.

A few samples from the 150 collected samples are demonstrated in Fig. 17.

Figure 17 shows an abnormal ECG signal. The highlighted area shows a signal known as premature ventricular beat arrhythmia in the medical world. It is characterized by the following:

– Absence of a p-wave.
– QRS is wide.
– T-wave is inverted.

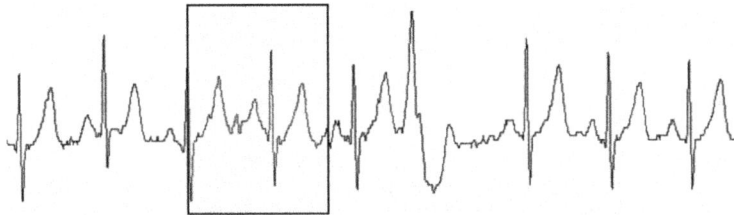

Fig. 17. Abnormal ECG signal.

At the gateway end, the data is analyzed using MATLAB. The data is fetched from a specific location and then based on that data some actions are taken by software that we design specifically for this application. This program has a Graphical User Interface that allows the nurses/doctors to interact with the system. The data is then saved in the server for future needs. It is worth mentioning that the collected data is very close to actual data collected by the ECG equipments. Using our developed system, we can test more than one patient simultaneously if needed. Each one of them will have to be connected with Alive ECG sensor and the rest is taken care of by the Teleosb sensor network.

Acknowledgement

The authors would like to thank King Fahd University of Petroleum and Minerals via projects IN080395 and IN090047 and University of Ottawa via WiSence project for their support. The authors would like to thank Research and Graduate Office of Acadia University and the Natural Sciences and Engineering Research Council of Canada (NSERC) for their support.

Conclusions and future work

An application with graphical user interfaces to control the Telosb motes is developed using TinyOS 2. Many interfaces were developed to build an application to carry out different experiments. Using this application, we were able to receive temperature readings from sensor nodes in single and multi hop fashions. Three modes of operations were developed to receive the temperature reading (periodic, threshold and query). Dissemination and Collection Tree protocols were implemented on top of sensor nodes. Collection Tree protocol was used for the uplink (from nodes to base-station), where nodes choose their parents based on the Link Quality Indicator. However, Dissemination protocol was used by the base-station to ensure that every node in the network receives a copy of the data. On the other hand, DYMO protocol was unstable and showed some unreliable behaviors.

An efficient power transmission algorithm was introduced to find out the optimum power of transmission in order to conserve energy. In addition, transmission power strength could be adjusted for each node at the base-station. Exact values of the received signal strength vs. distance, average throughput vs. the offered traffic and delay vs. number of transmitting nodes were measured and will be used in future simulation. Also, an exact shape of the radio power field was found with helical shape.

An ECG medical application was tested at a local clinic at KFUPM utilizing Telosb network and Alive ECG sensors. One hundred and fifty samples of health and abnormal patients were collected for duration of one year. The collected data was tested against the traditional ECG equipments and it was found that there was very small margin of error, due to calibration of the Alive ECG sensor. It is worth mentioning that our application received excellent remarks from cardiologists and nurses at KFUPM clinic.

References

[1]   A.B. Waluyo, B. Srinivasan and D. Taniar, *A Taxonomy of Broadcast Indexing Schemes for Multi Channel Data Dissemination in Mobile Database*, the 18th International Conference on Advanced Information Networking and Applications (AINA 2004), Volume 1, IEEE Computer Society, pp. 213–218, 2004.

[2]   A.B. Waluyo, B. Srinivasan and D. Taniar, *Optimal Broadcast Channel for Data Dissemination in Mobile Database Environment*, Proceedings of the 5th International Workshop on Advanced Parallel Programming Technologies (APPT 2003), Lecture Notes in Computer Science, Volume 2834, Springer, pp. 655–664 2003.

[3]   A.B. Waluyo, B. Srinivasan and D. Taniar, Research in mobile database query optimization and processing, *Mobile Information Systems* **1**(4) (2005), 225–252.

[4]   A. Durresi and M. Denko, Advances in mobile communications and computing, *Mobile Information Systems* **5**(2) (2009), 101–103.

[5]   A. Durresi, P. Zhang, M. Durresi and L. Barolli, Architecture for mobile Heterogeneous Multi Domain networks, *Mobile Information Systems* **6**(1) (2010), 49–63.

[6]   C. Li and L. Li, Energy efficient resource management in mobile grid, *Mobile Information Systems* **6**(2) (2010), 193–211.

[7]   Contiki, http://www.sics.se/contiki/.

[8]   D. Son, B. Krishnamachari and J. Heidemann, *Experimental Study of the Effects of Transmission Power Control and Blacklisting in Wireless Sensor Networks*, IEEE SECON, Santa Clara, USA, pp. 289–298 October 2004.

[9]   Dissemination, http://www.tinyos.net/tinyos-2.x/doc/html/tep118.html.

[10] E. Shakshuki, H. Malik and T. Sheltami, *Lessons Learned: Simulation Vs WSN Deployment*, The IEEE 23nd International Conference on Advanced Information Networking and Applications (AINA-09), pp. 580–587, Bradford, UK, May 26–29, 2009.

[11] H. Truong and S. Dustdar, A survey on context-aware web service systems, *International Journal of Web Information Systems, Emerald* **5**(1) (2009), 5–31.

[12] I. Chakeres, E. Belding-Royer and C. Perkins, *Dynamic MANET On-demand (DYMO) Routing*, draft version 04 IETF, Mar. 2006.

[13] I.D. Chakeres and J.P. Macker, Mobile ad hoc networking and the ietf, *SIGMOBILE Mob Comput Commun Rev* **10**(3) (2006), 79–81.

[14] I.F. Akyildiz and I.H. Kasimoglu, *Wireless Sensor and Actor Networks: Research Challenges*, Ad Hoc Networks, pp. 351–367, 2004.

[15] I.F. Akyildiz, W. Su, Y. Sankarasubramaniam and E. Cayirci, Wireless sensor networks: a survey, *Computer Networks* **38**(4) (2002), 393–422.

[16] J. Jeong, D.E. Cullar and J.H. Oh, *Empirical Analysis of Transmission Power Control Algorithms for Wireless Sensor Networks*, Technical Report No. UCB/EECS-2005-16.

[17] M. Kubisch, H. Karl, A. Wolisz, L.C. Zhong and J. Rabaey, *Distributed Algorithms for Transmission Power Control in Wireless Sensor Networks*, IEEE WCNC, New Orleans, Louisiana, pp. 558–563, 2003.

[18] Mao Ye, Chengfa Li, Guihai Chen and J. Wu, *EECS: An Energy Efficient Clustering Scheme in Wireless Sensor Network*, the 24th IEEE International Performance, Computing, and Communications Conference (IPCCC), Phoenix, Arizona, USA, pp. 535–540, 2005.

[19] R. Fonseca, O. Gnawali, K. Jamieson, S. Kim, P. Levis and A. Woo, *The Collection Tree Protocol (CTP)*, TEP 123, TinyOS Network Working Group, Aug. 2006.

[20] S. Lin, J. Zhang, G. Zhou, L. Gu, T. He and J.A. Stankovic, *ATPC: Adaptive Transmission Power Control for Wireless Sensor Networks*, the 4th International Conference on Embedded Networked Sensor Systems, Boulder, Colorado, pp. 223–236, November 2006.

[21] T. Bokareva, N. Bulusu and S. Jha, *A performance comparison of data dissemination protocols for wireless sensor networks*, in IEEE Globecom Wireless Ad Hoc and Sensor Networks Workshop, Texas, USA, pp. 85–89, 2004.

[22] Telosb, http://www.xbow.com/Products/Product_pdf_files/Wireless_pdf/TelosB_Datasheet.pdf.

[23] The Collection Tree Protocol (CTP), http://www.tinyos.net/tinyos-2.x/doc/html/tep123.html.

[24] TinyNode, http://www.tinynode.com/.

[25] TinyOs, http://www.tinyos.net/.

[26] tinyos-1.x, http://www.tinyos.net/tinyos-1.x/doc/.

[27] tinyos-2.x, http://www.tinyos.net/tinyos-2.x/doc.

[28] Vinh Pham, Erlend Larsen, Øivind Kure, Paal Engelstad, Routing of internal MANET traffic over external networks, *Mobile Information Systems* **5**(3) (2009), 291–311.

[29] Y. Yu, R. Govindan and D. Estrin, *Geographical and energy aware routing: a recursive data dissemination protocol for wireless sensor networks*, UCLA Computer Science Department Technical Report, Tech. Rep. UCLA/CSD-TR-01-0023, 2001.

[30] Zigbee, http://www.zigbee.org.

Tarek Sheltami is currently an associate professor at the Computer Engineering Department at King Fahd University of Petroleum and Minerals (KFUPM) Dhahran, Kingdom of Saudi Arabia. He joined the department on August 26, 2004. Before joining the KFUPM, Dr. Sheltami was a research associate professor at the School of Information Technology and Engineering (SITE), University of Ottawa, Ontario, Canada. He worked at GamaEng Inc. as a consultant on Wireless Networks (2002–2004). He was the Principle\Co Investigator of several research projects in the area of Ad hoc, Sensor Networks and Pervasive and Ubiquitous Computing.

Elhadi M. Shakshuki is a professor at the Jodrey School of Computer Science at Acadia University, Canada. He is the founder and the head of the Cooperative Intelligent Distributed Systems Group at the Computer Science Department, Acadia University. He received the BSc degree in computer engineering in 1984 from El-Fateh University, and the MSc and PhD degrees in systems design engineering respectively in 1994 and 2000, from the University of Waterloo, Canada. Dr. Shakshuki is an Adjunct Professor at Dalhousie University, Canada. He manages several research projects in his research expertise in the area of intelligent agent technology and its applications. He is a member of IEEE, ACM, AAAI and APENS.

Hussein T. Mouftah joined the School of Information Technology and Engineering (SITE) of the University of Ottawa in 2002 as a Tier 1 Canada Research Chair Professor, where he became a *University Distinguished Professor* in 2006. He has been with the ECE Dept. at Queen's University (1979–2002), where he was prior to his departure a Full Professor and the Department Associate Head. He has six years of industrial experience mainly at Bell Northern Research of Ottawa (became Nortel Networks). He served IEEE ComSoc as Editor-in-Chief of the IEEE Communications Magazine (1995–97), Director of Magazines (1998–99), Chair of the Awards Committee (2002–03), Director of Education (2006–07), and Member of the Board of Governors (1997–99 and 2006–07). He has been a Distinguished Speaker of the IEEE Communications Society (2000–07). Currently he serves IEEE Canada (Region 7) as Chair of the Awards and Recognition Committee. He is the author or coauthor of 6 books, 32 book chapters and more than 950 technical papers, 10 patents and 140 industrial reports. He is the joint holder of 12 Best Paper and/or Outstanding Paper Awards. He has received numerous prestigious awards, such as the 2008 ORION Leadership Award of Merit, the 2007 Royal Society of Canada Thomas W. Eadie Medal, the 2007–2008 University of Ottawa Award for Excellence in Research, the 2006 IEEE Canada McNaughton Gold Medal, the 2006 EIC Julian Smith Medal, the 2004 IEEE ComSoc Edwin Howard Armstrong Achievement Award, the 2004 George S. Glinski Award for Excellence in Research of the U of O Faculty of Engineering, the 1989 Engineering Medal for Research and Development of the Association of Professional Engineers of Ontario (PEO), and the Ontario Distinguished Researcher Award of the Ontario Innovation Trust. Dr. Mouftah is a Fellow of the IEEE (1990), the Canadian Academy of Engineering (2003), the Engineering Institute of Canada (2005) and the Royal Society of Canada RSC: The Academy of Science (2008).

12

A localization algorithm based on AOA for ad-hoc sensor networks

Yang Sun Lee[a], Jang Woo Park[b] and Leonard Barolli[c,*]

[a]*Department of Information Communication Engineering, Chosun University, Chosun, Korea*
[b]*Department of Information and Communication Engineering, Sunchon National University, Sunchon, Korea*
[c]*Department of Information and Communication Engineering, Fukuoka Institute of Technology (FIT), Fukuoka, Japan*

Abstract. Knowledge of positions of sensor nodes in Wireless Sensor Networks (WSNs) will make possible many applications such as asset monitoring, object tracking and routing. In WSNs, the errors may happen in the measurement of distances and angles between pairs of nodes in WSN and these errors will be propagated to different nodes, the estimation of positions of sensor nodes can be difficult and have huge errors. In this paper, we will propose localization algorithm based on both distance and angle to landmark. So, we introduce a method of incident angle to landmark and the algorithm to exchange physical data such as distances and incident angles and update the position of a node by utilizing multiple landmarks and multiple paths to landmarks.

Keywords: AoA, RSSI, wireless sensor network, localization

1. Introduction

Wireless sensor networks (WSNs) involve many different technologies such as communication, sensing and computing, which now becoming a very important research areas. The knowledge of the accurate locations of sensor nodes in WSN makes possible many attractive applications such as routing and tracking asset [1–4]. The position coordinates of sensor node in WSN can add significant meaning to information which are collected by sensors [5–7].

Sensor node location can be computed by Range based methods such as GPS (Global Positioning System) or Range-free method. The GPS-based method needs sensor nodes to be equipped by GPS which is very expensive and cannot be used in the case of blocking the GPS signal such as indoor environment.

The distance among sensor nodes can be determined by RSSI (Received Signal Strength Indicator), ToA/TDoA (Time of Arrival/ Time Difference of Arrival) and AOA (Angle of Arrival). If distances and angles between pairs of nodes can be measured accurately, the localization process will be very simple. But measurements can be with errors, so the localization algorithm can become complex and inaccurate.

There have been a lot of research on the positioning technologies [8–12]. Among these researches, DV-hop and DV-distance proposed by Niculescu [8,9] have been well known. In this paper, it is assumed

*Corresponding author: Leonard Barolli, Department of Information and Communication Engineering, Fukuoka Institute of Technology (FIT), 3-30-1 Wajiro-Higashi, Higashi-Ku, Fukuoka 811-0295, Japan. E-mail: barolli@fit.ac.jp.

that WSN consists of sensor nodes and landmarks. The coordinates of landmarks can be given by GPS or directly by humans. Also, it is also assumed that the sensors have ability to measure the distance and relative angles to their neighbors. Sensors except landmarks don't know their absolute reference bearing such as north so that they should infer the reference bearing from the bearings of landmarks. In [13], we have already introduced a method to obtain the bearings of sensor nodes and conformed the effectiveness of our method by comparing with DV-hop and DV-distance. In this paper, we will give more detail positioning algorithm and some results with node densities, measuring physical data and the number of landmarks.

This paper is organized as follows. In Section 2, we present the related works. Then, in Section 3, we introduce in details the method to measure the angles to neighbors [13]. In Section 4, we present the positioning algorithm proposed in this paper in more details. In Section 5 are shown the simulation results. Finally, in the last Section, we give the concluding remarks.

2. Related works

Several localization technologies using wireless radios have been developed and commercially available. The GPS is perhaps the most popular location system but GPS cannot be used indoor or when receivers cannot receive the wireless signal from the satellites. To complement the weakness of GPS, there have been many available localization systems such as Active badge, Active bat, Cricket, and Radar and their detail review is found in [14].

The Active badge localization system was developed at Olivetti Research Laboratory, now AT&T Cambridge and uses diffuse infrared technology. The Active badge system provides absolute location information and difficulty in locations with fluorescent lighting or direct sunlight. Because diffused infrared has an effective range of only several meters, Active badge system can be deployed in small or medium sized rooms.

Active bat localization system was developed by AT&T. This system uses ultrasound time of flight technique so that it can provide more accurate measurements than Active Badges. A bat sends an ultrasonic pulse to a ceiling mounted receivers and at the same time the controller sends not only the radio frequency request packet but also a synchronized reset signal to the ceiling sensors. Each ceiling sensors measures the time interval between RF signal reset arrival and ultrasonic pulse arrival and then the time difference will be converted distance. Active Bat can obtain accuracy within 9cm of the true position.

Cricket location support system was intended to complement the Active Bat system. Therefore, like the Active Bat system utilizes ultrasonic time-of-flight data and radio frequency synchronization signal. This system allows the objects to perform all their triangulation computations. Cricket uses the RF signal for synchronization as well as description of the available time domain. Cricket provides both the lateration and proximity techniques.

RADAR has been developed by Microsoft Research group and is based on the IEEE 802.11 WaveLAN wireless networking technology. RADAR can measure the signal strength and signal-to-noise ratio of signal which is used to compute 2D position within the building. This system needs only a few base stations and facilitates the same infrastructure providing the building's general purpose wireless networking. Wireless radio localization system operates by measuring radio signals between a set of landmarks (or beacons or fixed stations) and unknown nodes (or mobile stations) in WSN. In the localization system in WSN, nodes can calculate their own positions by using signals received from landmarks.

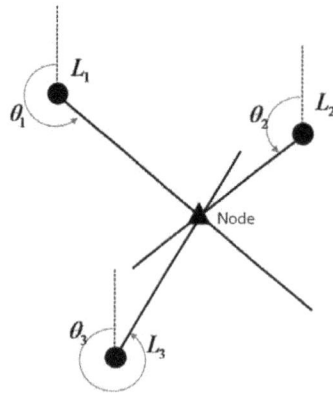

Fig. 1. Measuring the angles to determine the position of a node. Dots are landmarks which have already known their coordinates and triangle is a node to be located.

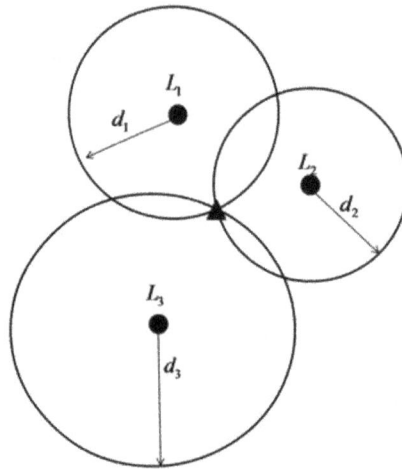

Fig. 2. Signal strength and ToA based localization system. Dots are landmarks which have already known their coordinates and triangle is a node to be located.

Wireless localization systems can be divided into angle based technique such as AoA, time based technique such as ToA and TDoA and signal strength based technique or their combinations. There is a brief review in [15].

Angle based techniques: AoA is the localization method based on Angle based techniques. Nodes in AoA techniques can estimate their locations by first measuring angles of received signals from several landmarks shown in Fig. 1. Measuring the AoA in nodes is available using directive antennas or antenna arrays. As shown in Fig. 1, finding the intersections of the lines-of-position allows the nodes to calculate their locations. AoA method needs a minimum of two landmarks to fix a position of a node.

Signal strength: The signal strength technique uses a known mathematical model describing the path loss attenuation with distance. The measured signal strength can convert the distance between a landmark and a node. A node to be located should be on a circle centered at the landmarks. As shown in Fig. 2, when measuring the signal strength from more than three landmarks, a node can find its location by using simple geometric relationships.

Time based localization: In ToA method, the distance between a landmark and a node can be found

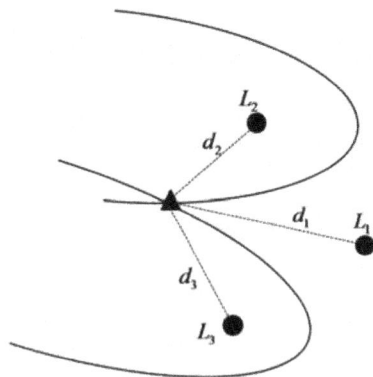

Fig. 3. TDoA based localization method. Dots are landmarks which have already known their coordinates and triangle is a node to be located.

by measuring one way propagation time between them assuming that the signals travel with the velocity of light (c). As shown in Fig. 2, if a node can know more than three distances to their neighboring landmarks, a node can locate its own position. The other hand, TDoA uses time differences of arrival signals. TDoA does not need to know the transmission time of a signal, which is important in ToA system. In TDoA technique, a line-of-position will be defined as hyperbola because a curve of constant time difference of arrivals for two landmarks is hyperbola as shown in Fig. 3.

3. The method to measure the angle to immediate nodes

WSN means the set of sensor nodes which are deployed ad-hoc. All nodes in WSN will be assumed to be possible to communicate with their neighboring nodes within their transmission range [16]. Especially the landmarks have already known their coordinates and they had their own reference bearings (for example, East). It will be assumed that all nodes have ability to measure the distance to their neighbors and relative angles based on their own axis (called heading). Figure 4 shows definitions of angles used in this paper [13]. The node's heading which is used for measuring the angle to neighbors is different among nodes. Node's headings are shown by thick arrows. In Fig. 4, $\hat{ab}$ is the measured angle between node A and node C. θ shows the incident angle from neighbors with reference of East. For example, θ_{AB} is the incident angle from node A measured at node B. Then, $\hat{a}$ is the azimuth of node A which means the angle of A's heading measured from East.

Because the landmarks have their azimuths, the calculations will be started from the nodes near landmarks. First of all, assume that node B has been aware of its own azimuth and the angle to node A. Node B is able to calculate the incident angle, θ_{BA} using its own azimuth and the angle to node A,

$$\theta_{BA} = \begin{cases} \hat{ba} - \hat{b} & \text{for } \hat{ba} \geqslant \hat{b} \\ 2\pi + (\hat{ba} - \hat{b}) & \text{for } \hat{ba} < \hat{b} \end{cases} \tag{1}$$

Figure 4 (a) is for $\hat{ba} - \hat{b}$ and Fig. 4 (b) is for $\hat{ba} \geqslant \hat{b}$. The obtained incident angle θ_{BA} will be transferred to node A and then will be used for calculating the angles related to node A. That is, node A will calculate the incident angle θ_{AB} from node B as follows,

$$\theta_{AB} = 2\pi - \theta_{BA}. \tag{2}$$

(a)   (b)

(c)

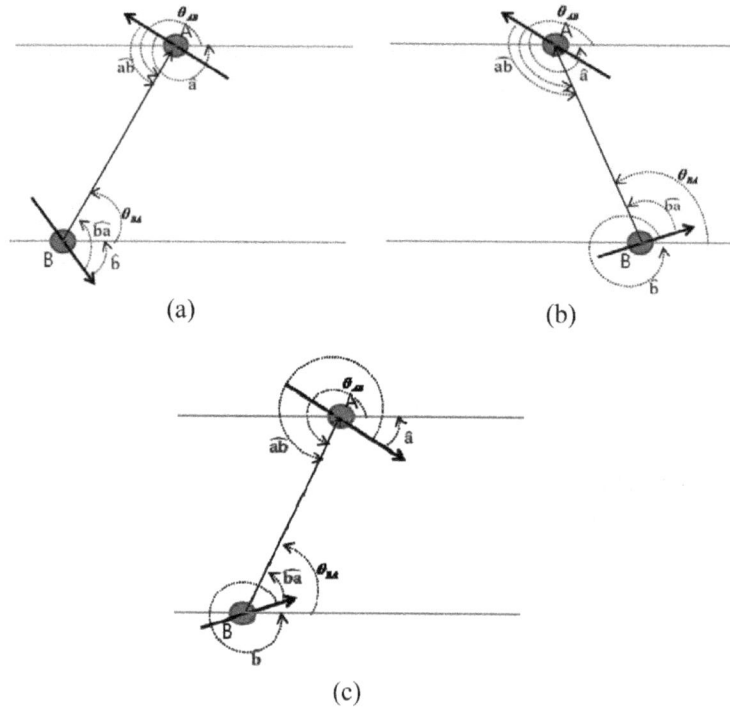

Fig. 4. Detail description for calculating useful angles between tow nodes (a) $\hat{ba} \geqslant \hat{b}$ and $\hat{ba} < 180$, (b) $\hat{ba} < \hat{b}$, (c) $\hat{ba} \geqslant \hat{b}$ and $\hat{ba} > 180$.

Also from θ_{AB} of Eq. (2) and the measured angle between node B and node A, $\hat{ab}$, the azimuth of node A can be obtained,

$$
\hat{a} = \begin{cases} \hat{ab} - \theta_{AB} & for \quad \hat{ab} \geqslant \theta_{AB} \\ 2\pi + \left(\hat{ab} - \theta_{AB}\right) & for \quad \hat{ab} \leqslant \theta_{AB} \end{cases} \tag{3}
$$

The same calculation will be performed for all nodes from the neighbors of landmarks so that all nodes with any connection to landmarks can have their own azimuths. And then nodes will also calculate all incident angles from their neighbors from their azimuths and the measured angles to their neighbors. Although nodes with at least one connection path to landmarks is possible to calculate the information related to angles, for nodes far from many hops from landmarks angle error accumulation will be profound. So, we will restrict the hop counts within 3 or 5.

4. Localization algorithm

The localization algorithm proposed in this paper consists of two phases. For distribution process, these two phases will be performed simultaneously. In the first phase, all sensor nodes in the sensor field will find the neighbor nodes within their own transmission ranges. And then they have to calculate the distances between them and their neighbors and the incident angles from their neighbors. This physical

id	distance	theta	L or S

Fig. 5. Table for physical information.

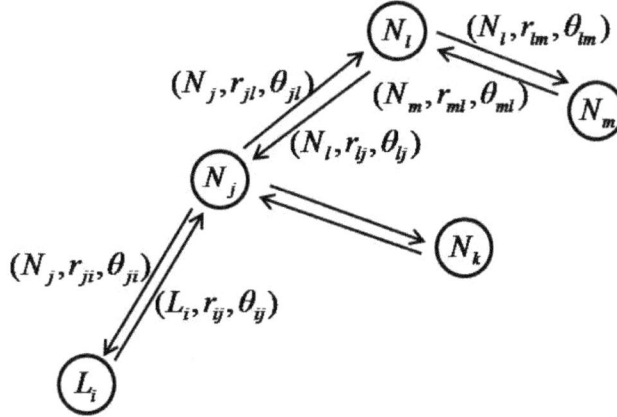

Fig. 6. The example for showing the data exchanges between nodes.

information will be recalculated and shared with their neighbors periodically or when the events happen such as adding new nodes. Each node will construct its physical information table as shown in Fig. 5.

In Fig. 5, the id is the identification numbers of neighbor nodes, distances means the distances between a node and its neighbors. These distances divide the distance calculated by a sensor node itself and the distance sent by its neighbors. 'Theta" is the calculated incident angles which are composed by the value calculated by a node itself and the values sent by its neighbors. To identify whether the neighbors are landmarks or ordinary nodes, "L or S" will be recorded.

Figure 6 shows the example of the process by which sensor nodes exchanges the physical data. Each node will send the message including physical data such as their identification number, the distance from itself to their proximities, the incident angle calculated by the method described in previous chapter. So, after finishing this process, all nodes can share all physical information to calculate their positions with their neighbors. For example, in Fig. 3, node N_j can construct its own physical table shown in Table 1.

A second phase will begin at landmarks, based on the physical data recoded in the physical data table. Firstly, since landmarks have known their position, they will broadcast their identification number and their coordinates. And then, the neighboring nodes of landmarks can calculate their positions from the stored physical data and the coordinates of their landmarks. For example, a node N_j adjacent to the landmark L_i in Fig. 6 can obtain its coordinate as follows

$$\begin{pmatrix} x_j \\ y_j \end{pmatrix} = \begin{pmatrix} x_i \\ y_i \end{pmatrix} + \begin{pmatrix} r_{ji} \cos(\theta_{ji}) \\ r_{ji} \sin(\theta_{ji}) \end{pmatrix} \tag{4}$$

After the nodes have calculated its coordinate, they construct their position table which includes their coordinates and the hop counts, the hop numbers far from the landmarks which used in calculating the positions. The nodes which have had their positions are going to broadcast their coordinates to their neighbors and then the nodes received the position information from their neighbors can calculate their position. The nodes that have calculated their coordinates can make their position tables. This process will continue until all nodes will have their positions and will also take place when their neighbors' positions update.

Table 1
The physical data table for node N_j

Node id	Distances		Incident angles		L or S
	From neighbors	By itself	From neighbors	By itself	
L_i	r_{ij}	r_{ji}	θ_{ij}	θ_{ji}	L
N_k	r_{lj}	r_{jl}	θ_{lj}	θ_{jl}	S
N_l	r_{kj}	r_{jk}	θ_{kj}	θ_{jk}	S

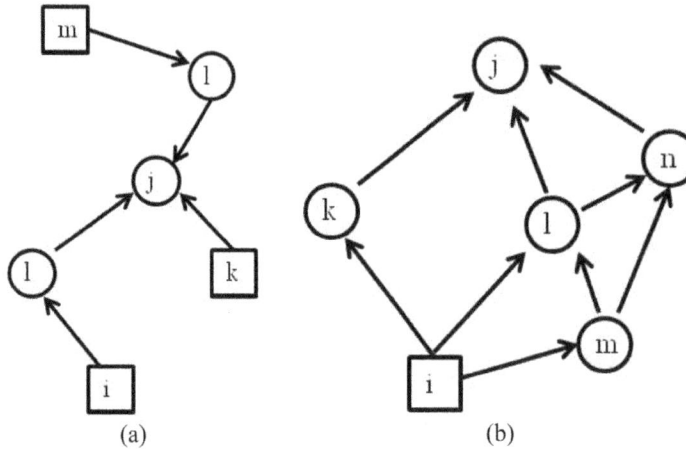

Fig. 7. In (a) sensor node j sees several landmarks; (b) there are several paths from landmark i to sensor node j.

4.1. Updating of coordinates of sensor nodes

As some nodes could have known several landmark's coordinates or there are many paths from landmarks to nodes which want to know their positions, a sensor node may have several coordinates. Figure 7 shows the situation that a node has linked several landmarks and had several paths to a landmark. In Fig. 7 squares are the landmarks and circles are sensor nodes. In the case of Fig. 7 (a), sensor node j knows 2 landmarks, so it is possible for this node to have three coordinates. Figure 7 (b) shows there are several paths from landmark i to sensor node j. In this case, the sensor node j will have several coordinates for itself. This means it is necessary for nodes to have a method to determine their positions among several coordinates.

This paper proposes four schemes to determine the sensor's coordinate. The first scheme is what we called "min-hop" method. This scheme can choose the landmark minimum hops away from a sensor node and then sensor node come to calculate its coordinate from the position of this landmark. However, there are multiple landmarks same hops away; the position of the sensor node will be averaged. The second scheme is called "min-distance" method. First of all, the sensor node estimates the distance to known landmarks and then choose the closest landmark in distance. Accordingly, the sensor node will determine its position calculated from the closest landmark.

The third method utilizes the several coordinates calculated from several landmarks. If there are several paths from a sensor node to a landmark, we choose the path with minimum hops. So, this method averages the coordinates from several landmarks to determine the node's coordinate. We can see that in Fig. 7 (a) the first method uses only landmark k to find the coordinate of node j but the third method utilizes all coordinates calculated from landmark i, k and m. We call this method "averaged min-hop" method.

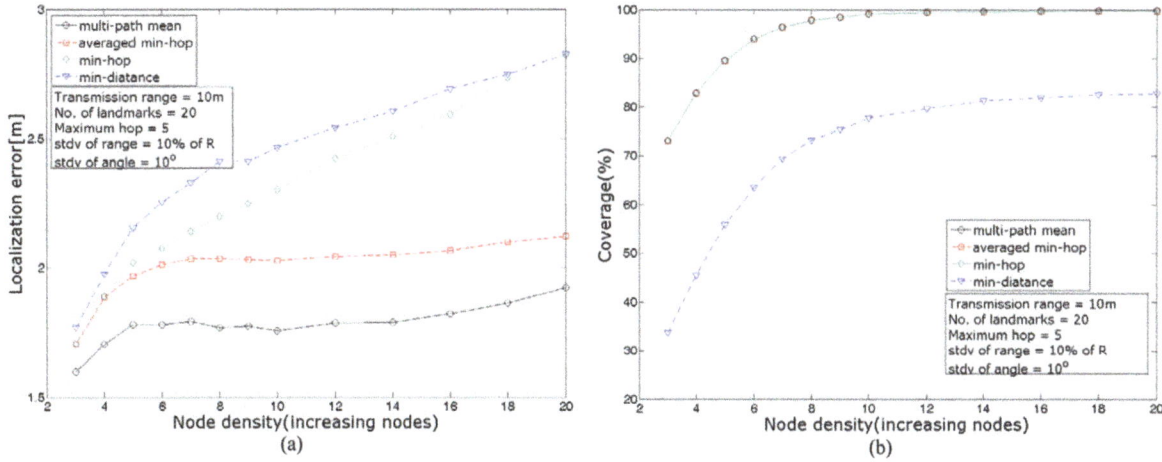

Fig. 8. Localization error and coverage with the variation of node density (node density is calculated by changing the number of nodes).

The final scheme is very simple. This method utilizes several paths from a landmark to a sensor node. As seen in Fig. 7 (b), there are multiple paths from a landmark to a node. This means a node can obtain several coordinates even when it knows only one landmark because several paths means several coordinates for a node. So, we will simply average the node's coordinates obtained from multiple path, so called "multi-path mean" method.

5. Results and discussion

The proposed method will be susceptible to angle error because the node's azimuth is calculated from a landmark far apart. This becomes severe when a landmark is far apart from. In this paper, the measured distance and angle are assumed to be Gaussian.

$$r_{meas} = r_{exact} \left(1 + \sigma_r N\left(0, 1\right)\right) \tag{5}$$

$$\theta_{meas} = \theta_{exact} \left(1 + \sigma_r N\left(0, 1\right)\right) \tag{6}$$

where r_{meas} (θ_{meas}) is the measured distance (angle), r_{exact} (θ_{exact}) is the true value of the distance(angle), σ_r (σ_θ) is a specific constant [17], and $N\left(0, 1\right)$ is a normally distributed random variable. Therefore, the noise error in measured values is modeled as additive and can be varied by changing the specific constants σ_r (σ_θ), where in Matlab® program, σ_r (σ_θ) in Eqs (5), (6) are the standard deviation of normal distribution, so we will simply call it a standard deviation.

Figure 8 shows the location error and coverage with the node density, where coverage means the ratio of nodes calculating the coordinate to all nodes. Node density [18–20] can be calculated as,

$$d = \frac{N\left(\pi R^2\right)}{A} \tag{7}$$

where N is the total number of sensor nodes deployed in sensor field, A is the area of sensor field, R means the transmission range of nodes which is the same among all nodes. Localization error obtained

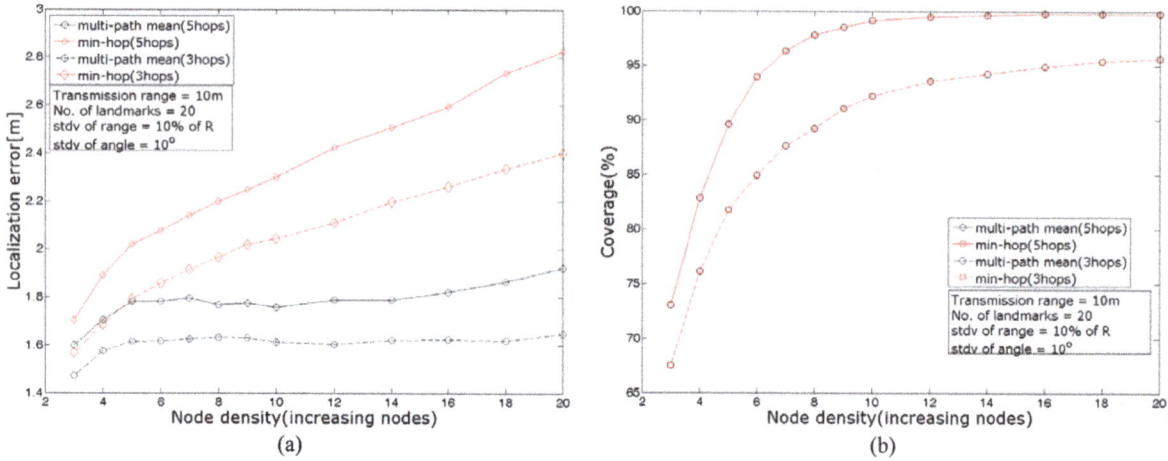

Fig. 9. Comparing the results between limited maximum hops of 3 and 5. (a) Accuracy (b) Coverage.

in the simulation is defined by:

$$L_{error} = \frac{\sum_{i=1}^{N} |r_{calc} - r_{real}|}{N} \tag{8}$$

where N is the total number of nodes, r_{calc} is the calculated coordinate of a node, and r_{real} is the real coordinate of a node.

In this simulation, sensor field is considered to be square of which length of each side is 100 m. The transmission range is 10 m. As shown in Eq. (7), the node density can be changed according to varying the number of nodes or transmission range when the area of sensor field is fixed. The results in Fig. 8 are obtained from varying the number of nodes. In this simulation, we restrict the hop count to landmarks within 5 in order to prevent the propagation of angle error. That is, when calculating the coordinates of nodes, only landmarks within 5 hops far from a node were considered. The larger in the hop counts, the larger coverage is but the larger localization error becomes. The "multi-path mean" method reveals the best result and the "averaged min-hop" method ranks second in the accuracy. Also, two methods are nearly independent upon the node density but the "min-hop" and "min-distance" method show that the accuracies increase with the increasing of the node density. In the aspect of the coverage, three methods except "min-distance" method show the same result. That is, the coverage obtained from our method reaches nearly 100% when the node density of about 10.

To see the effect of measuring errors in distance and angle on the accuracy and coverage, two results obtained with allowed maximum hops of 3 and 5 are compared in Fig. 9. In Fig. 9, the solid lines show the results for maximum hop limited to 5 and dashed line are the result for 3 hops. The accuracy in the case of allowed maximum hop of 3, in Fig. 9(a), is better than that of 5 hops in the expense of coverage shown in Fig. 9(b). The coverage of 5 hops reaches nearly 100% for node density of 12. However, in the case of maximum hop of 3, it reaches about 90%.

The effect of increasing the number of landmarks on the localization error and coverage is shown in Fig. 10. In general, increasing the number of landmarks can make hops to a node small so that the accumulation of angle error can be reduced, which allows the node to choose the landmark with the small error. When the number of landmarks are smaller than 40, the "multi-path mean" method shows the best accuracy, but with landmarks more than 40, the "averaged min-hop" method reveals better accuracy than

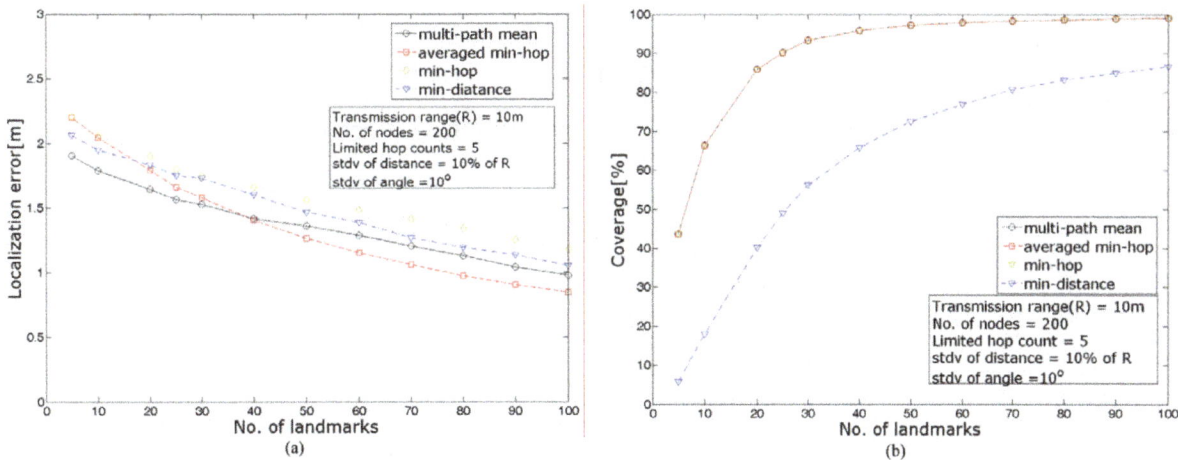

Fig. 10. Localization error and coverage with the number of landmarks.

other methods. In this simulation, because we restrict the hop counts to 5 to find landmarks, there are still isolated nodes from landmarks even though increasing the number of landmarks so that coverage will not reach 100%.

6. Conclusions

We introduced a localization algorithm to find the coordinates of nodes in WSN. This method assumed that the nodes have the ability to measure the mutual distance and relative angle to their neighbors within transmission range. The proposed algorithm starts at measuring the angle and distance at the landmark and their neighbors and then finding their azimuths and incident angles. Obtained incident angle and measured mutual distance allow a node to calculate its coordinates.

In this paper, we presented detailed description for utilizing multiple landmarks' coordinates and data from multiple paths. We presented four kinds of methods to calculate and update the coordinates of nodes. In "min-hop" method, the landmark is selected by minimum number of hops and the sensor node calculates its coordinates from the position of this landmark. The second method is called "min-distance" method where the sensor node estimates the distance to known landmarks and then chooses the closest landmark considering the distance. Therefore, the sensor node will determine its position calculated from the closest landmark. The third method utilizes several coordinates calculated from several landmarks, where the coordinates from several landmarks to determine the node's coordinate are simply averaged. This method is called "averaged min-hop" method. The last scheme, which is called "multi-path mean" method, utilizes several paths from a landmark to a sensor node when there are multiple paths from a landmark to a node. This method will simply average the node's coordinates obtained from multiple paths. The simulation results show that to utilize more data obtained from multiple paths can give more accurate position coordinates.

Acknowledgements

This work was supported by the National Research Foundation of Korea(NRF) grant funded by the Korea government(MEST) (No. 2010-0013411).

References

[1] A. Barolli, E. Spaho, L. Barolli, F. Xhafa and M. Takizawa, QoS Routing in Ad-hoc Networks Using GA and Multi-objective Optimization, *Mobile Information Systems* **7**(3) (2011), 169–188.

[2] W. Wu, X. Li, S. Xiang, H.B. Lim and K.L. Tan, Sensor Relocation for Emergent Data Acquisition in Sparse Mobile Sensor Networks, *Mobile Information Systems* **6**(2) (2010), 155–176.

[3] J. Jayaputera and D. Taniar, Data Retrieval for Location-dependent Queries in a Multi-cell Wireless Environment, *Mobile Information Systems* **1**(2) (2005), 91–108.

[4] A.B. Waluyo, B. Srinivasan and D. Taniar, Research on Location-dependent Queries in Mobile Databases, *International Journal of Computer Systems: Science and Engineering* **20**(2) (2005), 77–93.

[5] P. Sarkar and A. Saha, Security Enhanced Communication in Wireless Sensor Networks Using Reed-Muller Codes and Partially Balanced Incomplete Block Designs, *Future Technology Research Association International (FTRA), Journal of Convergence* **2**(1) (2011), 43–48.

[6] C.M. Huang, R.H. Cheng, S.R. Chen and C.I. Li, Enhancing Network Availability by Tolerance Control in Multi-Sink Wireless Sensor Networks, *FTRA Journal of Convergence* **1**(1) (2010), 15–22.

[7] D. Kumar, C. Trilok, R. Aseri and B. Patel, Multi-hop Communication Routing (MCR) Protocol for Heterogeneous Wireless Sensor Networks, *International Journal of Information Technology, Communications and Convergence* **1**(2) (2011), 130–145.

[8] D. Niculescu and B. Nath, Ad-hoc Positioning System (APS), *Proc. of IEEE Global Telecommunications Conference (GLOBECOM'01)* **5** (2001), 2926–2931.

[9] D. Niculescu and B. Nath, Ad Hoc Positioning System (APS) Using AOA, *Proc. of the 23-rd Conference of the IEEE Communications Society (INFOCOM'03)* (2003), 1734–1743.

[10] G.D. Stefano and A. Petricola, A Distributed AOA Based Localization Algorithm for Wireless Sensor Networks, *Journal of Computers* **3**(4) (2008), 1–8.

[11] A. Savvides, C. Han and M.B. Strivastava, Dynamic Fine-Grained Localization in Ad-hoc Networks of Sensors, *Proc. of MobCom-2001* (2001), 166–179.

[12] M. Boushaba, A. Hafid and A. Benslimane, High Accuracy Localization Method using AOA in Sensor Networks, *Computer Networks* **53**(18) (2009), 3076–3088.

[13] J. Park and D. Park, Ad-Hoc Localization Method Using Ranging and Bearing, *To appear in Proc. of International Conference on Information Technology Convergence and Services* (ITCS-2011) (2011).

[14] J. Hightower and G. Borriello, Location Systems for Ubiquitous Computing, *IEEE Computer* (2001), 57–66.

[15] J. Gibson, The Mobile Communication Handbook, *IEEE Press* (1999).

[16] B. Xie, A. Kumar, D. Zhao, R. Reddy and B. He, On Secure Communication in Integrated Heterogeneous Wireless Networks, *International Journal of Information Technology, Communications and Convergence* **1**(1) (2010), 4–23.

[17] P. Biswas, H. Aghajan and Y. Ye, Integration of Angle of Arrival Information for Multimodal Sensor Network Localization using Semidefinite Programming, *Proc. of 39-th Asilomar Conference on Signals, Systems and Computers* (2005), 1–9.

[18] K.K. Chintalapudi, A. Dhariwal, R. Govindan and G. Sukhatme, Ad-Hoc Localization Using Ranging and Sectoring, *Proc. of the 23-rd Conference of the IEEE Communications Society (INFOCOM-2004)* (2004), 2662–2672.

[19] K.K. Chintalapudi, A. Dhariwal, R. Govindan and G. Sukhatme, On the Feasibility of Ad-Hoc Localization Systems, *Tech. Report, Computer Science Department, University of Southern California, Los Angeles* (2003).

[20] K. Liu, S. Wang, F. Zhang, F. Hu and C. Xu, Efficient Localized Localization Algorithm for Wireless Sensor Networks, *Proc. of the Fifth International Conference on Computer and Information Technology (CIT-2005)* (2005), 517–523.

Yang Sun Lee received the B.S. and M.S. degrees in Electrical & Electronic Engineering from Dongshin University and Ph. D. degrees in Dept. of IT Engineering from Mokwon University in 2001, 2003 and 2007, respectively. He was a Senior Engineer at R&D Center, Fumate Co., Ltd. from 2007 to 2009. Since 2009, he has worked as a Research Professor in Dept. of Information Communication Engineering at Chosun University. His current research interests include UWB, multimedia communication, Network Transmission Scheme and Ubiquitous Sensor Networks. He is a member of the KICS, KIMICS, KEES, KIIT and KONI.

Jang Woo Park received the B.S., M.S., and Ph.D. degrees in Electronic engineering from Hanyang University, Seoul, Korea in 1987, 1989 and 1993 respectively. In 1995, he joined the faculty member of the Sunchon National University, where he is currently a professor in the Dept. of Computer & Communication Engineering. His research focuses on RFID circuits and system designs and RFID/USN technologies.

Leonard Barolli received B.S and Ph.D degrees from Tirana University and Yamagata University in 1989 and 1997, respectively. From April 1997 to March 1999, he was a JSPS Post Doctor Fellow Researcher at Department of Electrical and Information

Engineering, Yamagata University. From April 1999 to March 2002, he worked as a Research Associate at the Department of Public Policy and Social Studies, Yamagata University. From April 2002 to March 2003, he was an Assistant Professor at Department of Computer Science, Saitama Institute of Technology (SIT). From April 2003 to March 2005, he was an Associate Professor and presently is a Full Professor, at Department of Information and Communication Engineering, Fukuoka Institute of Technology (FIT). Dr. Barolli has published about 350 papers in referred Journals, Books and International Conference Proceedings. He was an Editor of the IPSJ Journal and has served as a Guest Editor for many International Journals. Dr. Barolli has been a PC Member of many International Conferences and was the PC Chair of IEEE AINA-2004 and IEEE ICPADS-2005. He was General Co-Chair of IEEE AINA-2006, AINA-2008, AINA-2010, CISIS-2009, CISIS-2010, and CISIS-2011, Workshops Chair of iiWAS-2006/MoMM-2006 and iiWAS-2007/MoMM-2007, Workshop Co-Chair of ARES-2007, ARES-2008, IEEE AINA-2007 and ICPP-2009. Dr. Barolli is the Steering Committee Chair of CISIS and BWCCA International Conferences and is serving as Steering Committee Co-Chair of IEEE AINA, IEEE INCoS, NBiS, 3PGCIC and EIDWT International Conferences. He is organizers of many International Workshops. Dr. Barolli has won many Awards for his scientific work and has received many research funds. He got the "Doctor Honoris Causa" Award from Polytechnic University of Tirana in 2009. His research interests include network traffic control, fuzzy control, genetic algorithms, P2P, wireless networks, ad-hoc networks and sensor networks. He is a member of SOFT, IPSJ, and IEEE.

Permissions

All chapters in this book were first published in MIS, by Hindawi Publishing Corporation; hereby published with permission under the Creative Commons Attribution License or equivalent. Every chapter published in this book has been scrutinized by our experts. Their significance has been extensively debated. The topics covered herein carry significant findings which will fuel the growth of the discipline. They may even be implemented as practical applications or may be referred to as a beginning point for another development.

The contributors of this book come from diverse backgrounds, making this book a truly international effort. This book will bring forth new frontiers with its revolutionizing research information and detailed analysis of the nascent developments around the world.

We would like to thank all the contributing authors for lending their expertise to make the book truly unique. They have played a crucial role in the development of this book. Without their invaluable contributions this book wouldn't have been possible. They have made vital efforts to compile up to date information on the varied aspects of this subject to make this book a valuable addition to the collection of many professionals and students.

This book was conceptualized with the vision of imparting up-to-date information and advanced data in this field. To ensure the same, a matchless editorial board was set up. Every individual on the board went through rigorous rounds of assessment to prove their worth. After which they invested a large part of their time researching and compiling the most relevant data for our readers.

The editorial board has been involved in producing this book since its inception. They have spent rigorous hours researching and exploring the diverse topics which have resulted in the successful publishing of this book. They have passed on their knowledge of decades through this book. To expedite this challenging task, the publisher supported the team at every step. A small team of assistant editors was also appointed to further simplify the editing procedure and attain best results for the readers.

Apart from the editorial board, the designing team has also invested a significant amount of their time in understanding the subject and creating the most relevant covers. They scrutinized every image to scout for the most suitable representation of the subject and create an appropriate cover for the book.

The publishing team has been an ardent support to the editorial, designing and production team. Their endless efforts to recruit the best for this project, has resulted in the accomplishment of this book. They are a veteran in the field of academics and their pool of knowledge is as vast as their experience in printing. Their expertise and guidance has proved useful at every step. Their uncompromising quality standards have made this book an exceptional effort. Their encouragement from time to time has been an inspiration for everyone.

The publisher and the editorial board hope that this book will prove to be a valuable piece of knowledge for researchers, students, practitioners and scholars across the globe.

List of Contributors

Francesco Palmieri
Dipartimento di Ingegneria dell'Informazione, Seconda Universit`a degli Studi di Napoli, Aversa (CE), Italy

Ugo Fiore
Universit`a degli Studi di Napoli "Federico II", Napoli, Italy

Aniello Castiglione
Dipartimento di Informatica "R. M. Capocelli", Universit`a degli Studi di Salerno, Via Ponte don Melillo, I-84084 Fisciano (SA), Italy

M. Goyal
Department of Computer Science, University of Wisconsin Milwaukee, Milwaukee WI 53201, USA

W. Xie
Department of Computer Science, University of Wisconsin Milwaukee, Milwaukee WI 53201, USA

H. Hosseini
Department of Computer Science, University of Wisconsin Milwaukee, Milwaukee WI 53201, USA

Dung T. Tran
Faculty of Information Technology, University of Science, Ho Chi Minh City, Vietnam

Trang T.M. Truong
Faculty of Information Technology, University of Science, Ho Chi Minh City, Vietnam

Thanh G. Le
Faculty of Information Technology, University of Science, Ho Chi Minh City, Vietnam

Yi-Fu Ciou
Department of Computer Science, Tunghai University, Tunghai, Taiwan

Fang-Yie Leu
Department of Computer Science, Tunghai University, Tunghai, Taiwan

Yi-LiHuang
Department of Computer Science, Tunghai University, Tunghai, Taiwan

Kangbin Yim
Soonchunhyang University, Asan, Republic of Korea

Hajar Mousannif
Cadi Ayyad University, Gu´eliz, Marrakech, Morocco

Ismail Khalil
Johannes Kepler University, Linz, Austria

Stephan Olariu
Old Dominion University, Norfolk, VA, USA

Fang-Yie Leu
Department of Computer Science, TungHai University, Tunghai, Taiwan

IlsunYou
School of Information Science, Korean Bible University, Nowon-gu, Seoul, South Korea

Feilong Tang
Department of Computer Science and Engineering, Shanghai Jiao Tong University, Shanghai, China

Jun-Ho Choi
Department of Computer Engineering, Chosun University, Gwangju, Korea

Chang Choi
Department of Computer Engineering, Chosun University, Gwangju, Korea

Byeong-Kyu Ko
Department of Computer Engineering, Chosun University, Gwangju, Korea

Pan-Koo Kim
Department of Computer Engineering, Chosun University, Gwangju, Korea

Keisuke Goto
Department of Multimedia Engineering, Graduate School of Information Science and Technology, Osaka University, Osaka, Japan

Yuya Sasaki
Department of Multimedia Engineering, Graduate School of Information Science and Technology, Osaka University, Osaka, Japan

Takahiro Hara
Department of Multimedia Engineering, Graduate School of Information Science and Technology, Osaka University, Osaka, Japan

Shojiro Nishio
Department of Multimedia Engineering, Graduate School of Information Science and Technology, Osaka University, Osaka, Japan

Tetsuya Oda
Graduate School of Engineering, Fukuoka Institute of Technology (FIT), Higashi-Ku, Fukuoka, Japan

Admir Barolli
Department of Computers and Information Science, Seikei University, Musashino-Shi, Tokyo, Japan

Fatos Xhafa
Department of Languages and Informatics Systems, Technical University of Catalonia, Barcelona, Spain

Leonard Barolli
Department of Information and Communication Engineering, Fukuoka Institute of Technology (FIT), Higashi-Ku, Fukuoka, Japan

Makoto Ikeda
Department of Information and Communication Engineering, Fukuoka Institute of Technology (FIT), Higashi-Ku, Fukuoka, Japan

Makoto Takizawa
Department of Computers and Information Science, Seikei University, Musashino-Shi, Tokyo, Japan

Yongjun Li
School of Computer Science & Engineering, South China University of Technology, Guangzhou, 510640, China

Hu Chen
School of Computer Science & Engineering, South China University of Technology, Guangzhou, 510640, China

Rong Xie
School of Computer Science & Engineering, South China University of Technology, Guangzhou, 510640, China

James Z. Wang
School of Computing, Clemson University, Clemson, SC 29634, USA

Tarek R. Sheltami
Computer Engineering Department, King Fahd University of Petroleum and Minerals, Dhahran, Saudi Arabia

Elhadi M. Shakshuki
Jodrey School of Computer Science, Acadia University, Wolfville, Nova Scotia, B4P 2R6, Canada

Hussein T. Mouftah
School of Information Technology and Engineering, University of Ottawa, ON, Canada

Yang Sun Lee
Department of Information Communication Engineering, Chosun University, Chosun, Korea

Jang Woo Park
Department of Information and Communication Engineering, Sunchon National University, Sunchon, Korea

Leonard Barolli
Department of Information and Communication Engineering, Fukuoka Institute of Technology (FIT), Fukuoka, Japan

www.ingramcontent.com/pod-product-compliance
Lightning Source LLC
Chambersburg PA
CBHW080651200326
41458CB00013B/4809